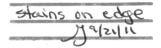

Flight Discipline

Tony Kern

McGraw-Hill

New York   San Francisco   Washington, D.C.   Auckland   Bogotá
Caracas   Lisbon   London   Madrid   Mexico City   Milan
Montreal   New Delhi   San Juan   Singapore
Sydney   Tokyo   Toronto

Library of Congress Cataloging-in-Publication Data

Kern, Tony.
    Flight discipline / by Tony Kern.
      p. cm.
    Includes index.
    ISBN 0-07-034371-3 (hardcover)
      1. Air pilots—Psychology. 2. Airplanes—Piloting—Human factors.
    3. Self-control. I. Title.
    TL555.K47   1998
    629.132'52—dc21   *629.1325 2*   97-49042
                                                         CIP

McGraw-Hill

A Division of The McGraw·Hill Companies

Copyright© 1998 by The McGraw-Hill Companies, Inc. All rights reserved. Printed in the United States of America. Except as permitted under the United States Copyright Act of 1976, no part of this publication may be reproduced or distributed in any form or by any means, or stored in a data base or retrieval system, without the prior written permission of the publisher.

6 7 8 9 0 DOC/DOC 0 2

ISBN 0-07-034371-3

The sponsoring editor for this book was Shelley Carr, the editing supervisor was Sally Glover, and the production supervisor was Pamela Pelton. It was set in Garamond by Michele Bettermann and Michele Pridmore of McGraw-Hill's desktop publishing department, Hightstown, N.J.

Printed and bound by R. R. Donnelley & Sons Company.

McGraw-Hill books are available at special quantity discounts to use as premiums and sales promotions, or for use in corporate training programs. For more information, please write to the Director of Special Sales, McGraw-Hill, Professional Publishing, Two Penn Plaza, New York, NY 10121-2298. Or contact your local bookstore.

 This book is printed on recycled, acid-free paper containing a minimum of 50% recycled, de-inked fiber.

GEN1-AV

Contents

Acknowledgments *xxi*

List of contributors *xxiii*

Preface *xxv*

Introduction *xxxi*

**Part One The Problem of Poor Discipline
in Aviation** *1*

1 Discipline: The foundation of airmanship *3*
The critical importance of flight discipline *3*
Why study to be a better flyer? *4*
Flight discipline as a personal decision *5*
Expertise and discipline *6*
The airmanship-discipline connection *6*
   Discipline must be the first step *8*
Objectives of this book *9*
   Defining flight discipline *9*
   Recognizing hazardous scenarios *9*
   Developing airmanship after discipline *10*
The many faces of flight discipline *10*
Case study: Unforced errors *11*
   Fifteen seconds to impact *11*
   Background to a tragedy *12*
   The players *13*

Organizational failures of discipline: A culture of
    non-compliance *15*
  Organizational failure 1: Failing to implement CRM
    training mandate *16*
  Organizational failure 2: Failing to enforce the
    prohibition against flying unapproved instrument
    approaches *17*
  Organizational failure 3: Failing to take "no" for
    an answer *18*
Aircrew failures of discipline *20*
External pressures and organizational influences *21*
  External factor 1: High operations tempo and culture of
    noncompliance *21*
  External factor 2: VIP passengers *22*
  External factor 3: Multiple mission changes *23*
Internal factors *24*
  Internal factor 1: Fatigue *24*
  Internal factor 2: Pressing *26*
  Internal factor 3: Distraction *27*
Final approach *28*
An analysis of aircrew actions *29*
The deadly chain of failed discipline *30*
Good intentions and discipline *30*
A final perspective on discipline: The tough questions *31*
Chapter review questions *32*
References *32*

2 The costs of poor flight discipline *35*

Captain "Jepp" *36*
  The flight discipline cost-benefit equation *38*
The scope of the discipline problem *38*
  Statistical evidence *39*
    Mishap data *39*
  Commercial costs *39*
  Military costs *40*
  Self-reported statistics *42*
  Media representative gets the ride of his life *43*
  Air taxi risks collision to keep passenger dry *45*
The cost to others *46*
  Case Study: A simple checklist oversight *46*

The slippery slope of poor flight discipline   48
"The best B-52 pilot I've ever known"   49
   A standard setter   50
   Two faces of poor discipline emerge   50
   Situation one: Fairchild AFB Air Show: 19
      May 1991   51
   Situation two: 325th Bomb Squadron (BMS) Change of
      Command "Fly-Over": 12 July 1991   51
   Situation three: Fairchild Air Show: 17 May 1992   52
   Situation four: Global Power Mission: 14-15
      April 1993   52
   Situation five: Fairchild Air Show: 8 August 1993   52
   Effects on other aviators   52
   Situation six: Yakima Bombing Range: 10
      March 1994   53
   Situation seven: Air Show Practice: 17 June 1994   54
A final perspective   55
Chapter review questions   56
References   57

3 The letter of the law: Regulatory discipline   59

Regulations are made to be broken, right?   59
Three types of regulatory deviations   63
   Regulatory deviation #1: "I didn't know."   63
      Self-assessment: Finding out what you need
         to know   64
      Regulatory deviation #2: "It's not my fault, I was a
         victim of circumstances."   66
   Case study: I shouldn't have trusted those weather
      forecasters   66
      Situation awareness (SA): The key to "unavoidable"
         violations   68
   Regulatory deviation #3: Willful noncompliance   69
   Case study: Unqualified for the task at hand   69
   Case study: A situation that got out of hand   70
Combating noncompliance   72
Good pilots follow the rules   74
Chapter review questions   75
References   75

4 The problem with shortcuts: Procedural discipline 77

What is procedural error? 78
  Four keys to procedural excellence 78
  Procedural knowledge 79
    Procedure versus technique 79
  Case study: The lights were out, but someone was home 81
    Analysis 82
    Role of air traffic control 83
    Applications for flight discipline 85
    Skill and proficiency 85
  Case study: Final flight 86
  Attention management 89
    Task saturation and channalized attention 90
  Case study: The single-focus pilot 91
    Habit patterns: The key to consistency 93
    Negative transfer of training 93
  Case study: One too many sets of procedures to remember 93
    A final caution on habits 94
  Case study: Checklist delay and confusion 95
  Techniques (not procedures!) for improving procedural discipline with habit patterns 96
Some final words on procedure 97
Chapter review questions 97
References 98

5 Organizational issues for flight discipline 99

Organizational impacts 99
Climate verses culture 100
  Shaping a climate 102
  Chaos and confusion 104
  Economic pressures 106
Rogue practitioners 108
  Typical patterns: Rogues cross all disciplines 109
    Rogues on the rails 109
    Rogues at sea 110

Medical rogues   *111*
Rogues in business: Nick Leeson   *113*
Common characteristics   *114*
Don't kill creativity   *115*
Establishing a non-receptive culture: Hero or villain?   *116*
Taking action: Bad news doesn't improve with age   *116*
What action to take?   *116*
Seven principles for organizational flight discipline   *117*
Chapter review questions   *118*
References   *119*

Part Two   The Anatomy of Flight Discipline   *121*

6 Personality factors and flight discipline   123

The early picture of pilot personality   *124*
New demands on the pilot   *127*
Personalities are permanent   *128*
How personality shapes performance   *128*
Personality factors and communications   *128*
Personality factors and stress   *131*
Personality factors and decision making   *132*
The "top gun mentality"   *132*
Self-assessment is the key   *135*
Chapter review questions   *135*
Additional resources on personality types for self-assessment and understanding   *136*
Books   *136*
Internet sites   *136*
References   *137*

7 Hazardous attitudes   *139*

Case study: Pushing the limits   *139*
Hazardous attitudes impacting flight discipline   *142*
Pressing (also known as get-home-itis)   *143*
Case study: Pressing for the customer   *143*
Avoiding the "pressing " trap   *144*
Let's "take a look" syndrome   *145*
Resisting the "take a look" syndrome   *145*

Antiauthority   *145*
    Avoiding the antiauthority impulse   *146*
Machismo   *146*
    Avoiding the machismo trap   *146*
Invulnerability   *147*
    Fighting the invulnerability syndrome   *147*
Impulsiveness   *148*
    Overcoming impulsiveness   *148*
Resignation   *149*
    Fighting resignation   *150*
Complacency   *150*
    Combating complacency   *150*
Airshow syndrome   *151*
    Case study: An impromptu airshow   *152*
    Defending against the airshow syndrome   *153*
Emotional "jet lag"   *153*
    Bouncing back: Fighting emotional jet lag   *154*
Excessive deference   *154*
    Assertiveness and specifics: The answers to excessive
        deference   *155*
A final perspective on hazardous attitudes   *155*
Chapter review questions   *156*
References   *156*

8 Peer pressure   *157*

Peer pressure defined   *157*
Sources of peer pressure   *158*
How peer pressure works   *159*
Pride of performance   *159*
    Case study: "If they can hack it, so can I."   *160*
    Case study: Peer pressure prevents effective crew
        coordination   *161*
Mission weighting   *163*
    Case study: Freezing rain, gusty winds,
        broken dreams   *163*
    Overzealous instructors   *165*
Active advocacy   *166*
    Case study: Meeting the quota   *166*
Competing with yourself   *169*

Pressure from above   *171*
    Case study: The captain is always right?   *171*
    Case study: A company out of control   *172*
Resisting peer pressure   *174*
    Principled decision making   *174*
Peer pressure as a positive aspect of flight
    discipline   *175*
Chapter review questions   *176*
References   *176*

Part Three   Practical Issues for Flight Discipline   *179*

**9 Guiding lights: The critical role of
instructors and mentoring to flight
discipline   *181***
    On losing a student—and a friend   *182*
    The proficiency challenge   *185*
    Case study: Proficiency and pride—When do you say
        "this won't work?"   *185*
    Case study: Multiple stressors   *187*
    Case study: Disciplined instruction means expecting
        the unexpected   *188*
    Human stops   *191*
Flight discipline and role models: Students
    never forget   *192*
    Role modeling is tough work   *195*
    Flight discipline is preventative medicine   *196*
    Confrontation   *196*
    Flight discipline is consistency   *197*
    A final point for supervisors   *198*
Chapter review questions   *199*
References   *199*

10 Communications discipline   *201*
    The complex airspace environment   *203*
    The pilot-controller team   *204*
    Read back, hear back   *205*

Causes of communications breakdowns   *205*
   Read-back problems   *206*
   Hear-back problems   *206*
   Digging deeper   *207*
   Summary and recommendations from the
     ASRS study   *208*
Lost communications   *209*
   How pilots lose communications   *210*
   Phase of flight   *210*
   Low experience increases risk   *211*
Recommendations   *211*
Some finer points of communications discipline   *212*
   Phraseology   *212*
   Written and automated communications   *213*
   Clearing   *213*
   Subtle messages   *214*
Crew leadership and communications discipline   *214*
   Ask the right questions at the right time   *215*
   Frankly state opinions   *215*
   Work out differences   *215*
Ground communications   *216*
   Case study: "I thought you said..."   *217*
Communications discipline demands precision   *218*
A final perspective   *220*
Chapter review questions   *221*
References   *221*

11 Automation discipline   *223*

Tales of two disasters   *223*
   A moonless night in Texas   *224*
   A windy day in New York   *225*
   Have we overautomated?   *226*
Automation strengths and weaknesses   *227*
Automation philosophies, policies, and procedures   *229*
   Human-centered automation   *231*
   Pilot views on automation   *232*
   Pilot decisions   *232*
Operations philosophies   *233*
     Delta   *234*
     Cathay Pacific Airways   *235*

United   *235*
U.S. Air Force C-17 operating philosophy   *236*
Automation pitfalls   *237*
Workload   *238*
Captain's role   *239*
Complacency   *240*
Mode confusion   *240*
Automation in general aviation: The limits of
the GPS   *243*
Limits of hand-held GPS   *244*
GPS water rescue   *245*
Building personal discipline for automation use   *246*
Chapter review questions   *247*
References   *247*

12 Disciplined attention: The next best thing to a crystal ball   *249*

Addressing the now   *250*
Monitor   *250*
Evaluate   *251*
The future *251*
Anticipate   *252*
Consider contingencies   *252*
Plan   *252*
Traps   *253*
Focus on the right information at the right time   *253*
If something doesn't look or feel right, it
probably isn't   *253*
Watch out when you're busy or bored   *254*
Habits are hard to break   *254*
Expectation can reduce awareness   *254*
Things that take longer are less likely to get
done right   *255*
Reliable systems aren't always reliable   *255*
It's hard to detect something that isn't there   *256*
Automation keeps secrets   *256*
Distractions come in many forms   *257*
How to build a crystal ball   *258*
Manage crew awareness   *258*
What do they know that I need to know?   *259*
What do I know that they need to know?   *260*

What do none of us know that we need to know?   *260*
  Create reminders   *260*
Summary   *261*
Chapter review questions   *261*
References   *262*

13 Killing conditions: Common scenarios for breakdowns of flight discipline   *263*

Lethal scenario 1: The unexpected weather encounter   *264*
  Case study: My "hail-lacious" low level   *264*
  Case study: What you see isn't always what you get   *266*
Lethal scenario 2: Spatial disorientation—transitioning between instrument and visual conditions   *267*
  Some "autogravic" illusion thing   *267*
Lethal scenario 3: Flying low   *270*
  Case study: Low-altitude antics lead to tragedy   *270*
  Special skills are required   *271*
Lethal scenario 4: Midmission changes   *272*
  Case study: A few minutes to play   *272*
  Case study: Descent into disaster   *273*
  The "4C" approach to changing situations   *276*
A final perspective on lethal setups   *276*
Chapter review questions   *277*
References   *278*

14 Flight Planning: Discipline at ground speed zero   *279*

Knowns and unknowns   *279*
Case study: A tanker with a view   *280*
Flight planning as an art   *281*
Case study: Short planning—long landing   *282*
Disciplined planning means knowing yourself   *285*
  Medical airworthiness   *285*
  Psychological airworthiness   *285*
  Dangerous minds   *286*
  Subtle implications and personal minimums   *286*
  What scares you?   *287*

Disciplined planning means knowing your
  aircraft *287*
    General aircraft knowledge *287*
    Specifics on the individual aircraft *287*
    Fundamental planning criteria *288*
Disciplined planning means knowing your team *288*
    Team formation and participation *289*
    Briefing *289*
    Interpersonal relationships *290*
    The positive side of conflict *291*
Disciplined planning means knowing your
  environment *291*
    Workload management *292*
Case study: A failure to prepare *292*
Disciplined planning means knowing your risk *296*
    Risk management *296*
    Identifying risk *297*
    Controlling risk *298*
    Summary of risk management *298*
    Formal risk management tools *299*
A final perspective on flight planning *299*
Chapter review questions *299*
References *300*

15 Chaos theory?: Structuring change in the cockpit *301*

When things change *301*
Two reasons why we react poorly to change *302*
Small inputs, big outputs *303*
Case study: Coincidence, chaos—or both? *303*
Reacting to change *308*
The hurry-up syndrome *309*
    Target seen but not hit *310*
    Doing something wrong—or maybe not at all *311*
    Errors of commission *311*
    Errors of omission *312*
    What led to the error? *312*
    Time compression leads to poor discipline *313*
Managing change: Operating in the fourth
  dimension *313*

The critical importance of time and tempo: Four critical
    questions *313*
        Do I need to act immediately? *313*
        Will it help—or hurt—to delay this decision? *314*
        What help do I need? *314*
        What is the worst-case scenario? *314*
        Tempo: The final element *314*
    More hints for managing change *315*
        Don't forget the basic four *315*
        Use the checklist *316*
        Keep an eye on capabilities *316*
A final perspective on in-flight change *316*
Chapter review questions *317*
References *317*

16 Flight discipline in action *319*

    Case study: Prepared to use all available resources *319*
Operation Eagle Claw *322*
    The plan *323*
    Execution *324*
    The end of the plan *327*
Case study: Black Knight: A mission that couldn't afford
    to fail *329*
    The aircraft *331*
A final perspective *335*
Chapter review questions *336*
References *336*

17 Flight insurance: A personal program for improving flight discipline *337*

Denial—Not just a river in Egypt *338*
    Withdrawing from the discussion *338*
    Procrastination *339*
        Selective perception 339
        Rationalization 340
        Egotism 341
    A final word on denial *342*
Professionalism *342*

Becoming a more disciplined flyer   *347*
   Step 1: Licensing yourself for change   *348*
   Step 2: Flight planning for better discipline   *348*
   Step 3: Recording a baseline   *349*
   Step 4: Double check your flight plan   *350*
   Step 5: Establishing waypoints   *351*
   Step 6: Record and reward your progress   *352*
   Step 7: Arrive intact, enjoy your success, and set a
     new goal   *352*
A final perspective: Some practical reminders   *353*
   Understand and use your checklist   *354*
   Know your resources and be able to tap into them under
     stress   *354*
   Overlearn emergency procedure, regulations, and
     systems   *355*
   Don't bite off more than you can chew   *355*
Chapter review questions   *355*
References   *356*

Appendix A   *357*

Automation-related aircraft accidents and incidents   *357*

Appendix B   *385*

Controlled flight into terrain (CFIT)   *385*

Glossary   *391*

Bibliography   *395*

Index   *401*

Acknowledgments

I wish to express my sincere appreciation to all those who have assisted me in completing this project. A particular debt of gratitude is owed to those who contributed to the work through active writing or allowing earlier work to be utilized here, especially Charles Billings, Sherry Chappel, Ron Westrum, Corby Martin, Pat Barker, Randy Gibb, and Dave Wilson of Hughes Training Division, all of whom contributed major pieces to the project. I would also like to thank Neil Krey, Vince Mancuso and all of the members of the CRM Developers Group for insights and ideas, which added greatly to the depth of the book.

I would also like to publicly acknowledge the steadfast support and assistance of Shelly Carr, my acquisitions editor at McGraw-Hill, who routinely goes beyond the call of duty to help a struggling and often confused author.

This project could not have been completed without the support of my employers, the United States Air Force, whom I continue to serve with pride and enthusiasm, and specifically Colonel Carl Reddel and Colonel Mark Wells, my history and airpower mentors, respectively. In a wider sense, I could not have arrived at a station from which this work could be completed without the direct support of the Air Force in helping me achieve both the education and practical experience which were the foundations for this book.

Finally, I thank my wife and best friend Shari, for hours of editing, word processing, and much, much more. My boys Jacob and Trent were also instrumental in allowing this work to be completed, primarily by catching enough trout in the spring to allow me the summer "off" to finish. As always and in all things, I thank the Lord for the creative energy and sense of purpose to pursue productive activities.

List of contributors

Patrick Barker
United States Air Force

Sherry L. Chappell
Program Manager for Human Factors Services,
Delta Air Lines

Randall Gibb
United States Air Force

Corby Martin
United States Air Force

Ron Westrum
Eastern Michigan Universtiy/Aeroconcept

Dedication

To my fallen comrades:

Mike Moynahan

Tim Cookson

Scott Genal

Paul Ziemba

Zen Goc

Clay Smith

Dennis Rando

Glenn Comeaux

Pace Weber

Glenn Profitt

I'm tired of going to funerals.

Preface

The Mad Hatter's perspective: A parable on aviation professionalism

It had been many years since I had read Lewis Carroll's *Alice's Adventures in Wonderland,* but as I had burned out on reading professional journals and flight manuals, I thought a bit of fantasy might be just the trick for an aging mind. As I came to Chapter VII, you know it as the story of the Mad Hatter's Tea Party, I had a sudden lucid moment, rare these days. You might even call it an "ah-ha" type of phenomenon; the light bulb came on. Many of the grave issues in aviation training came together in a single, crystal-clear insight. I can relate it to you in a single sentence. You can't fix the broken watch no matter how good the butter is.

Now before you accuse me of sampling some of the smoking Caterpillar's hookah-pipe stash, let me explain. You may recall the story of the Mad Hatter's Tea Party, a festive affair with all the finest character's in attendance, March Hare, the Dormouse, Alice, and, of course, the Hatter himself. During the conversation, the March Hare is deeply disturbed about his broken watch, which he had sought to repair by applying butter to the inside. The Mad Hatter is quite perplexed by the Hare's action, and forthrightly tells the babbling bunny, "I told you butter wouldn't suit the works!"

The Hare is upset, "But it is the best butter!" confused as to why such a high-quality product had not repaired his watch. And then it hit me.

We have been trying to fix our watch with the wrong product, and we don't understand why it won't work. The watch of which I am speaking, of course, is aviation human factors, that toughest of watches to repair. We have applied all kinds of butter, ointments, education and training programs, and whatnot to the challenge of reducing human-error accidents, and while we have had some modest improvements in human performance, the percentage of mishaps caused by human error remains relatively stable. We haven't fixed the watch. What is needed, I believe, is something different.

The good news is that we have many highly educated and capable experts looking at every individual component of our watch. We have cleaned and polished each part, and we have even begun to understand what makes the watch work. Consider the following description of the evolution of aviation human factors training from Neil Krey, a human factors expert with Hughes Training Inc. He is speaking to the professional training community, but his vision of the future is valuable to us all.

> It seems to me that there are three eras of training that we are progressing through in our industry (aviation human factors). The first era came as we realized that training was important and we made it mandatory. To define the requirement, we said that you must train every X months, and the training must consist of X hours covering X topics. This was the event-based training era. Many of us are living in this era today.
>
> More recently, we recognized that event-based training didn't necessarily provide any assurance that crew members were maintaining proficiency over the long term. To address that, we required a formal definition of proficiency standards and developed a process to go with it. This Advanced Qualification Program (AQP) is designed to ensure continuing proficiency throughout the career of an airman. Some of us are now in the middle of the proficiency-based training era.
>
> The proficiency-based training era shares a weakness with the event-based training era, however. The weakness is that a single standard is used to evaluate proficiency regardless of whether you are on your very first qualification check for a new aircraft or the hundredth recurring check. Neither era provides opportunities for, or expects, professional growth during an airman's career. Such expectations rely on our professionalism, not the regulatory requirements.
>
> So I propose that we need to move toward a new era in training—the growth-based training era. In this era, we will

not encourage professional growth. We will require it. And this brings us full circle back to the basic premise of Tony Kern's book Redefining Airmanship. (Redefining Airmanship (Kern 1997) argues for a shared set of fundamental criteria to define professional airmanship. His ten elements of airmanship are derived from a historical analysis of success in aviation.) For now, each airman must create their own growth plan, but in the future, those of us in the training department may be tasked with providing support for those efforts.

This insight does an outstanding job of prescribing the characteristics of the watch we need to build, and cuts to the heart of the subject at hand—personal discipline to grow. The future of aviation human factors lies primarily with the discipline of individual flyers, not high-powered training programs, 3D simulation, or advanced technology aircraft. Although it would be unjustified to fully expand the metaphor to say that everything we have done to date was as useless as butter in a broken watch, we have missed the point that individual discipline underscores all professional growth. And that is what this book is all about.

Words mean things

I'm not quite ready to abandon the Mad Hatter's perspective just yet. Later at the tea party, the Mad Hatter chastises Alice for her imprecise use of words. The Hare joins in and tells her to "say what you mean." Alice's reply infuriates her table mates. "At least I mean what I say—and that's the same thing."

"Not the same thing a bit!" cries the Hatter. "Why you might just as well say that 'I see what I eat,' is the same thing as `I eat what I see'."

"You might just as well say," added the March Hare "that 'I like what I get,' is the same thing as 'I get what I like.'"

"You might as well say," added the Dormouse, which seemed to be talking in its sleep, "that, 'I breathe when I sleep,' is the same as 'I sleep when I breathe."

Click—light bulb number two came on in my head. In the constant struggle with hazy concepts in aviation, the need for clearly communicating what we do know is essential. Clear definitions of standards for terms such as airmanship, judgment, and flight discipline are far more than exercises in semantics. Without a common understanding of these critical terms, we may not be able to move forward in human factors at all. For if we know not what we seek, then all roads lead...to where?

Movement is not always progress

Alice had finished a rather odd conversation with the Red Queen in the forest, when much to Alice's surprise, they suddenly began to run. Alice relates, "The curious part of the thing was, that the trees and all other things round them never changed their places at all: however fast they went." Still the Queen kept crying "Faster! Faster!...Don't try to talk!"

"Are we nearly there?" panted Alice. "Nearly there!" the Queen repeated, "Why we passed it ten minutes ago. Faster!"

When the exhausting gallop was finished, Alice found herself next to the same tree from whence the run had began. "Why, I do believe we've been under this tree the whole time! Everything is just as it was!"

Movement is not necessarily progress, at least not on this side of the Looking Glass. Personal growth activities should not be mistaken for personal growth. You must begin with a clear notion of where you are and where you want to go. This book is written to help you identify a personal path of improvement by establishing an unmistakable starting point.

Time for the tea party to end

In perfect harmony with Alice's version of time in Wonderland, let me end at the beginning. When Alice first sat down at the table with her new acquaintances, the Mad Hatter looked up and asked the following riddle. "Why is a raven like a writing desk?" For the remainder of the discussion, Alice is preoccupied with finding the answer to the elusive riddle. In the end, the Hatter asked, "Have you guessed the riddle yet?"

"No, I give it up," Alice replied. "What's the answer?"

"I haven't the slightest idea," said the Hatter. "Nor I," said the March Hare.

The point here is that we all fall victim to being distracted by areas of personal interest, as opposed to more fundamentally important questions for our aviation safety and growth. Although individual agendas may not be as insignificant as the riddle with no answer posed by the Mad Hatter, it brings me to my most fundamental point, flight discipline is the cornerstone for all airmanship development, and as such should—better said must—receive appropriate attention. In preparing to fly, as in flying itself, distraction from the truly important can be, as the famous poster says "extremely unforgiving of any carelessness, incapacity or neglect." This is true at both the organizational and individual levels.

In summary, if we view human-factors training as a "not quite broken" but "not quite fixed" watch—which I will ask you to do for just a moment—then flight discipline is the mainspring (or battery if you prefer a more modern analogy) of the watch. Nothing runs well without it, no matter how bright and shiny the parts are. However, a skilled pilot without discipline is far more than a broken watch; he or she is a walking time bomb. We are never certain of his or her next decision, and no level of stick and rudder expertise or systems knowledge will make up for this missing piece. As we seek to improve our personal abilities across the spectrum, we must start with clear understanding of both the concept, and the need for—uncompromising flight discipline.

Introduction

There is only one kind of discipline—perfect discipline.
General George Patton

General Patton had it right, because discipline is by its very nature either self-sustaining or self-destructive. Nowhere is this more evident than in aviation, where even single small failures of discipline often play out in the most dramatic and tragic endings. At other times, it seems as if an aviator can literally "cheat death" by practicing poor flight discipline on multiple occasions without ever suffering a negative repercussion. Stephen Coonts, a former naval aviator and the author of *Flight of the Intruder* and *The Minotaur*, has written a piece called "The Philosophy of Luck" in which he addresses this apparent incongruity.

"I've never thought much of the old saying, "I'd rather be lucky than good." I think the good are lucky...There is no substitute for sound, thorough preparation to avoid or cope with foreseeable misfortune. People who drive straddling the centerline can get around a few curves, but sooner or later they are going to meet a Kenworth coming the other way. That's not just predictable; it's inevitable."

For far too long, we have talked around the central issues in aviation—personal responsibility and accountability. According to John Shaud, the former commander of the Air Force Air Training Command and the president of the Air Force Association, "in our zeal to improve, we have dissected aviation almost beyond recognition," referring to the mountains of research available on aviation human factors. But information does not always equal improvement.

I believe most of us can demand—and get—more out of ourselves than we are currently getting, and that all of us can improve our flight discipline. But simply wanting it is not enough. Working together, the community of aviators must begin to define the critical components of airmanship, and further to define standards of excellence in these areas. But even the combination of information and motivation is still not enough.

We must establish individual methods for accomplishing these objectives of greater knowledge, expertise, and discipline. We need a roadmap, which brings me to the point of this book. Flight discipline is the foundation of airmanship. It is where the journey towards excellence must begin, and this book is designed to help get you there.

Plan and overview of the book

In these chapters we will travel over a great deal of terrain, stopping to view the landscape where it appears most beneficial for illuminating the many characteristics of this phenomenon known as flight discipline. We will see the importance of this topic to organizations as well as individuals, and how even small lapses of discipline can have tragic effects. We will see that the path of poor discipline is a downhill slope, where any false step can be lethal.

Part One of the book establishes flight discipline as the foundation for all airmanship development and details research results which identify the severity of the problem of poor flight discipline. This analysis scrutinizes military, commercial, and general-aviation accident databases to demonstrate that poor flight discipline is a leading cause of mishaps, and that even so-called "good" aviators can fall victim to temptation, often with tragic results. Using several case studies, the section details the origins of poor flight discipline from a historical perspective and poses the question, "What causes a good aviator to forsake good judgment and give in to temptation to deviate from the proven path of compliance?" It is seen that there are three principle types of flight discipline deviations: regulatory, procedural, and deviations from organizational policy. A thorough discussion on each of these areas fleshes out the theoretical perspective of the intricate ideal of flight discipline.

Part Two looks at internal and external factors which often lead to problems with flight discipline. Separate chapters dealing with personality traits, hazardous attitudes, and external factors such as peer pressure are addressed.

Part Three of the book focuses on practical issues of flight discipline, from the critical role of instruction and mentoring, to issues of planning, communication, automation discipline, and attention management. Each of these areas is discussed as it relates to sound cockpit decision making. In addition, one chapter develops a set of scenarios which research reveals as the most likely to cause sudden breakdowns in discipline and judgment in individual aviators. Throughout the chapters, case studies will provide the opportunity for each reader to project themselves into the scenario and ask themselves what they might have done to counteract the pressures towards poor judgment. Each chapter begins with a short introduction to the specific aspect of flight discipline and is followed by one or more case studies that illustrate either positive or negative applications of the aspect in question. An evaluative summary provides guidelines for self-improvement.

The book concludes with a personal plan for safer and more effective flying, addressing an often overlooked aspect of aviation improvement—personal willpower and, if necessary, behavior modification and training for flight discipline. By breaking down the complex and often mystifying ideal of flight discipline into sound research findings and case studies, the book establishes clear checkpoints for individual accountability and improvement. Let's take a closer look at each of these areas.

Part One: The problem of poor discipline in aviation

In Chapter 1—Discipline: The foundation of airmanship—the author identifies the critical nature of discipline as the foundation for all airmanship. He reviews his findings on the structure of airmanship, as discovered through a historical analysis of successful airmen over the past 90 years. A case study of the US Air Force CT-43 crash, which took the life of Secretary of Commerce Ron Brown and 34 others, is used to illustrate the importance of discipline at all levels.

In Chapter 2—The costs of poor flight discipline—the author outlines the significance of violations of flight discipline, using research results and case studies to create a comprehensive picture of the severity of the problems associated with deviations from regulatory, procedural, and policy guidance. This chapter introduces the need for absolute personal discipline in flight operations.

Chapter 3—The letter of the law: Regulatory discipline—looks at regulatory deviations as a subset of poor flight discipline, pointing out which regulations are most often intentionally violated, and using case studies to point out the potential consequences. The chapter concludes with an evaluative summary of the cost-benefit ratio associated with poor flight discipline, and the argument for better discipline becomes clear.

In Chapter 4—The problem with shortcuts: Procedural discipline— the author highlights the importance of the basic procedural knowledge, skill, and proficiency in overcoming the temptation to deviate from established procedures. Using case studies which illustrate the risks involved with poor checklist discipline, the chapter concludes with an evaluative summary which suggests ways to obtain and maintain procedural discipline.

Chapter 5—Organizational issues for flight discipline—illustrates the conflicting nature of organizational policies, starting with the FAA and working down to the individual organization under which a pilot operates. Issues such as organizational culture and climate are discussed, as well as recommendations for creating a healthy climate for disciplined behavior. Organizations are seen as confronting opposing agendas of safety and mission accomplishment, which often place the pilot in the middle of a time-constrained and high-risk situation, with life-and-death consequences.

Part Two: The anatomy of flight discipline

Chapter 6—Personality factors and flight discipline—looks at personality types and traits and how they impact the behavioral characteristics of the pilot. Personalities are shown as fixed and unchangeable, so the emphasis is centered on self-awareness as the key to good flight discipline and points the reader to excellent outside sources for further study and development.

In Chapter 7—Hazardous attitudes—the author reviews the literature and research on internal factors that can be modified, pointing out several well-documented attitudes that can start the pilot down a path from which he or she may not recover. Antidotes for the attitudes are identified, and key identification characteristics are presented to assist the pilot in identifying hazardous attitudes in others.

Chapter 8—Peer pressure—examines one of the key causes of lost discipline, identifying sources of peer pressure as well as techniques for resisting it. Case studies include the tragic story of seven-year-old Jessica Dubroff, and other vignettes which illustrate how powerful a force peer pressure can become.

Part Three: Practical issues for flight discipline

Chapter 9—Guiding lights: The critical role of mentoring and instruction to flight discipline—looks at the roles of instructors and mentors in the makeup of an individual's flight discipline. Using gripping and often emotional testimonials from experienced instructor pilots, this chapter shows how even a single deviation by a respected mentor can set the stage for disaster from others who see the mentor's actions as a template for their own. Instruction and mentoring of flight discipline provide the link from experienced aviators to future generations of pilots. Its critical nature is highlighted through case studies, and the chapter concludes with a lessons-learned section.

Chapter 10—Communications discipline—illustrates the crucial importance of effective cockpit communication, demonstrating along the way how poor communication can lead to breakdowns of discipline in the cockpit. In addition to inter- and intra-cockpit communication, special attention is given to the link between pilots and air traffic controllers; a thorough discussion of lost communication—perhaps the single most significant breach of communication discipline—rounds out the chapter.

Chapter 11—Automation discipline—goes into great detail in discussing one of the most important topics in modern aviation, the human-machine interface. For perhaps the first time, a user-friendly discussion of automation philosophies combines the perspectives of the designers, the organizations who use the technology, and the pilots themselves. This chapter covers the breadth of new technologies, focusing on everything from mode confusion on an Airbus 320, to tips for using a Global Positioning System receiver in a Piper Cub.

Chapter 12—Disciplined attention: The next best thing to a crystal ball—provides an in-depth discussion of attention discipline from multiple perspectives. Tips for avoiding traps and improving focus and concentration are interlaced with illuminating case studies. The chapter concludes with a pilot-specific set of recommendations on "how to build a crystal ball."

Chapter 13—Killing conditions: Common scenarios for breakdowns of flight discipline—takes a critical view of prevalent conditions which precipitate sudden losses of judgment. These discussions include the role of experience and the fatal character flaw of exhibitionism in the aircraft. In addition, distraction and cockpit prioritization are considered. Accident and incident analysis show that pilots often place too much emphasis on things that are far less significant than returning to terra firma in one piece.

Chapter 14—Flight planning: Discipline at ground speed zero—examines the problem behind most violations of flight discipline—the failure of the pilot to adequately understand and prepare for flight. A discussion of how prepared you need to be is accompanied by a Controlled Flight Into Terrain (how not to) checklist which is included as an appendix.

Chapter 15—Chaos theory: Structuring change in the cockpit—is an illuminating look at managing the rapid pace of change in the cockpit. Techniques for safety and efficiency are offered and the chapter concludes with a discussion of how aviators can use time to their advantage, instead of having it work against them.

Chapter 16—Flight discipline in action—presents three gripping case studies illustrating the big picture of flight discipline, where time-critical human decision-making tips the scales between success and failure, victory and defeat, and often between life and death.

Chapter 17—Flight insurance: A personal program for improving flight discipline—brings the book to its logical conclusion with a personal plan for action, outlining several principles for avoiding the traps associated with poor flight discipline. It starts with a "zero tolerance" policy for any intentional deviations, unless emergency conditions exist. It follows up this basic tenet of airmanship with techniques for modifying your behaviors to bring them in line with a new and improved understanding of flight discipline. By establishing clear guidelines for preparation, recognition of hazardous attitudes and scenarios, and in-flight decision making, this chapter summarizes proven approaches for a rock-solid foundation of airmanship—uncompromising flight discipline.

How to read this book

This book is intended to be utilized as a personal improvement manual or academic text in a classroom setting. As such, key words and concepts are italicized throughout, and chapter review questions are provided to stimulate discussion, gauge the depth of your understanding, and provide a summary of the main points of the chapters. It is not necessary to read this book in the order presented, although the first chapter provides the definition and follow-on case study which sets the stage for the more detailed aspects of flight discipline which follow.

Author's intent and disclaimer

This book represents the private views of the author and does not represent any official U.S. Government position. I make no claim to practicing—or even fully understanding—aviation perfection. However, for the past decade, I have made it my life's work to try and grasp the meaning of aviation excellence. As a career aviator, I had the experience to ask the right questions. As a historian, I possessed the tools with which to dig for the answers. Finally, as a trainer and educator, I am duty-bound to provide these findings in as clear and usable a format as God has given me the talent to provide. I seek then only to provide the means for personal improvement for any and all who wish to begin down this path towards personal growth and excellence.

Although I have logged nearly all of my flying hours in military aircraft, this book is not intended for a strictly military audience. The principles and case studies found inside these covers apply to all who fly, regardless of their aviation environment. Some seasoned aviators may find the discussions contained herein oversimplified; others may struggle with the psychological aspects inherent in any discussion of the human equation. But there is something here for all.

Finally, I challenge any and all who read this book to join in the effort towards greater understanding of the human factor as it relates to professionalism in the air. This can occur in many ways, from attending local safety events, to joining aviation associations and participating in the continuing dialogue of flyers, researchers, educators, and trainers at national and international conferences. None of us is as smart as all of us. Reach for the top; fly your best!

Part One

The problem of poor discipline in aviation

1

Discipline: The foundation of airmanship

Greater prudence is needed rather than greater skill.
Wilbur Wright, 1901

The critical importance of flight discipline

Discipline is the foundation of airmanship. With it, a flyer or an organization can safely and systematically build towards excellence. Without discipline, we cannot hope to mature to our full potential as aviators or aviation organizations. In fact, without a solid foundation of flight discipline, we are always on thin ice, consistently flirting with tragedy. Failures of flight discipline can—in a single instant—overcome years of skill development, in-depth systems knowledge, and thousands of hours of experience. Without discipline, none of these attributes can protect us against a sudden loss of judgment.

Failures of flight discipline cross all boundaries of aviation. They can be found in general aviation, the military, and even within the cockpits of major commercial carriers. Experience alone is not a guarantee against a failure of flight discipline. Discipline failures are often found in the most unlikely places—within the ranks of our most skilled and proficient aviators. High-time, experienced pilots fall victim to failures of discipline at almost the same rate as their younger counterparts, but often for different reasons.

Tragic flaws and unfulfilled potential are almost a proverb in aviation. How often have we heard of the "best pilot" someone knew going down unnecessarily—often due to a simple oversight or single-point failure of judgment? We shake our heads in disbelief and mutter something like, "If it could happen to ol' Bob, it could happen to any of us." We are right when we say this, but only to a point. Discipline is a learned skill, just like landing in a crosswind. If appropriately developed it can tip the scales well in our favor when Murphy strikes. Unfortunately, we seldom get the same kind of step-by-step training in formulating effective flight discipline as we are given in crosswind landings. This is a serious and potentially deadly oversight, which this book seeks to remedy. We will accomplish this by explaining what flight discipline is in considerable detail, why it is so critical to airmanship development, and how it can be learned and practiced on a daily basis. But before we delve too deeply into the intricacies of airmanship and flight discipline, perhaps we should briefly discuss the reasons behind educating the mind to fly more effectively.

Why study to be a better flyer?

Unlike the hard sciences, expertise in aviation cannot be dissected in the same way as, say, the chemical structure of a hydrogen molecule. Airmanship is at once an art and a science, and a complex one at that. While there is little doubt that experience is often the most effective way to develop an art, aviation must be viewed differently than painting or poetry for a variety of reasons.

First, of course, is the safety factor. An old instructor pilot of mine once said, "The college of hard knocks is a great learning institution, but the price of tuition is often an aircraft and a funeral." Point made. While the hard sciences have made tremendous progress over the past several centuries by starting where the last genius left off and moving forward, this process is much more difficult in a less precise science/art such as flying an aircraft. More difficult, but not impossible. We, too, can stand on the shoulders of those who have gone before to reach the next level of aviation expertise, but first we must discover and validate what the useful lessons really are, and then be willing to expend the time and effort to learn them.

The second reason for systematically building upon the lessons of others is the cost factor. Flying aircraft is damned expensive. To maximize the efficiency factor and squeeze the most out of every flying hour, we need to establish a condition of "readiness"—meaning mentally prepared to learn. This preconditioning is accomplished through such techniques as study and chair flying.

A final reason for looking to the past for establishing standards and guidance for present-day and future growth is the level of the challenge that aviators from bygone days faced. Although most of us hate to admit it, flyers who mastered the air with the technologies of yesteryear had to develop qualities that most of us do not have today. Trying to land with hot motor oil spraying back on your goggles, or navigating from Ireland to the Canadian Maritimes with only a bubble sextant or pressure navigation for course guidance required powers of expertise and concentration that we can only imagine today in our high-tech aircraft. Of course, occasionally—in serious emergencies—we must still call upon the sum of all of our abilities in modern aviation. Tragically, that call is often unanswered at the moment of truth. So it is appropriate—and the time is right—for us to take time with the aviators of earlier years and learn their secrets so that our new technologies can take us as individuals—like our predecessors—to the pinnacle of our human potential as flyers.

Flight discipline as a personal decision

It is critically important that aviators—as individuals—come to grips with exactly what is meant by this often-misunderstood term—flight discipline. This is not to say that organizations have no role in developing discipline. On the contrary, organizations must motivate and provide resources for greater individual discipline. Traditionally, poor discipline has been associated with "hot-dogging," "scud-running," or taking other intentional and unnecessary risks. But the full meaning of flight discipline goes well beyond this narrow view. On an individual level, we must internalize and assess our own big-picture approach to discipline. What tempts us? What do we overlook or take for granted? Where are our own personal quicksand bogs? What factors would it take to push us over the line into a region of poor judgment? How safe are we...really? The pages which follow seek to help each aviator answer these questions by defining and conceptualizing the complex idea of flight discipline, using case studies to point out the severity of the problem and providing research-based insights for recognition of high-risk environments. Recommendations for prevention and self-improvement are made throughout.

Flight discipline deals with the individual motivations of the pilot, and the purpose of this book will be to bring each of us face to face with circumstances that often compromise thorough planning and preparation, degrade situational awareness, and negatively impact good judgment. We can—and must—prearm and prepare ourselves to deal with the temptation to violate principles of good flight discipline when and where they occur—both on the ground and in flight. To effectively integrate the multiple talents required to become an expert aviator, we must first understand where discipline fits into the overall picture of expertise and airmanship.

Expertise and discipline

Expertise in aviation is often difficult to understand, assess, and therefore—to develop. Dreyfus and Dreyfus (1986) found that "experts" from many fields often have to rely on intuitive judgment to interpret guidance from multiple sources in making decisions. This is a strong hint that a rational and disciplined approach is critical to sound decision making. In a more focused study into expertise in aviation, aviation psychologists Helmreich and Foushee (1986) found the following characteristics, among others, present in their definition of an expert pilot/crew manager:

1 Recognizes his or her personal limitations.
2 Recognizes diminished decision-making capacity in emergencies.
3 Discusses personal limitations.
4 Encourages others to question decisions.

Each of these studies highlights the importance of individuals' abilities to self-regulate—to discipline—their own actions and approaches to decision making. From the historical perspective, the findings run closely parallel.

The airmanship-discipline connection

This section is a brief overview from the author's previous book and study on the nature and structure of airmanship. For a detailed and instructional discussion of airmanship as a whole, see *Redefining Airmanship* (McGraw-Hill 1997).

Good airmen share common traits. In a four-year study of the elements found in 153 successful aviation events over the past 90 years, several clues to the nature of good airmanship emerged. Perhaps the most significant finding was that better integration of multiple skills is the key to successful aviation outcomes. The questions "What is airmanship?" and "How do we develop it?" can now begin to be answered by looking back at clues from the past, as well as forward at the likely demands of future aviation. Interestingly, both lenses produce similar views of airmanship. Historically, great aviators tend to possess certain common qualities and characteristics, and a glimpse into the crystal ball of future technology suggests little change in the necessary attributes of the successful airman. The changes that have occurred over time appear to be changes of degree only and not fundamental shifts in the nature of what constitutes superior airmanship.

This historical analysis revealed three fundamental principles of expert airmanship, regardless of the time frame analyzed: skill, proficiency, and the discipline to apply them in a safe and efficient manner. Beyond these basic principles, five areas of expertise were identified as common among expert airmen. Expert airmen have a thorough understanding of their aircraft, their team, their environment, their risks, and themselves. When all of these elements are in place—the superior aviator exercises consistently good judgment and maintains a high state of situational awareness. All of these areas are illustrated together at Figure 1-1 to show the conceptual relationships. From this model, it is readily apparent that discipline is clearly the cornerstone of effective airmanship.

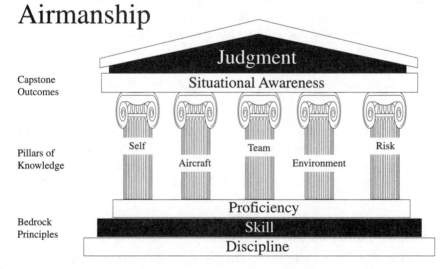

1-1 *The airmanship model. This historically developed model shows that flight discipline is the cornerstone for all systemic development. Failure to inculcate flight discipline into your personal development could result in a poor foundation of airmanship.*

An expert aviator combines many factors into a comprehensive whole and further understands that all of these factors are dynamic— requiring constant and calculated attention. Flyers with the right stuff understand the capabilities and limitations of themselves, their team, their aircraft, the physical, regulatory, and organizational environment, and the multiple risks associated with the flight. An expert flyer builds upon a bedrock of flight discipline, skills, and proficiency. No single-focus flyer approaches excellence. True expertise and continued growth demand mastery of a special set of skills that have come to be known as human factors. Total airmanship blends technical and tactical expertise, proficiency, and a variety of human factors to smoothly and effectively integrate the capabilities of the pilot and the machine. Total airmanship leads to improved situational awareness, fewer mistakes, increased operational effectiveness, improved training, and safer flying operations. By eliminating gaps in airmanship, a flyer is better able to handle the rapidly changing, dynamic, and often hostile environment of flight.

To be certain, there are other factors involved in airmanship. Opportunity, external situational factors, and just plain luck figure into the success or failure of nearly every decision that an aviator makes. This model does not attempt to explain all success in flight with a simple graphic representation. Nor does it suggest that good airmanship has always been a "balanced" approach, as suggested by the model. Many extremely successful flyers have compensated for a lack in one area with an abundance in another. History shows that this "single-focus" approach may work on occasion, but more often than not results in failure rather than success. Murphy is patient, and if there is a weakness to be found, he will find it sooner or later.

While all elements of airmanship may not be possessed by each and every aviator, in general terms—consistently successful outcomes rely on a mix of attributes. The ideal of a solid foundation of discipline, skill, and proficiency, combined with multiple bases of knowledge from which to make decisions, is sound guidance for today's flyers.

Discipline must be the first step

But developing total airmanship is not a simple learning task. Reaching this level of expertise must start from within and begin with a disciplined motivation to improve. Many have suggested that judgment and situational awareness are innate qualities, something one either has or does not have. Don't buy into this rationalization for laziness or complacency. Research and training experts agree that there are specific underpinnings to good airmanship that can be systematically developed through study and practice—and it all begins with discipline.

Objectives of this book

This book proposes to contribute to reducing flight discipline mishaps in three primary ways. The first purpose is by defining and conceptualizing the often foggy ideal of flight discipline. The second reason is to help the reader to appreciate and recognize likely scenarios for failed flight discipline. The third purpose is to provide a set of techniques for self-analysis and self-instruction for improving individual flight discipline, first in ourselves—and then by example, in others.

Defining flight discipline

For all of the reasons stated on the previous pages, flight discipline has taken on an expanded definition from the narrow view of poor in-flight decision making.

Flight discipline is the strength of will required to systematically develop all areas of airmanship and execute sound judgment in the presence of temptations to do otherwise, as well as to safely plan and employ an aircraft within all operational, regulatory, organizational, and common sense guidelines.

That's a mouthful, but an achievable mouthful if we break it down into bite-sized chunks. This will be accomplished and explained in the following chapters. This definition seeks to improve personal accountability and reliability, and aviation safety as a whole, by putting the emphasis for compliance and prevention where it belongs—with each individual aviator.

Recognizing hazardous scenarios

The second purpose is to provide an architecture for systemic recognition of likely scenarios and environments where flight discipline violations occur. This will be accomplished through research findings which show where, when, and by whom, flight discipline mishaps occur in the real world of aviation, coupled with case studies from the commercial, general, and military aviation environments.

Developing airmanship after discipline

Finally, the book provides a framework with which training programs can tie into the big picture of airmanship by showing flight discipline as the lowest common denominator—or the bedrock—of good airmanship. This will provide the "why" as well as the "what" to students who often question the relevance of current training or the rationale behind reading a self-improvement book. In the chapters that follow, a set of principles for self-analysis and daily operations are suggested for possible incorporation into a pilot's kit bag.

The many faces of flight discipline

Flight discipline exists in many forms and at many levels. It is found in the study habits or checklist adherence of each individual flyer, as well as in the decisions made in "top-brass" offices of large organizations such as the FAA, commercial airlines, or within the military. It is at once an attitude and a behavior. As an attitude, the disciplined mind harbors no room for complacency, failures of preparation, or unnecessary risk-taking. As a behavior, this zero-tolerance attitude manifests itself in our everyday decisions and actions. In one form or another, flight discipline exists within nearly every organizational entity we deal with in aviation, including the likes of weather forecasters, Flight Service personnel, and air traffic controllers.

Organizations are key in the development of flight discipline, as they establish a culture and climate within which individual flight discipline can either flourish—or perish. The following case study illustrates the many faces of flight discipline at both the organizational and individual levels. Unfortunately, most of the examples highlighted in this study are negative. These failures include organizational compliance with governing directives, failing to train aircrews sufficiently, and putting mission accomplishment ahead of peacetime safety on the institutional priority list. On the individual level, we see both external and internal factors at work upon the minds and bodies of the aircrew members, resulting in a failure to perform the most basic in-flight tasks. There is evidence of failed flight discipline during planning as well as in the air. In short, the following case study is illustrative of breakdowns across the entire spectrum of flight discipline, and for that reason, it is an ideal illustration of the need for us all to take a fresh look at flight discipline. These errors are identified and discussed through the analysis of a much-publicized accident in the mountains of Croatia that took the life of

35 people. The purpose of this example is not to cast blame or disparage individuals or organizations, but rather to show how single failures of discipline can lead to a deadly chain of events. The details that follow provide ample evidence of the critical importance of inculcating flight discipline at all levels of an organization.

Case study: Unforced errors

The case study of the factors leading up to the crash of IFO 21—a US Air Force CT-43, are illustrative and full of examples of failures of discipline at both the organizational and individual levels. It is presented to allow us to see the linked factors associated with flight discipline that we will discuss in detail in later chapters.

Fifteen seconds to impact

Captain A.J. Douglas (a pseudonym) must have felt something was wrong. Perhaps it was a glimpse of rising terrain through a break in the cloud cover. Maybe it was just a sense that the crew must have overflown the missed approach point by now, which they were having great difficulty identifying. Or perhaps it was a verbal prod from the copilot, Captain Tim Simmons (pseudonym)—something like, "Hey pilot, something's not right here; let's go missed approach." Although we will never know what actually occurred in that final fifteen seconds, we do know that for some reason, Captain Douglas added power and began a shallow turn to the right (Coolidge 1996 p. 30). While this intuitive correction was indeed appropriate, it was far too late.

At 2:47 P.M. local time on the third of April 1996, a United States Air Force CT-43A (Boeing 737-200), call sign IFO 21, slammed into the rocky slope of a mountain nearly two miles north of the intended airport at Dubrovnik, Croatia. All aboard were killed, including six Air Force crewmembers and 29 passengers—among which was the United States Secretary of Commerce, the Honorable Ronald H. Brown.

The significance of this accident lies not in the fact that a high-ranking U.S. cabinet member was killed, or that a critical error was made at the moment of truth, although there are lessons for us there as well, but rather in the series of unforced errors—or poor flight discipline—which put the crew in the position for the lethal mistake. Only the final error was of the split-second, time-constrained type we train for in our emergency-procedure simulators. The remainder of the errors—the ones that built the labyrinth with only one exit—were made out of inattention, complacency, or convenience. In short, they were failures of discipline.

To aid in the analysis of these failures, we must view the event through several lenses, both organizational as well as individual. Throughout the analysis, possibilities are explored that may or may not have actually had a direct bearing on the crash itself. More importantly for our learning purposes, is the fact that these events could have contributed to this mishap. We study all possibilities to gain the maximum learning potential from this tragedy, in hopes of preventing the next. Factual analysis of provable cause-and-effect relationships has already been expertly accomplished by a hand-picked team of investigators. While their task was to look for the absolutes—as learners our task is to look at the maybes and analyze the what-ifs. As with all of the case-study analyses in this book, there is no intent here to focus blame, but rather to learn from these errors to become better airmen and more responsible aviation organizations.

Background to a tragedy

The story of IFO 21 actually begins with a tale of two wars. Following the fall of the Berlin Wall and the end of the cold war, dozens of airfields formerly considered primarily as targets by American military aircraft suddenly became open to western traffic. Since western aviation personnel had little or no access to these formerly hostile airdromes, there were no instrument approach procedures that had been officially "approved" by western aviation standards. This led to some confusion as to what requirements had to be met for U.S. aircraft to fly into these newly opened countries and airfields. More on that later.

The flare-up in the former Yugoslavia only made matters worse. As the war raged on, certain pieces of critical terrain changed hands several times. One of these was the Cilipi airport in Dubrovnik, Croatia. Since the days of Alexander the Great, it has been common practice for a withdrawing force to take a few goodies—the "spoils of war"— with it as it occupies or retreats from enemy territories. Such was the case with the precision approach capability at the Cilipi airport, the primary airdrome serving the coastal city of Dubrovnik, through which hundreds of would-be peacemakers and various negotiators made their entrance into the still relatively unstable region. During the period of conflict from 1992 to 1995, the Instrument Landing System (ILS) and Very High Frequency Omnidirectional Range (VOR), and a third NDB were all stolen (Coolidge 1996 p. 41).

The end result of all of this was that the critical crossroads in this hot region were only serviced by one instrument approach procedure, a nondirectional beacon (NDB) approach—the least accurate of instrument approach systems currently in use at major airports. Furthermore, the approach that IFO 21 would be required to execute used two different NDBs to complete the approach and missed-approach procedure. The NDB to runway 12 at Cilipi approach is depicted at Figure 1-2. As can be seen from this depiction, an aircraft flying this approach would require two Automatic Direction Finding (ADF) receivers to complete both the approach and missed approach—since regulations prohibit "cross-tuning" a single receiver after the final approach fix. The CT-43A has only one.

While not exceedingly difficult, NDB approaches require constant attention and good approach planning to fly effectively and have a high potential and margin for both pilot and equipment error (Kelly 1996 p. 1). Theoretically, adequate crew and staff planning should include a review of the equipment required to fly this approach, which would have quickly identified the CT-43 incompatibility. It did not. We will look into that failure of discipline more deeply in a moment.

So the composite backdrop of the accident reveals a relatively confusing and somewhat uncertain playing field, primarily due to the recent opening of the eastern block countries as well as the ongoing war in the former Yugoslavia. These circumstances presented unique challenges for all levels of the organization tasked with ensuring safety while simultaneously completing the normal peacetime military missions of the United States Air Forces in Europe (USAFE) as well as the peace-keeping operations in the former Yugoslavia. In order to properly analyze this scenario, we need to first understand the relationships between the major organizational players.

The players

The hierarchical command structure between the U.S. Air Forces in Europe (USAFE) military headquarters and the 76th Airlift Squadron is fairly straightforward, although there are always several side players in

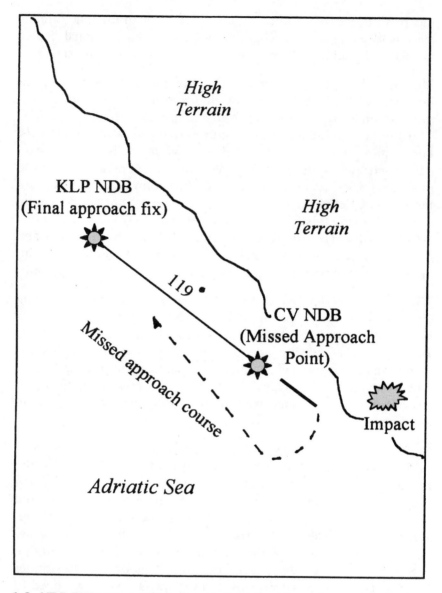

1-2 *NDB RWY 12 at Cilipi airport. It is clear from the approach depiction that two ADF receivers were required to accomplish this approach.*

any mission as complex as the one assigned to USAFE. Figure 1-3 shows the principle players in the decision process relating to airfield and instrument approach suitability for the newly opened countries which were not serviced by an approved Department of Defense (DOD) instrument approach procedure.

Organizational failures of discipline: A culture of noncompliance

At multiple levels of command and supervision, there were several opportunities to break the chain of events which led to this tragedy. It is appropriate and illuminating to start at the upper echelons of command and work our way down to the aircrew level in this analysis. We will see that a willingness to accept less than full regulatory

Organizational Players in the Crash of IFO 21

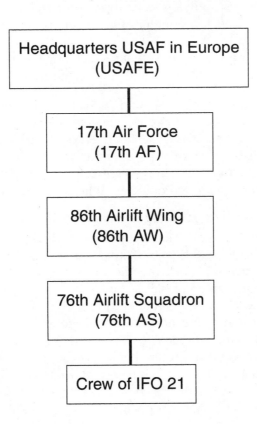

1-3

compliance occurs at all levels, and that an organizational culture of noncompliance may well have set the stage for the crash of IFO 21.

Although the formal investigation of the mishap focused a great deal of attention on the fact that the crew flew an instrument approach that did not meet U.S. Department of Defense standards, there were other factors that may have been equally—if not more significant. A myopic or single-focus analysis does not tell the entire story. Accidents are seldom this simple, and this one is no exception. We will limit our analysis to two organizational failures at echelons above the Airlift Wing. The first was a training failure—the failure to implement an effective Cockpit/Crew Resource Management (CRM) program as required by a regulation nearly two years old at the time of the accident. This was surely a CRM mishap if ever there was one. The second breakdown was one of enforcement, the inability of the command and supervisory positions to enforce their orders, which told aircrews to stop flying to unapproved airfields. It is clear from the investigation that the intent of all echelons above the wing was for the 86th Airlift Wing (AW) to stop flying to unapproved fields immediately upon notification, yet several months of noncompliance were allowed to occur before the cows came home on April third.

The remaining organizational and individual breakdowns of discipline will be discussed relative to elements internal to the airlift wing, concluding with the crew of IFO 21. Throughout the process there were opportunities to fix the problems and break the accident chain. Critical organizational decisions were made at each of these junctures, and these decisions turned out to have life-and-death consequences.

Organizational failure 1: Failing to implement CRM training mandate

Cockpit/Crew Resource Management is the flight crew's insurance policy against multiple human failures within the hostile flight environment. It teaches aircrew members to identify, access, and utilize all available resources to safely and effectively complete mission objectives and has been credited with documented and significant accident reduction wherever it has been thoroughly and systematically implemented. Although Air Force Instruction 36-2243, Cockpit/Crew Resource Management required aircraft and mission-specific CRM training for all Air Force crewmembers and had been in force for nearly two years, there was no major command CRM program in place at the time of the crash of IFO 21 (Coolidge 1996 p. 61). This failure to provide the required training resulted in the 76th airlift squadron attempting to develop CRM training on its own. Even though this was a noble effort by hardworking visionaries at the squadron level, this program clearly did not meet the requirements set forth by the existing regulatory guidance (USAF AFI 36-2243 1994 p. 6).

This curious disregard for an existing training requirement is only the first verse in a sad song of an organization that appeared incapable of fully implementing and enforcing governing regulations and policy. Should it have been a great surprise that subordinate organizations took the same cavalier attitude towards regulatory guidance?

Organizational failure 2: Failing to enforce the prohibition against flying unapproved instrument approaches

Pilots do not question the reliability of printed instrument approaches—at least not until recently. They rely implicitly on the accuracy of the depicted approach plate to provide the required margin of safety above obstacles and terrain. But as discussed earlier, the opening of new airports in previously hostile areas caused the military to question the reliability of these new published approaches and to require a comprehensive safety review. "Air Force Instruction 11-206, paragraph 8.4.1 requires that any instrument approach procedure not published in a Department of Defense or National Oceanic Atmospheric Administration flight information publication be reviewed by the major command Terminal Instrument Procedures (TERPS) specialist before it can be flown by Air Force crews" (Coolidge 1996 p. 51)—unless the weather is good enough for a visual approach. These distinctions are made to ensure that approaches developed by sources unfamiliar to the U.S. aircrews meet or surpass western safety requirements. To say that this new restriction would negatively impact on the mission of the 86th Airlift Wing—who serviced the entire European region—would be a huge understatement.

Before we go any further into this discussion, the reader needs to know that if the safety review process had been accomplished for the NDB Rwy 12 approach at Cilipi airport in Dubrovnik, Croatia, the TERPs specialist would have found an error of at least 400 feet on the minimum descent altitude (MDA) for the approach (Coolidge 1996 p. 45). If the directives had been followed, either the crew would not have flown the approach at all, or they would have flown it with correct altitudes that guaranteed adequate terrain clearance. In either case, they would likely be alive today.

The new approach guidance went into effect in November of 1995, and the 86th AW Operations Group Commander immediately recognized its adverse effect on his units' ability to perform its mission. The CT-43 had historically flown into many airfields where the only published approach is a "Jeppesen." Jeppesen refers to a company which merely publishes approaches given to them by host nations, and they are exceedingly clear on this point. In fact, Jeppesen publishes a disclaimer specifically stating that they "do not review or approve the adequacy, reliability, accuracy or safety of the approach procedures they publish" (Coolidge 53). Yet there seemed to be some confusion on this point by several organizational supervisors, as the following e-mail from a senior official clearly points out.

> *"The implications ($$ & manpower) of this 'new' guidance is significant...especially with all the...countries opening up!! What's the matter with Jeppesens...we've used them for years...don't our airline bro's use them all the time?"(Coolidge 1996 p. 58).*

Obviously, this official did not understand—or choose to accept—the concept that Jeppesen was merely a reproducer of approach plates of sometimes dubious reliability. Somehow he equated the term "Jeppesen" with "safe" or "reliable"—a point the company itself goes to great lengths to avoid.

In spite of the impact and confusion surrounding the new regulation, the major command headquarters had little choice but to notify their subordinate commands, including the 86th AW, of the new requirements. As soon as he received and reviewed the new regulation, the 86th AW Operations Group Commander (OG) requested a blanket waiver to allow crews to fly Jeppesen approaches to minimum weather criteria without the required review.

Organizational failure 3: Failing to take "no" for an answer

On 2 January 1996, the request to fly Jeppesen approaches without a TERPs review was denied by headquarters Air Force after a review by the Air Force Flight Standards Agency, and the rationale for the denial provided as follows.

> *"The MAJCOM TERPS review of non-DOD/NOAA FLIP products before they are authorized for USAF aircrew use provides a reasonable and prudent balance between operational flexibility and instrument approach development requirements....Some country/regions have approach design/flight*

*check procedures similar to those used by the USAF and prob-
ably require little in the way of "hands on" review. Other parts
of the world use less reliable practices when applying their
approach building procedures and would warrant a closer
look....Proper approach development is one factor aircrews
take for granted every time they fly an instrument approach.
When planning an approach, our aviators assume that if
they fly the approach as depicted, they will have adequate
obstacle/terrain clearance. The requirements outlined in
11-206 will help us maintain that high level of confidence—
we should keep them as they are" (Coolidge 1996 p. 53).*

On 23 Jan 96, a Major from the USAFE Operations training division
delivered the news directly to the 86th OG Commander (via e-mail).

*"HQ AFFSA (Flight Standards Agency) has denied the...waiver
request to AFI 11-206, submitted to authorize use of Jeppesen
approaches without MAJCOM review. The result of this is two-
fold. 86th AW FCIF (flight crew information file) 95-20, as
presently written, which authorizes the continued use of
Jeppesen approaches, will have to be rescinded....Presently,
with the waiver being denied, and upon rescinding the
FCIF, 86th AW aircrews will have no authorized Jeppesen
approaches to fly."*

Case closed? Not quite. About two hours after receiving the e-mail,
the commander at the Operations Group crafted his own message to
subordinate squadron commanders' offices, as well as information
copies to both echelons of supervision above him, stating in part,

*"This is a start—will await further guidance...on fields we've
never flown into. My view on this: Safety is not compromised
if we continue flying ops normal until approaches are
reviewed—then we rescind the FCIF" (Coolidge 1996 p. 55).*

Although the 86th AW commander initially expressed some reserva-
tions about the response, he did not countermand his subordinate's
order. In fact, the Accident Investigation Board Report determined
that "credible testimony shows that the OG commander's action to
not rescind the FCIF was taken with the concurrence of the 86th AW
commander" (Coolidge 1996 p. 55). Later, at an Operations Group
staff meeting "the consensus from the squadron commanders and
the chief of standardization and evaluation was that safety was not
compromised, and Jeppesen approaches could continue to be flown

pending...review" (Coolidge 1996 p. 55). In relatively short order, the 86th had thumbed its nose at higher headquarters. The FCIF that was directed to be rescinded on 23 January, was still in use on 3 April—the day of the mishap. It was rescinded on 4 April.

To quote country singer Lorry Morgan, "What part of no don't you understand?" The clear and timely dissemination of the waiver denial was categorically dismissed with the concurrence of all relevant commanders in the 86th AW. Although it is highly likely that the lower-echelon squadron commanders and chief of standardization/ evaluation felt considerable pressure to go along with the boss's recommendation to ignore the waiver denial and continue "Ops normal." What would prompt professional military officers—especially ones who had shown the mettle to climb through the ranks to command—to ignore clear directives from higher headquarters? In a word—the mission.

However, in the final analysis none of these officers flew the aircraft into the ground, and no number of organizational failures can adequately explain the breakdowns of discipline at the aircrew and individual levels. Yet the multiple breakdowns of the organization surely had some impact on the individual aviators who flew within them by establishing a climate of noncompliance, and this perhaps begins to explain the apparent disregard of directives which occurred at the aircrew level.

Aircrew failures of discipline

From the aircrew perspective, the portrait of the final flight of IFO 21 can be viewed as a collage of external pressures, busted crew rest, poor planning, mismanaged resources, violations of regulations, ignored or misapplied checklists and tech order procedures, distractions, lost situational awareness, and extremely poor judgment. For professional pilots, these guys had a real bad day. The accident investigation calls this collection of errors "uncharacteristic mistakes" which included "misplanning the flight...flying outside of a protected corridor...excessive speed and not having the aircraft configured by the final approach fix...beginning the approach without approval (clearance) and without a way to identify the missed approach point" (Coolidge 1996 p. 60).

When you add to these errors the fact that the copilot willingly broke crew rest on two occasions, that the crew was apparently unaware of the active airspace restrictions contained in the special mission instructions (SPINS), that they did not have the required equipment to fly the approach into Dubrovnik, that they had scheduled less than the regulation minimum ground time at Dubrovnik, and that they did not properly manifest the passengers on board the aircraft—it seems that the accident investigation report may have understated the issue when it found "behaviors indicative of a reduced capacity to cope with the normal demands of the mission" were present in the crew of IFO 21 (Coolidge 1996 p. 60, 61). In documenting the crew failures, it is not enough to make a laundry list of errors, or to wag a condescending finger of blame at the dead pilots—who were certainly trying to do their best. We must look into the potential causes of these multiple failures.

External pressures and organizational influences

The flight to Dubrovnik was clearly not a "normal" mission, but crew had flown many similar to it. The accident investigation states "external pressures to successfully fly the mission were present, but testimony revealed a crew that would have been resistant to this pressure and would not have allowed it to push them beyond what they believed to be safe limits" (Coolidge 59). What external pressures could have been responsible for such deviations from normal performance?

External factor 1: High operations tempo and culture of noncompliance

The 76th Airlift Squadron, and in fact all of USAFE, had been operating at a fever pitch for months, if not years, prior to the mishap. The demands of the mission, coupled with the military draw down that has impacted negatively on all branches of the service, simply left too few people to do too many things. Like all good soldiers, each echelon, from the senior commanders to the lowest ranking of the enlisted forces, leaned forward to get the job done. This operations tempo may well have been partially responsible for other accidents and mishaps as well. As we have seen, this combination of high ops tempo and mission-oriented commanders began to create an atmosphere of "can-do at all costs" and caused some to blatantly ignore regulatory guidance in the sacred name of the mission.

It is impossible to isolate aircrews from this command atmosphere. In fact, the 76th squadron commander had been recently relieved from command by the 86th AW commander for "a loss of faith in his leadership abilities." The relieved commander felt that he was relieved because "of his concern about flying General Officers and on allowing...missions to fly into potentially hostile fire areas" (Coolidge 1996 p. 59). At least one aircrew member was clear in his opinion that the firing of his squadron commander did have an impact on how he approached the mission. "It does force you to find...more ways to get a mission done. I don't know if that is good or bad, but it will get you to thinking of how to preclude those problems as quickly as you can" (Coolidge 1996 Tab EE-1/12). The views of the former squadron commander stressed safety over mission accomplishment.

The wing commander may have felt that such views were interfering too strongly with the ability to accomplish the wing mission. The human factors representative on the accident investigation board stated:

> "...there were indirect messages from the...wing that even though safety was properly acknowledged and advocated in the formal sense, mission accomplishment...was foremost. Examples include (1) that when there was a safety stand-down day in October 1995, the (squadrons) continued to fly scheduled missions, (2) the day following the mishap, the 76th did not stand down because missions had to be flown, (3) testimony that there was a constant struggle (with the wing) to lessen the flying per day so that the crews could train or obtain rest for the crews; and (4) they could not have a safety down day because there were too many missions to fly" (Coolidge 1996 Tab EE-1/10).

Although the accident investigation report stated that "the replacement of the squadron commander and its timing (four days prior to the mishap) were coincidental to this accident" (Coolidge 59), it seems difficult to believe that squadron crewmembers would not perceive the firing of their boss for stressing safety over mission accomplishment as anything but a clear message to get the job done.

External factor 2: VIP passengers

Although this unit regularly carried distinguished and high-ranking passengers, the combination of a presidential cabinet member and the flight into a recent combat zone carried certain pressures that were sure to affect the crew. On one previous documented occasion, the Commerce party had attempted to pressure a C-20 pilot to take a potentially unsafe course of action when scheduling difficulties were encountered (Coolidge 1996 p. 59). The pilot of IFO 21 had demonstrated his capacity to stand up to pressures such as these on previous occasions, including a recent flight where he had transported the presidents of Croatia, Bosnia-Herzegovina, and Serbia and had to divert from the intended destination of Sarajevo. It is unknown what pressures—if any—might have been generated by the Commerce party on this flight, but it is unlikely that it would have been enough to convince the crew to forsake safety for mission accomplishment by itself. As a contributor to the overall stress level on the crew, however, it could well have been a factor.

External factor 3: Multiple mission changes

A third external stressor that may have been more contributory was the multiple and late arriving mission changes to which the aircrew had to respond. Aviators are *controllers* by nature, and as such they abhor feelings of unpreparedness. It can be stated with some degree of certainty that the crew was agitated—more likely damned mad—about the last-minute changes to this high-profile mission. The accident investigation states "frequent changes to the mission itinerary contributed to the possibility of inadequate mission planning" (Coolidge 1996 p. 59). Once again, this may be a significant understatement, and the mission changes may have had implications which go beyond mere planning considerations. The multiple changes may have affected the basic physiological capabilities of the crew by contributing to broken crew rest by the copilot—for certain—and perhaps other crew members as well. We will discuss the implications of that in a moment.

Although many, if not most, military missions experience changes prior to and even during the mission, this flight experienced four separate major changes to the original itinerary, the last of which occurred on 2 April, the day after the crew had "completed" their official mission planning at Ramstein AFB. This may well have created a situation where the crew had to make difficult planning choices related to planning adequacy, thoroughness, and even regulatory compliance. In fact, it is quite clear that these multiple changes forced the crew to do some mission planning well into the night prior to a 3:30 A.M. mission show time on April 3rd.

Apparently, busting crew rest was almost commonplace in the airlift squadron. The accident investigation revealed multiple cases of crews who felt the need to violate crew rest minimums to get the mission accomplished. Although the former squadron commander had tried to discourage this practice, he stated "every now and then I hear a trip report come back in and the crew—the aircraft commander will write how they made it happen, four hours away from crew rest. I know some of the guys are still doing that" (Coolidge 1996 Tab EE 1/11).

Internal factors

While many of the factors that affected the crew of IFO 21 were beyond their control (such as the organizational climate of noncompliance, the potential pressure of flying a presidential cabinet member, and the multiple mission changes), there were also internal factors at play. The internal drive for success often found in high achievers, like the aircraft commander, can often manifest itself in negative ways. A hesitancy—or even inability—to say "no" to a tasking from above is one such hazard. Another phenomenon that may have been occurring was the fact that the pilot was on a rapid career upswing after a less-than-spectacular start in the airlift squadron. He may have viewed this "second chance" as something he wasn't about to mess up by failing to get this high-profile mission accomplished. Although each of these internal factors may have played a small role in the crew's sudden inability to cope with the mission demands, the most serious, and likely internal contributory factor, was self-induced fatigue by the copilot.

Internal factor 1: Fatigue

Fatigue can severely impair an individual's performance, and in the cockpit of an aircraft it can have lethal implications. Something caused multiple breakdowns on the crew of IFO 21, and based on the analysis of the copilot's sleep pattern the night before the accident, fatigue must be considered as a likely contributory—if not outright causal—factor.

A former Air Force wing safety officer who specialized in training for the night environment points out the seriousness of fatigue to military pilots.

> *"Fatigue is potentially the most serious human factor problem associated with...flying. Fatigue, fatigue recognition, quality sleep and fatigue management techniques should be a priority concern for everyone involved with...flying operations"* (Hoey 1992).

Hoey goes on to list several typical aircrew errors caused by fatigued pilots, including despair, short temper, reduction in the will to work, loss of appetite (which can lead to hypoglycemia), *loss of the desire to interact with others* (emphasis added), mental depression, a defeatist attitude, and loss of memory (Hoey 1992). This list of symptoms identifies several areas critical to successful and safe air operations and may reflect possible causes behind the "uncharacteristic mistakes" aboard IFO 21 cited by the accident investigation report. Perhaps the most significant is the finding of a reduced desire to interact with others, which could have been critical during those last few minutes.

Beyond the seriousness of impaired performance lies a more insidious effect of fatigue. Curt Graeber (1990) of the NASA-Ames Research Center states that fatigue not only contributes to serious performance errors, but that crew members often cannot accurately assess their own fatigue levels, thereby rendering them less capable of self-regulation. Graeber and Hoey's conclusions demonstrate that fatigue may well have a serious impact on the interaction required for effective crew resource management, exactly what was missing on the flight deck of IFO 21. Let's take a moment to review the copilot's actions the evening of April 2, the night before the fateful mission.

At 10:00 P.M. the night before the mission, the copilot made a call to the European Operations Center—a controlling agency for the flight—and requested "the latest mission change." He was verbally briefed on the change, which to his surprise added a whole new segment to the preplanned mission. After his phone call, he was faxed a copy of Change 4, but only the cover sheet survived transmission (Coolidge 1996 p. 11). This indicates that the copilot did not even begin to plan this added segment of the flight until less than six hours before show time for the mission—a clear violation of minimum crew rest periods. Pilot crew rest in the U.S. Air Force requires 12 hours off duty and eight hours of uninterrupted rest prior to showing for a mission. Anyone who has ever waited on technology to deliver a critical piece of information can just imagine the young captain's attitude as he watched the FAX stop after a single cover sheet rolled off the hotel machine. Should he call back for a retransmission? Should he wake the pilot and tell him they were being pushed too far, too late and recommend a safety-of-flight delay in the mission? Or did he say to himself, "The hell with it. I'll suck it up and get the job done. I'll hack the mission?"

There is no information available to indicate what the copilot did with the information after he received it, but a likely scenario would be that he proceeded to plan out the new segment of the mission. After all, why would he not have waited until morning to make the call if he did not plan on using the information that evening? If he did begin to plan the new mission segment, it would have taken a minimum of 45 minutes to an hour to put the new information together, meaning that he would have hit the sack sometime around 11:00 P.M. Assuming he went to sleep immediately, which is doubtful with the worries of the changed flight fresh on his mind, he would have had the opportunity for about four hours of sleep. But even this short period would not pass without interruption.

Sometime between midnight and one o'clock, a pilot who had recently arrived from Cairo called upon the IFO 21 copilot to return some personal items he had brought back from his trip. They talked for a few moments and the visiting pilot also gave the copilot some mission planning materials the crew had prepared, trying to help out because they knew the crew IFO 21 was receiving late changes to the mission. Now if he went to sleep immediately, he would now be able to add perhaps two hours of sleep to the less than one hour he had already gotten. Although the accident report states that "it could not be determined if the copilot had sufficient sleep," it is clear from the testimony that he did not—at least in terms of Air Force regulations. How significant was this? After all, it was only a single crewmember and it was just one night.

Wilkinson (1965), a noted sleep researcher who studied human performance degradation following periods of sleep debt, noted that effects of sleep loss vary widely between individuals from essentially no effect to an almost complete breakdown of performance. In short, the aircraft commander of IFO 21 might well have been flying solo and not known it. On a normal mission, he might have hacked it, but the pressure was on and the number of distractions present on this approach would have challenged a well-rested and fully functional crew.

Internal factor 2: Pressing

If the aircraft commander of IFO 21 may have been flying with a partially impaired copilot due to fatigue, he may also have been competing with himself. He had recently seen a rapid upturn in his flying career progression and may well have been trying to demonstrate that he deserved it.

It hadn't always been so. After Capt. Douglass's arrival at Ramstein in 1994, the squadron commander had noted that he "did not display adequate procedural knowledge for upgrade to aircraft commander" (Coolidge 1996 p. 35). In fact, the commander did not approve his upgrade during his entire eight months in command. However, about five months later, in October of 1995, the young captain finally upgraded to aircraft commander. Less than three months following this upgrade, he was granted a waiver by the operations group commander (for insufficient flying hours as an aircraft commander) and was upgraded to instructor pilot (IP). He completed his instructor checkout on 15 February 1996, and less than *one week later*, the operations group commander approved another waiver of requirements for his upgrade to flight examiner status—the military equivalent of a check airman. So while this aviator had labored in obscurity for nearly 13 months as a copilot, his fortunes had changed dramatically recently. In less than four months, he had upgraded sequentially to aircraft commander, instructor pilot, and evaluator pilot. Although he may have been a late bloomer, Capt. Douglas was on his way now. As the lone evaluator pilot in his squadron, he knew he was viewed as the guy who could get things done. He had come a long way in a short time, perhaps too short. This scenario of rapid advancement may well have set the stage for a hazardous attitude known as *pressing*.

Pressing is defined as an unwarranted—and occasionally obsessive—drive to accomplish flight objectives. It has also been called a "can-do" attitude, get-home-itis, get-there-itis, or mission-itis. By any name, it can lead to unsafe conditions associated with poor risk management. When a pilot presses, he or she places more emphasis on mission accomplishment and less on safety. The implications are obvious. But would this scenario drive a normally good pilot into a region of bad judgment, one that could lead to the incredible series of "uncharacteristic mistakes?" Perhaps not on its own, but when coupled with a fatigued copilot, and a few unexpected distractions...

Internal factor 3: Distraction

One of the greatest enemies of the aviator is channelized attention, or the inability to rapidly scan and process multiple inputs, commonly referred to by pilots as *crosscheck*. There can be many reasons for channelized attention and lost situation awareness, but the most common is simple distraction—a phenomenon which the crew of IFO 21 was about to deal with in abundance. The accident investigation report explains the source of these distractions. "During the flight from Tuzla to Dubrovnik, the mishap crews misplanning of the route caused a fifteen minute delay in the planned arrival time (an unpardonable sin when transporting VIPs). Pressure may have begun to mount for the crew to make the scheduled arrival time, especially because responsibility for the delay now rested with the crew. As IFO 21 neared the final approach fix, there were two additional distractions: a delay in clearance to descend from 10,000 feet and external communication with a Croatian aircraft, 9A CRO" (Coolidge 1996 p. 60).

Testimony indicates that as IFO 21 approached the final approach fix, the pilot of 9A CRO asked them to switch frequencies and proceeded to explain an "unpublished circling procedure" that he had used to get the U.S. ambassador to Croatia and the prime minister to Croatia on the ground only an hour earlier. It appears that the aircraft commander was hand flying (autopilot off) the aircraft and simultaneously talking to the Croatian pilot. The copilot was talking on the tower frequency and most likely running the checklists. Neither were adequately preparing to fly an NDB approach to Cilipi airport in bad weather.

Final approach

The post-accident analysis of radar tapes and aircraft wreckage indicates that the following sequence of events took place as IFO 21 passed the final approach fix. The aircraft crossed the final approach fix without clearance and approximately 80 knots above the flight-manual final-approach airspeed of 133 knots. In addition to being hot, they began tracking approximately nine degrees left of the final approach ground track. The copilot was not backing up the aircraft commander with his navigation instrument settings, and neither pilot had any way of identifying the missed approach point.

At this point, still four minutes from mountain impact, the crew was clearly well behind the aircraft. In addition, the high airspeed was limiting their time to fix the problems and salvage the approach. A well-disciplined and normally functioning aircrew should have realized the danger and executed some version of a missed approach at this time—but the crew of IFO 21 pressed on. The crew eventually slowed the aircraft to 150 knots and descended to the minimum descent altitude (MDA) of 2,150 feet. But these actions were being taken at the expense of accurate course guidance. The aircraft was still tracking nine degrees left of course, the weather was poor, and Murphy was waiting patiently on a 2,300-foot peak less than four miles away at their twelve o'clock position. Simply stated, the crew had broken down as a team entity, and the pilot's individual crosscheck was failing.

The missed approach procedure for Dubrovnik requires a right turn and a climb to 4,000 feet and is identified and executed at the "CV" NDB locator. Post-accident analysis found that the single ADF receiver on board IFO 21 was tuned to the KLP beacon—which was required for course guidance. In the absence of a second ADF receiver, the crew was unable to identify the missed approach point, and as a result overflew it without executing the required procedure. Although there are several unauthorized procedures that the crew might have been attempting to use to identify the missed approach point, including timing, inertial navigation system coordinates, cross-tuning the single ADF receiver, and visual identification—whatever procedure they used, if any—failed them. The final failure of discipline had occurred, and the crew impacted the rocky mountainside more than one nautical mile past the published missed approach point, killing all aboard.

An analysis of aircrew actions

The crew of IFO 21 got behind the aircraft and never caught up. The lack of a complete crosscheck at regular intervals has been responsible for a multitude of pilot-error accidents (Nance, 1986). Fatigue and distraction appears to exacerbate this tendency. (Alluisi, 1967, 1972) found weighted tasks, those with high priority, caused the fatigued operator to attempt to maintain his performance on the task deemed most important at the expense of secondary, or less-important tasks. This is especially dangerous in aviation, where "less-important" tasks are just as potentially lethal as those considered "primary." In this case, the need to get the secretary of commerce on the ground by a certain time may well have been deemed the most important task. By focusing on it, the crew—quite probably degraded by the copilot's fatigued state—was unable to function up to normal standards. More significantly, they did not realize the danger of their degraded performance in time to save their passengers or themselves. A combination of a late descent, poor planning which added the pressure of a late arrival, and a relatively difficult approach set the stage for a breakdown of the basic crosscheck and checklist discipline required to fly a safe instrument approach. All of this could have been solved with a single trip around the holding pattern.

The deadly chain of failed discipline

From the moment that higher headquarters decided not to implement CRM training, they were making a decision to operate at a higher-than-necessary risk level. When the 86th Airlift Wing decided not to comply with the directive to stop flying Jeppesen approaches that had not been reviewed by DOD instrument specialists, they too made a decision which put all of their aircrews in a region of increased risk. The failure of several levels of oversight to ensure compliance on both of these decisions indicate that adequate checks were not in place at multiple levels of supervision.

On an individual level, the aircraft commander of IFO 21 allowed his crew to be pushed into a very small corner by accepting mission changes that they did not have time to adequately plan. This resulted in a failure to identify the fact that the CT-43 did not have the required equipment (two ADF receivers) to fly an instrument approach into Dubrovnik. There may have been considerable external and internal pressures at play, but as always in aviation, the buck stops with the pilot in command.

The copilot failed as a team member by not pointing these items out to his aircraft commander, and by violating clearly established crew rest criteria. As a result, he was not as sharp as he needed to be at the moment of truth. He did not adequately back up the aircraft commander on the approach, failed to accomplish required check-lists in a timely manner, and failed to advise the aircraft commander to go missed approach as the situation deteriorated and the crew lost situation awareness.

Good intentions and discipline

All of these decisions were made with good intentions. At the most senior level, CRM training was just not a high priority. Staffing was down, operations tempo was up, and there were just too few resources to go around. At the wing level, the mission came first. Each tasking was important, and the new restrictions got in the way of priority one—getting the job done. The pilots of IFO 21 were clearly aware of the heavy emphasis on the mission, especially in the wake of their squadron commander being relieved of command. They knew the importance of the commerce secretary's mission, and were just trying their best to be "can-do" team players. But good intentions are not sufficient rationale for poor discipline, and that is why this case study is so effective at introducing the multiple perspectives of flight discipline. As aviators—or those responsible for aviation policy—we must clearly understand and follow established guidance. We must practice sound flight discipline. The road to hell is paved with good intentions.

A final perspective on discipline: The tough questions

If Secretary Brown had been delivered in one piece, would we still view this event as a case of misplaced priorities—or as a positive demonstration of a "can-do" attitude? Would the aircrew have professionally benefited from a couple of letters of appreciation from the Secretary Brown's office, or gotten a wink and a pat on the back from the senior staff if they had successfully flown the "special circling procedure" given them by the pilot of 9A CRO? Would all be forgiven and forgotten if they had hacked the mission? Simply put, does the result of a decision—or string of decisions—determine the legitimacy of the process used to get there? Have we reached the point in our aviation decision making where the end truly justifies any means of achievement? Has the unwritten motto in aviation at all levels become "don't get caught?" Before you trivialize these questions, ask yourself how many "small infractions" you have witnessed— or perhaps been a part of—during your flying career that were necessary to "get the job done." This may be as small an infraction as a little "scud-running" to get back to the home airport or through to your destination—or perhaps something more severe. After you analyze your own flying history, ask yourself how you would sleep at night—if you were still around to attempt it—if the result had turned out like the one described in this case study.

One final point that should be made is that this case study represents *caricatures* of poor judgment. That is to say, the failures in this case were obvious, large and easy to recognize—they stood out like a cartoonist's rendering of W.C. Field's nose. But most failures of discipline are much more difficult to analyze—or even to identify, and it is these failures which are the most prevalent and dangerous to our day-to-day operations. Sometimes just acquiescing to a bad idea can be a fatal link. The British novelist and philosopher C.S. Lewis makes this point clearly when he states in *The Screwtape Letters*, "indeed the safest road to Hell is the gradual one—the gentle slope, soft underfoot, without sudden turnings, without milestones, without signposts" (43).

Thus it becomes the purpose of this book to provide milestones and signposts with which to see and correct our path towards greater discipline—the foundation of airmanship.

Chapter review questions

1 How does flight discipline fit into the overall picture of airmanship? Why?

2 Do you agree with the author's definition of flight discipline? Why or why not?

3 What do organizations have to do with an individual's flight discipline?

4 How many failures of discipline were pointed out in the case study?

5 Does the end result of a flight decision (either by the organization or the aircrew) justify the means of accomplishment in your flying organization?

6 Are there times when it is necessary to bend the rules to get the job done?

References

Alluisi, E. A. 1967. Methodology in the use of synthetic tasks to assess complex performance. *Human Factors.* 9, 375–384.

Alluisi, E. A. 1972. Influence of work-rest scheduling and sleep loss on sustained performance. In W. P. Colquhoun (Ed.), *Aspects of Human Efficiency*, pp. 199–214. London: The English Universities Press.

Coolidge, C. H. Jr. 1996. AFI 51-503 *Report of Aircraft Accident Investigation on USAF CT-43 73-1149*, vol 1., United States Air Force (USAF).

Diehl, A. 1992. Does cockpit management training reduce aircrew error? *ISASI Forum*, 24 (4).

Dreyfus, H. L. and Dreyfus, S. E. 1986. *Mind over machine: The power of human intuition and expertise in the era of the computer.* New York: The Free Press.

Graeber, C. 1990, February. The tired pilot. *Aerospace*, pp. 45–49.

Hoey, R. 1992, November. Fit to fly? Fatigue in the cockpit. *Flying Safety*, pp. 8–13.

Kelly, Bill. 1996. NDB's High Margin of Error. *Aviation Safety.* June 15.

Kern, T. 1997. *Redefining Airmanship.* New York: McGraw-Hill.

Lewis, C. S. 1961. *The Screwtape Letters.* New York: Touchstone.

Nance, J. J. 1986. *Blind Trust*. New York: Morrow Publishers.

Wilkinson, R. T. 1964. Effects of up to 60 hours of sleep deprivation of different types of work. *Ergonomics*. 1 175–186.

Wilkinson, R. T. 1965. Sleep deprivation. In O. G. Edholm and A. Bacharach (Eds.). *The Physiology of Human Survival*, pp. 399–430. New York: Academic Press.

2

The costs of poor
flight discipline

Self-control is the highest form of rulership.
Apocrypha: Aristeas, 222

The costs associated with human flight have always been high, but throughout the past century many dedicated professionals have sought to reduce it. It is ironic that, in spite of the Herculean efforts by many to make aviation as risk-free as humanly possible for modern day flyers, there remains an element within our community that routinely violates technical manuals, regulatory guidance, and organizational policy in the name of expediency, ignorance, ego, or simply for the thrill and the adrenaline rush. The costs associated with these failures of flight discipline are unacceptably high.

By yesterday's standards, flying today is a piece of cake. For decades, even the most disciplined and professional flyers literally risked their lives every time they took off. This was due in large part to the technology of the day, as well as the as-yet undeveloped airspace and instrument approach structure that regulates current flying activities. Yet in spite of—or perhaps because of—this extremely high risk factor, successful flyers from yesteryear took very few unnecessary chances. To put this in perspective, let's look at the approach taken by one of the giants of aviation, a man who saw the costs of operations as simply too high. But rather than accept the risks, he sought to reduce them. After reading the story of Captain Jepp, ask yourself why you would take an unnecessary risk for mere convenience. Where can we reduce the risks and costs of operations today?

Captain "Jepp"

This section is quoted and paraphrased with permission from Babette André's online article "Captain Jepp, the gentile air-mapping pioneer, passes away" from General Aviation News and Flyer found at http://www1.drive.net/evird.acgi$pass*222.../ ganflyer/dec201996/capt._jepp_dies.html

Elrey B. Jeppesen was born Jan 28, 1907, little more than three years after the Wright brothers flew at Kitty Hawk. He was issued a pilot certificate in Oregon in 1929, signed by Orville Wright. In 1930 he began his primary aviation career with a job flying mail between Salt Lake City; Cheyenne, Wyoming; and Oakland, California. This was not the most forgiving terrain to be flying over, especially at night, without the aid of reliable weather forecasts, accurate terrain maps, or radio aids to navigation. But the mail had to get through, and this was the highest-paying route—$50 per week and 7 cents per mile (14 cents at night)! "That winter began with 18 pilots flying between Oakland and Cheyenne. Before the season was out, Jepp attended the funerals of four of them (André 1996 p. 1)." It wasn't that these pilots were foolish or undisciplined. They were not attempting to take navigational or procedural shortcuts. It was simply a matter of a lack of—as yet uninvented—instrumentation and no reliable or complete terrain maps for pilots to use when determining their next course of action. The costs were simply too high, and Jepp set out to remedy this situation.

"Captain Jepp started recording field lengths, slopes, drainages, lights, and obstacles. He illustrated airport layouts and the terrain, and noted local farmers who had phones and would provide weather reports (André 1996 p. 1)." He quite literally established the first METRO weather system and developed the first set of "approach plates" in an attempt to protect his friends and himself. He knew that to survive they must fly in a more disciplined and systematic manner. Captain Jepp took no unnecessary risks, his disciplined approach is clearly evidenced by his description of how he mapped the elevation of Blythe Mountain, a particularly dangerous peak just outside of Salt Lake City.

"I climbed it with three altimeters strapped to my back, and I took the temperature when I went up. I took the readings to

the physics department at the University of Utah and had them figure the elevation. Then I added 500 feet to it to be safe. (Later developed technology showed that) I was off by 200 feet, so with the 500 added...I was 300 feet in the clear." *(André 1996 p. 1).*

There is much for us to learn from the methodical approach taken by this pioneer of disciplined flight operations. For Captain Jepp, it wasn't enough to climb the peak with a single altimeter—he took three. Although he was quite good at mathematics, and the equations for temperature conversion were available to him, Captain Jepp took his readings to the experts at the physics department to provide a level of certainty beyond which he himself felt adequate to provide. Finally, because he recognized the limits of the technology of the day, he added a 500-foot margin for error— "just to be safe."

Of course the rest of the story is the stuff of aviation legend. His mapping business—which he once offered to United Airlines for $5,000—grew from selling his "little black book" to other pilots for $10 a pop, to a multimillion-dollar profit margin per year. Today, more than 2,000 charts are updated every month at Jeppesen corporate headquarters in Englewood, Colorado.

Captain Jepp was laid to rest on December 2, 1996. Although the wind chill was below zero, three open-cockpit biplanes flew a missing man formation over the cemetery as a tribute to what he stood for and accomplished during his lifetime. But the legacy of Captain Jepp—and other early pioneers—is perhaps best shared by this letter from a young Elrey Jeppesen to his parents, who were worried about his future when he began flying the mail.

"Dear Folks,

I am studying hard and long most of the time and intend to do that as long as I can, but under no circumstances am I ever going to stop flying. I like it too well and when the time comes when I can no longer fly, I might just as well start looking for the next world.

So far as I can see, the most important thing in this world is to learn the art of living and that is what I have been trying to do for some time. I will not get very much out of life if I cannot fly."

Captain Jepp mastered the art of living—and flying—with a methodical discipline that may be lacking in many modern flyers operating in a far more forgiving environment."

The flight discipline cost-benefit equation

This chapter looks at the phenomenon of flight discipline from both the individual and organizational perspectives, and seeks to drive home two essential lessons. First, violations of flight discipline have both obvious and unseen costs to the organization within which an aviator flies, and to individuals themselves. Second, the potential costs invariably outweigh the potential gains from an event of poor discipline. The elementary conclusion to this equation is that it's not worth it to tempt fate with an act of poor discipline, either in the preparation or execution phases of flight.

Violations of flight discipline have an insidious, creeping effect on an aviator's good judgment, often affecting future courses of action. A single step down the slippery slope to ethical compromise—one small intentional infraction—can deflect an aviator's future judgment and willpower. Further, the full consequences of a single compromise may not be seen for months, or even years. Aviators who get away with bending a rule in one situation are far more likely to try it again in a tighter box of conditions, only to find themselves simultaneously out of airspeed, altitude, and ideas, wondering why they ever took such a foolish chance. If they could reflect from the grave, they would likely find the answer in a small "no-harm/no-foul" infraction earlier in their flying career.

Flight discipline violations are contagious. Whenever we see someone breaking a regulation or exercising poor judgment, many of us will begin—perhaps unconsciously—to lean towards an accommodating attitude towards this type of behavior. Some flyers will openly admire a daredevil maneuver and begin to wonder how they might find an opportunity to accomplish such a feat of daring. Because of the infectious nature of discipline violations, we share a common bond of trust to uphold standards, lest we drag others down with us.

The scope of the discipline problem

There are two ways in which we can look at the depth and breadth of the flight discipline challenge. The first is the most obvious and involves a statistical analysis of how many failures of discipline occur, to whom, and how often. Self-reported incidents and accident and mishap databases only hint at the magnitude of discipline violations, but they can help us see the tip of the iceberg. The second method for looking at the scope of the flight discipline problem is to look at how many ways it can affect other areas of our airmanship development. We will do this by discussing multiple ways that poor discipline impacts upon our training, knowledge growth, and situational awareness. Let's begin with a look at the statistics.

Statistical evidence

A host of statistical data suggests that human factor failures such as poor flight discipline remain responsible for a large percentage of mishaps, many of which are fatal. The following statistics seem to indicate that the scope of the problem is indeed large and crosses all boundaries of aviation.

Mishap data (Aviation Monthly Safety Summary and Report 1996, p.1)

According to the National Transportation Safety Board, total fatalities involving U.S. registered aircraft show a mixed set of results. On an encouraging note, total deaths from airline mishaps fell in 1995 to 175 from a total of 264 in 1994. The great majority of these—160—came in the American Airlines B-757 crash in Colombia in December. This accident was attributed to human error related to automation. Other fatalities from the commercial sector included 9 deaths attributed to mishaps on scheduled commuter airlines, and 52 fatalities from 76 accidents which occurred in on-demand air taxis. On the whole, the picture is improving for commercial air traffic, but there is obviously still work to do.

However, general aviation trends are disheartening. Fatalities rose from 723 in 1994 to 732 in 1995. There were 2,066 general aviation mishaps in 1995, including 408 fatal accidents. The general aviation accident rate per 100,000 hours flown was 10.33 in 1995, up from 9.09 the year before. To put all of these numbers in perspective, the general aviation accident rate is about 40 times the rate for scheduled air carriers operating under Part 121 of the FARs.

Commercial costs

In addition to the obvious cost to an airlines of replacing an aircraft and a crew, there are other, less obvious costs resulting from mishaps. Of course, all air carriers are covered by vast amounts of liability insurance, but you may be surprised at the uninsured costs that can accumulate as a result of a mishap. Figure 2-1 lists the multiple costs that can be associated with accidents.

Military costs

Historical trends in military mishap data also point to poor discipline and other human factors as the primary accident cause. The General Accounting Office (GAO) reported to the U.S. House of Representatives Committee on National Security in 1996 that an analysis of mishap data from 1975 to 1996 showed "human error was reported as a contributing cause in 71% of Class A flight mishaps (Air Force)...76% in Army mishaps...and a causal factor in 80% of Navy and Marine Corps mishaps" (GAO 1996 p. 25). Class A mishaps refer to mishaps where the damage is greater than one million dollars, the aircraft is destroyed, or there is a fatality or permanent total disability that results from the mishap.

Taxpayers shell out big bucks for military hardware, and the cost in hard-earned dollars is often overlooked in the aftermath of a media-hyped mishap. No one gets down to brass tacks like the GAO.

> *Since fiscal year 1975, the services report the cost of Class A flight mishaps at about $21 billion. The value of Class A losses has been fairly constant over the last six years, ranging from a high of approximately $1.6 billion in fiscal year 1993 to a low of $1.2 billion in fiscal year 1994. Even given that fiscal years 1994 and 1995 had generally low mishap rates (the measure usually reported for public consumption), the value of...losses still exceeded $2.5 billion during that time (GAO 1996 p. 17).*

As the shepherd of a considerable portion of the public treasury, the military must continually seek new and better ways in which to counter any acts of poor flight discipline. Although these dollar amounts do not correspond directly with acts of poor flight discipline, a large percentage have clear links to human error—many of which are failures of discipline. As military critics will be quick to point out, the costs of poor personal and organizational discipline could buy a lot of school lunches. While these data do not suggest that all fatalities or Class A mishaps are caused by poor flight discipline, a deeper analysis shows that many are—a fact supported by the much larger database of self-reports.

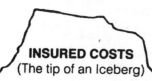

INSURED COSTS
(The tip of an Iceberg)

INJURIES

- Compensation for lost earnings
- Medical and hospital cost
- Awards for permanent disabilities
- Rehabilitation costs
- Funeral charges
- Pensions for dependents

PROPERTY DAMAGE

Insurance premiums or charges for
- Fire
- Loss and damage
- Use and occupancy
- Public & Liability

UNINSURED COSTS

INJURIES

- First aid expenses
- Transportation costs
- Cost of investigations
- Cost of processing reports

WAGE LOSSES

- Idle time of workers while work is interrupted
- Man hours spent in cleaning up accident area
- Time spent repairing damaged equipment
- Time lost by workers receiving first aid

PRODUCTION LOSSES

- Product spoiled by accident
- Loss of skill and experience
- Lowered production of worker replacement
- Idle machine time

ASSOCIATED COSTS

- Difference between losses and amount recovered
- Rental of equipment to replace damaged equipment
- Surplus workers for replacement of injured employees
- Wages or other benefits paid to disabled worker
- Overhead costs while production is stopped
- Loss of bonus or payment of forfeiture for delays

OFF THE JOB ACCIDENTS

- Cost of medical services
- Time spent on injured workers welfare
- Loss of skill and experience
- Training replacement worker
- Decreased production of replacement
- Benefits paid to injured worker or dependents

INTANGIBLES

- Lowered employee morale
- Increased labor conflict
- Unfavorable public relations
- Loss of goodwill

By permission of QANTAS Airways

2-1 *Insured versus uninsured costs of an aircraft mishap.*

Self-reported statistics

If self-reported statistics are any indication, we have a long way to go to cure the problem of poor flight discipline. The NASA aircrew safety reporting system (ASRS) is an extremely effective, nonattribution system whereby pilots, air traffic controllers, and anyone else associated with commercial or general aviation can report a significant event related to safety without fear of retribution. Although the system is available for use by military pilots, it is seldom utilized for their purposes. Nonetheless, these self-reports form the basis of a huge database of incidents through which aviation analysts can sift looking for clues as to the nature—and potential cures—of safety issues. Several statistical analyses hint strongly at the need for a new and improved approach towards flight discipline.

From 1988 to 1994, the ASRS database shows 126,225 reported incidents of nonadherence to rules, far and away the largest single category reported (ASRS Directline No. 8 1996 p. 11). In 1994 alone, there were 20,889 reported incidents of nonadherence to rules—or about 57 reported incidents per day. Since it is likely that less than one in five violations are actually reported, we can safely assume that several hundred pilots or flight crews break the rules *every day.* This kind of data makes you want to keep your eyes outside the cockpit a bit more, doesn't it?

A study of self-reported critical incidents from the Gulf War of 1991 illustrates that the military is not immune to the problem of poor flight discipline. In a study titled "A historical analysis of U.S. Air Force tactical aircrew error in Operations Desert Shield/Storm," at least 36% of self-reported incidents were failures of discipline related di-

Nonadherence to . . .	1994 Incidents	1994 % of Incident Base	1988–1994 Incidents	1988–1994 % of Incident Base
ATC Clearances	14,075	53%	90,280	53%
FAA Regulations	6,201	23%	35,290	21%
Published Procedures	4,448	17%	22,593	13%
Company Policy and Others	578	2%	2,556	1%
Total Unique Relevant	20,889	79%	128,225	75%
Irrelevant or unknown	5,524	21%	42,573	25%
Incident Base	26,413	100%	170,798	100%

2-2 *ASRS reported incidents. The sheer magnitude of reported incidents is cause for a pilot to sit up and take notice. Deviations from regulations, procedures, and policy are clear indicators of a need for greater attention to flight discipline.*

rectly to decision making. (Kern 1994 p. 58) The following example was typical.

> BACKGROUND: *Although no minimum altitude limits had been placed on missions, squadron leadership had stressed remaining above 8,000 to 10,000 feet and limiting reattacks on targets (to lesson the exposure to Iraqi surface-to-air missiles). On this mission the wing commander launched a maverick (missile) at approximately 10,000 feet, but coming off of the target he saw an enemy-armored personnel carrier with troops getting out.*
>
> BEHAVIOR: *The wing commander immediately reversed and rolled in to gun the target, pulling out at approximately 6,000 feet.*
>
> CONSEQUENCE: *While pulling out after the gun attack, the aircraft was hit by a surface to air missile (SAM), resulting in serious battle damage to the aircraft. The pilot was able to safely recover.*

So who are these wild-eyed crazies breaking the regulations with such reckless abandon? The statistics show that they come from all sectors of aviation—commercial, military, and general aviation. They occur on the flight decks of major airline jumbo jets, and in the tight cockpits of military F-16s or civilian Piper Cubs. In short, the typical rule-buster is just that—typical. In a humorous analogy from the movie *The Addams Family*, Wednesday Addams was asked why she hadn't dressed up for a Halloween party. She replied, "I'm dressed like a homicidal maniac. They look just like everybody else." The same can be said of violators of flight discipline.

Statistics can only tell us so much. The second method for understanding the scope of the flight discipline problem is by considering real-world scenarios. The following examples demonstrate how poor flight discipline impacts airmanship development, training, knowledge growth, and situation awareness.

Media representative gets the ride of his life (Hughes 1995)

The day began full of excitement for a local TV media representative. He was the envy of his colleagues because he had been selected to accompany the local fighter wing on a training mission that would include, he was sure, a great deal of excitement. He listened carefully as he was briefed on the specifics of the mission and the unlikely possibility of an emergency. He was strapped into the back seat of an A-7K—a two-seat training version of the high-speed, low-altitude, all-weather attack aircraft.

The flight lead and number four were part of a four-ship air refueling and air combat training mission. The A-7K (two-seater) was the flight leader with a local media representative as a passenger. The refueling was pretty interesting but uneventful, and the flight progressed to 2v2 air-to-air practice engagement. The media representative felt certain that this would be the highlight of the ride—he didn't know how right he was. On the fourth engagement, numbers one and four collided. Both pilots and the passenger ejected successfully. The aircraft were destroyed. The accident investigation revealed a series of failures of discipline at multiple levels.

The commander failed to establish and maintain a formal air-to-air combat training program in accordance with command directives. The preflight briefings were not conducted as required by both command and local directives and failed to define roles, establish specific objectives, set engagement criteria, and identify "knock-it-off/terminate" criteria. Formal element briefings were not held, in violation of local and command directives. On the fourth engagement, three and four (bomber and defender element) ingressed the target area undetected. After egressing undetected they "jumped" one and two (the fighter and attacker element). Lead condoned three and four's unbriefed and unanticipated actions and did not intervene to knock-it-off. Four engaged one, and their efforts degenerated into a "scissors," a dangerous and often no-win condition. After "killing" number two (simulated, of course), number three engaged the flight leader without directing four to disengage—another clear violation of regulation and procedure. At this point, the flight leader was attempting to defend against both three and four, all the while trying to give the media rep a good show. It was about to get better.

During this close maneuvering, number four made the final critical error of discipline—he diverted his attention from the other aircraft into his cockpit to reset the roll axis of his automated flight control system. Though the flight lead recognized the dangerous situation developing, he once again failed in his leadership role to call "knock-it-off" or take positive action to avoid the collision. The number four pilot reacquired the flight leader a moment too late to take successful evasive action and the aircraft collided.

There is no information available as to whether or not the media representative requested another ride so he could film a landing.

In the following ASRS report, a corporate aircraft copilot describes an incident illustrating that the problem of poor discipline extends into the commercial environment as well.

Air taxi risks collision to keep passenger dry (ASRS 1995)

We were taxiing eastbound across the main terminal ramp when ground control pointed out a conflict off to our left that was on a converging course with us at the intersection of the southeast portion of the main terminal ramp and the parallel taxiway to the FBO. The captain, who was taxiing and handling the radios, told ground control that we would give way to the other aircraft. Ground control acknowledged the transmission and instructed us to follow the other aircraft to the FBO. As we approached the intersection, I could clearly see that we were on a collision course and that we were just a bit too fast if we were going to give way. Fearing that the captain might have misunderstood the instructions from ground control, I looked over at him and could see him closely watching the other aircraft off to our left. I had my feet on the brakes but didn't use them since the captain had the plane. (I was told in flight training never to have two pilots on the brakes simultaneously, due to the possibility of losing all brake pressure.) I started to warn the captain verbally, but I didn't since I would have had to raise my voice considerably in order to be heard through his headset. (This plane didn't have an intercom.) I chose not to do this since the wife of a VIP was in the back, and I did not want to alarm her. Just as I was about to step on the brakes at the last possible minute to avoid collision, the captain shoved the throttles up and cut in front of the other aircraft, barely missing it. Needless to say, both ground control and the pilot of the other aircraft were very upset.

The captain's excuse for this inexcusable near accident was that he had a VIP on board. It was raining at the time and he didn't want the other aircraft to take the parking place closest to the FBO. Ironically, the VIP's wife was met at the plane with a car and didn't need to ever go inside. I finally

*said, "I thought you were going to follow him?" The captain
replied, "I changed my mind, sorry."*

In this incident the copilot was nearly as much at fault as the captain—
who was clearly way out of bounds. By failing to take assertive action
either by word or deed because "the wife of a VIP was on board," he
failed in his role as a part of the team. In this case, the only cost of
failed discipline was a little embarrassment and a tarnished reputation.
Others have not been so fortunate. Consider the cost in the next case
study, which also involved VIPs.

The cost to others

Failures of flight discipline often result in effects that go well beyond the perpetrator or his or her aircraft. For this reason alone, we
have a moral responsibility to each other—as well as the public at
large, to eliminate practices of questionable discipline in our own
flying habits and maintain a watchful posture on others. Pilots enjoy considerable freedom of action with which to address inflight
events, and the ability to operate safely and efficiently within this
flexible environment is an essential prerequisite of good airmanship. Airmen are given a sacred trust not to take advantage of a system designed to allow flexibility and creativity. Yet, unfortunately,
some still do. Consider the cost of poor discipline in this example.

Case study: A simple checklist oversight

On Tuesday 8 November 1994, a Beech P58 Baron rolled down Runway 21 at Great Falls airport in Montana. On board were three local-
area businessmen, executives from a local investment company who
were headed to eastern Montana on company business. The aircraft
was flown by an experienced 50-year-old pilot from a local aviation
company who specialized in corporate shuttle flights. Shortly after
liftoff, the pilot radioed the tower that "we need to go around and
land." He did not declare an emergency. The tower controller saw
the aircraft take off, climb to about 50 feet AGL, and begin a slow left
turn. As the Baron crossed over the frontage road for the airport, it
suddenly pitched nose down and crashed in a deserted field about
one mile from the point where it lifted off.

The accident investigation is illuminating as to the likely cause of the
mishap. *Aviation Safety*, a journal which summarizes NTSB accident
reports for its readers, described the sequence of events.

"Although the pilot did not state the nature of his emergency, it appears that the nose baggage door may have opened on takeoff. Witnesses saw something fall from the airplane as it began its turn, and investigators found a garment bag belonging to one of the occupants lying on the departure end of the runway." (Aviation Safety No. 22 1996, p. 10).

Further analysis revealed that the baggage door was most likely left in the unlocked position, although aircraft crash damage made it impossible to know for certain. On the Baron, it is difficult to tell if the baggage door is latched or just resting in the closed position, without physically checking the baggage door on the exterior inspection checklist. Imagine the predicament for the corporate shuttle pilot when he saw the first bag fly out of the baggage door just as he broke ground. He was most likely embarrassed and angry at himself, a mindset which may have led to the decision to make the low-altitude, low-airspeed turn. It was a fatal decision fueled by a simple checklist oversight. Flight tests indicate that the Baron series can fly with the baggage door open and suffer "no adverse control of performance problems" (*Aviation Safety* No. 22 1996 p. 11).

We are often desensitized to the human costs of accidents unless we happen to know someone who was involved. Mishap reports refer to "crewmembers" and "passengers," and we forget about the impact of their loss on the families and community from which they came. Occasionally, the news media will cover a mishap in a way which can bring some of this pain to the surface. But more often than not, mishap casualties become merely statistics to be briefed at the next safety meeting, obscuring the human costs. To give some human measure to the price that was paid for an overlooked checklist item, let's take a closer look at those who were lost in this mishap. Consider this list of victims and the associated extended human costs of the tragedy.

Eugene Lewis, 46, was the president of a large division within the company, but his loss was felt across the state of Montana. He was a volunteer advisor to many of the state business schools, the Chairman of the Montana council on economic education, and was a member of the Montana Ambassadors—an economic promotion group. He had been married for 28 years and had two children.

Don Knutson, 49, was a senior vice president and trust officer in the company. The chairman of the company said that "He frankly brought in more trust business than anybody else in the area. He was a fixture in the corporation." He was also an active leader in the local Boy Scouts and had three athletic sons. Dad hardly ever missed a game.

Bob Bragg, 54, was also a high roller in local business as the president and CEO of another investment entity. Like the other victims, his loss extended well beyond the workplace. He was an active volunteer with the Great Falls symphony, a Rotary director, and had been named Montana Ambassador of the Year in 1993. He also left behind a family with three children.

Finally, the pilot Harold Graff, 50, was known throughout the community as a "nice, clean-cut guy" who left behind five children and a wife who was just recovering from an automobile accident. Although three of these individuals held key management positions in a vital and dynamic corporation, the president and CEO of the parent organization summed up the relative costs to the company in this manner.

> *"Money and jobs are secondary. These people were all in key management posts so we are left with a business dilemma that is totally overshadowed by the personal loss. It's (the considerable loss to the company) totally secondary. They are not just people we work with, they are our friends." (Great Falls Tribune 9 Nov 1994 p. 5a.)*

It is clear from these short biographies that all accidents have incalculable human costs associated with them. When the human loss is placed alongside the financial cost of an aircraft and trained pilot—plus any litigation which might come about as a result of the mishap, and the lost public confidence and trust in the case of commercial or military mishaps—it becomes obvious that the potential downside of poor flight discipline is enormous. Yet aviators continue to take unnecessary risks by taking planning and procedural shortcuts, flying into conditions that they are not qualified to handle, and worst of all, merely for convenience. This is an unbalanced equation. The potential costs are simply not worth the little that might be gained through a flight discipline error. All of the costs we have detailed so far are only applied if the gamble fails—if the shortcut results in a crash. But there are other costs, and these affect flyers even when the outcome of the dice roll does not result in a mishap—this time.

The slippery slope of poor flight discipline

What makes a good flyer go bad? There are many stories of solid aviators who somehow get on the wrong track with regard to flight discipline. The following case study details three years of documented failures of discipline and the results of what occurs when neither the individual nor the organization can stop it. It illustrates the point that poor flight discipline feeds upon itself in a way that may overpower an individual's ability to counteract the downward spiral towards disaster. It is the story of a military pilot with all the tools but one—discipline.

"The best B-52 pilot I've ever known"

For a more thorough discussion of the organizational issues associated with this case study, see Chapter 2 of *Redefining Airmanship* (Kern, McGraw-Hill, 1997). The study is synopsized here to show the progressive nature of flight discipline failures which go unchecked. All quotations and information in this case study are taken from the Air Force Regulation 110-14 Report on the B-52 Mishap, 24 Jun 1994 at Fairchild Air Force Base, Washington. The entire report is available through the Freedom of Information Act.

While all aircraft accidents that result in loss of life are tragic, those that could have been prevented are especially so. The crash of Czar 52, a B-52 flying an airshow practice mission at Fairchild AFB in June of 1994, was primarily the result of actions taken by a singularly outstanding stick-and-rudder pilot, but one who, ironically, practiced incredibly poor flight discipline. Of equal or greater significance was the fact that supervision and leadership facilitated the accident through failed policies of selective enforcement of regulations, as well as failing to heed the desperate warning signals raised by peers and subordinates over a period of three years prior to the accident.

As the process of devolving flight discipline progressed, other young pilots began to see this rogue aviator as a role model, and only good fortune prevented other accidents from occurring. At the time of the accident, there was considerable evidence of Lt Col Hammond's poor discipline spanning a period of over three years.

This case study is worth our analysis and contemplation, not because it was a unique aberration from what occurs in other flying organizations, but rather because it is a compilation of tendencies that are seen across the spectrum. Many aviators report that rules and regulations are bent on occasion, and some individuals seem to be "Teflon-coated" because their mistakes are ignored or overlooked by their supervisors. Routinely, these aviators become role models— even heroes—for impressionable younger pilots, illustrating the contagious nature of poor flight discipline.

A standard setter

Lt Col Arthur Bob Hammond (pseudonym) was the chief of the Wing Standardization and Evaluation Section at Fairchild Air Force Base, Washington. This position made him responsible for the knowledge and enforcement of academic and in-flight standards for the wing's flying operations. By nearly any measuring stick, Bob Hammond was a gifted stick-and-rudder pilot. With over 5,200 hours of flying time and a perfect 31–0 record on checkrides, he had flown the B-52G and H Models since the beginning of his flying career in March of 1971. He was regarded by many as an outstanding pilot, perhaps the best in the entire B-52 fleet. But between 1991 and June of 1994, a pattern of poor discipline began to surface. Perhaps his reputation as a gifted pilot influenced the command staff, who allowed this pattern of behavior to continue.

Two faces of poor discipline emerge

Many flyers who are able to get away with poor discipline over prolonged periods are socially very adept, able to present themselves differently depending on the image they wish to convey. The following were typical comments from Lt Col Hammond's superiors:

"Bob is as good as a B-52 aviator as I have seen."

"Bob was...very at ease in the airplane...a situational awareness type of guy...among the most knowledgeable guys I've flown with in the B-52."

"Bob was probably the best B-52 pilot that I know in the wing and probably one of the best, if not the best within the command. He also has a lot of experience in the CEVG, which was the Command Stan Eval...and he was very well aware of the regulations and the capabilities of the airplane."

A far different perspective on Lt Col Hammond's flying is seen in statements by more junior crewmembers, who were required to fly with him on a regular basis.

"There was already some talk of maybe trying some other ridiculous maneuvers...his lifetime goal was to roll the B-52."

"I was thinking that he was going to try something again, ridiculous maybe, at this airshow and possibly kill thousands of people."

"I'm not going to fly with him, I think he's dangerous. He's going to kill somebody someday and it's not going to be me."

"(Lt) Col Hammond made a joke out of it when I said I would not fly with him. He came to me repeatedly after that and said, 'Hey, we're going flying Mikie; you want to come with us?' And every time I would just smile and say, 'No. I'm not going to fly with you.'"

"Lt Col Hammond broke the regulations or exceeded the limits... virtually every time he flew."

The reasons for these conflicting views may never be entirely known, but hint at a sophisticated approach to breaking the rules that became a pattern in Hammond's flying activities, and is often apparent in those who manipulate the flexible flight environment for their own purposes. Keep in mind as you read through this sequence of events, that prior to this time, there is no evidence that Lt Col Hammond had violated regulations or other guidance. In fact, he was known as a by-the-book instructor and evaluator. But he apparently fed on the adulation he received from his first few dalliances with poor discipline. Perhaps the fact that the B-52—also known as the "BUFF" (Big Ugly Fat Fellow)—suffered from an image problem in the era of the sleek B-1, combined with the closing of the bomb wing at his base, prompted his first violations. After that it was all downhill, and no one seemed capable or willing to stop the descent. To use an appropriate analogy—Bob was in a death spiral towards disaster.

Situation one: Fairchild AFB Air Show: 19 May 1991

Lt Col Hammond was the pilot and aircraft commander for the B-52 exhibition in the 1991 Fairchild AFB air show. During this exhibition, Lt Col Hammond violated several regulations and tech order limits of the B-52 by exceeding bank and pitch limits and flying directly over the airshow crowd in violation of Federal Aviation Regulation (FAR) Part 91. In addition, a review of a videotape of the maneuvers leaves one with the distinct impression that the aircraft may have violated FAR altitude restrictions as well.

Situation two: 325th Bomb Squadron (BMS) Change of Command "Fly Over": 12 July 1991

Lt Col Hammond was the aircraft commander and pilot for a "fly over" for a Change of Command ceremony. During the "practice" and actual fly over, Lt Col Hammond accomplished passes that were estimated to be "as low as 100 to 200 feet." Additionally, Lt Col Hammond flew steep bank turns (greater than 45 degrees) and extremely high pitch angles, in violation of the Dash 11 Tech Order, as well as a "wingover"—a maneuver where the pilot rolls the aircraft onto its side and allows the nose of the aircraft to fall "through the horizon" to regain airspeed. The Dash 11 recommends against wingover-type maneuvers because the sideslip may cause damage to the aircraft.

Situation three: Fairchild Air Show: 17 May 1992

Lt Col Hammond flew the B-52 exhibition at the Fairchild Air Show. The profile flown included several low-altitude, steep turns in excess of 45 degrees of bank, and a high-speed pass down the runway. At the completion of the high-speed pass, Lt Col Hammond accomplished a high-pitch-angle climb, estimated at over 60 degrees nose high. At the top of the climb, the B-52 leveled off using a wingover maneuver.

Situation four: Global Power Mission: 14–15 April 1993

Lt Col Hammond was the mission commander of a two-ship GLOBAL POWER mission to the bombing range in the Medina de Farallons, a small island chain off the coast of Guam in the Pacific Ocean. While in command of this mission, Lt Col Hammond flew a close visual formation with another B-52 in order to take closeup pictures, a maneuver prohibited by regulations. Later in the mission, Lt Col Hammond asked a member of his crew to leave the main crew compartment and work his way back to the bomb bay to take a video of live munitions being released from the aircraft. This was also in violation of current regulations.

Situation five: Fairchild Air Show: 8 August 1993

Lt Col Hammond flew the B-52 exhibition for the 1993 Fairchild Air Show. The profile included steep turns of greater than 45 degrees of bank, low-altitude passes, and a high-pitch maneuver which one crewmember estimated to be 80 degrees nose high—10 degrees shy of completely vertical. Each of these three maneuvers grossly exceeded technical order and regulatory guidance.

Effects on other aviators

By now, the crewmembers of the local bomb squadron had grown accustomed to Lt Col Hammond's air show routine. But a more insidious effect of his ability to consistently break the rules with apparent impunity was manifested in younger, less-skilled crewmembers. In one example, a B-52 Aircraft Commander who had seen several of Lt Col Hammond's performances attempted to copy the "pitch-up" maneuver at an air show in Kamloops, Canada—with near disastrous results. The navigator on this flight said "we got down to seventy knots and...felt buffeting" during the recovery from the pitch up. At 70 knots, the B-52 is in an aerodynamically stalled condition and is no longer flying. Only good fortune or divine intervention prevented a catastrophic occurrence in front of the Canadian audience.

A second example occurred at Roswell, New Mexico, when a new aircraft commander was administratively grounded for accomplishing a maneuver he had seen Bob Hammond do at an air show. "It was a flaps-down, turning maneuver in excess of 60 degrees of bank, close to the ground," his former instructor said of the event. "I was appalled to hear that somebody I otherwise respected would attempt that." The site commander was also appalled and sat the man down and administered corrective training." The bad example set by Col Hammond had begun to be emulated by junior and impressionable officers and had resulted in one near disaster and an administrative action against a junior officer.

Situation six: Yakima Bombing Range: 10 March 1994

Lt Col Hammond was the aircraft commander on a single ship mission to the Yakima Bombing Range to drop practice munitions and provide an authorized photographer an opportunity to shoot pictures of the B-52 from the ground as it conducted its bomb runs. Lt Col Hammond flew the aircraft well below the established 400-foot minimum altitude for the low-level training route. In fact, one crossover was photographed at less than 30 feet, and another crewmember estimated that the final ridge line crossover was "somewhere in the neighborhood of about three feet" above the ground, and that the aircraft would have impacted the ridge if he had not intervened and pulled back on the yoke to increase the aircraft's altitude. The photographers stopped filming because "they thought we were going to impact...and they were ducking out of the way." Lt Col Hammond also joined an unbriefed formation of A-10 fighter aircraft to accomplish a flyby over the photographer. This mission violated regulations regarding minimum altitudes, FAR Part 91 and Air Force Regulation (AFR) 60-16 regarding overflight of people on the ground. There were several occasions during the flight where other crewmembers verbally voiced their opposition to the actions being taken by Lt Col Hammond. The following is one crewmember's recollection of the events that took place on one low-altitude pass:

> *"We came around and (Lt) Col Hammond took us down to 50 feet. I told him that this was well below the clearance plane and that we needed to climb. He ignored me. I told him*

(again) as we approached the ridge line. I told him in three quick bursts 'climb-climb-climb.'...I didn't see any clearance that we were going to clear the top of that mountain...It appeared to me that he had target fixation. I said 'climb-climb-climb.' Again, he did not do it. I grabbed ahold of the yoke and I pulled it back pretty abruptly...I'd estimate we had a crossover around 15 feet...The radar navigator and the navigator were verbally yelling or screaming, reprimanding (Lt) Col Hammond and saying that there was no need to fly that low...his reaction to that input was he was laughing—I mean a good belly laugh."

It is obvious that by this point, the problem of poor flight discipline had gotten pathological. In spite of these facts, Lt Col Hammond was selected to perform the 1994 air show. "It was a nonissue," the commander stated, "Bob was Mr. Air Show." The final link in the accident chain was in place.

Situation seven: Air Show Practice: 17 June 1994

Lt Col Hammond and the crew flew the first of two scheduled practice missions for the 1994 air show. The profile was exactly the same as the mishap mission except that two profiles were flown. Once again they included large bank angles and high-pitch climbs in violation of regulations and technical order guidance. The wing commander had directed that the bank angles be limited to 45 degrees and the pitch to 25 degrees. These were still in excess of regulations and technical order guidance. Both profiles flown during this practice exceeded the wing commander's stated guidance. The next mission would be Hammond's last. At 1335 Pacific Daylight Time (PDT) 24 Jun 94, Czar 52 taxied to runway 23 for departure. Following the completion of the profile, the tower directed a low approach because a KC-135 was still on the runway. After applying full power, Hammond banked the aircraft past 90 degrees in a slow-speed, low-altitude turn around the control tower, a maneuver he had accomplished several times before. But this time he went just a bit too far, and the aircraft impacted the ground, killing all aboard.

The crash of Czar 52, like most accidents, was part of a chain of events. These events were facilitated through the failed flight discipline of a single aviator, who may have influenced scores of young and impressionable aviators who were influenced by watching a rogue aviator as their role model for over three years. These aviators remain on active flying status in various Air Force wings, or are now employed within the ranks of commercial air carriers, passing along what they have learned. Because of this, the final domino in this chain of events may not yet have fallen.

Before we write off Bob Hammond as an isolated case, let me share with you a letter I received from an aviation safety colleague in Australia who was responding to me after reading my initial report on the Hammond accident. It seems as if one of their pilots had recently died in a similar accident involving poor flight discipline.

> *"Two pilots. I couldn't believe it when I read your report...It was like they were the same person! So I put Hammond (BH) in, alongside my bloke, John F. (JF). My interpretation of...(multiple factors in) his life was that it bespoke contempt for societal rules and values...Same thing really, refusing to be contained within society's strictures...JF had, like BH, exhibited a long line of violations, in front of senior officers, typically going too close to cloud, or even briefly into it, while flying display manoeuvres, going on with the show when the cloud base was really too low. (I didn't find this out till after the investigation, years later—from 'the boys' club.) I call the phenomenon 'upwards intimidation' and analyse its origins as being in how others perceive the target's flying skills. If the fellow is thought to be a 'top gun', the senior officers don't feel they can rein him in. The manner in which both JF and BH had achieved this was identical—in my view...Finally, adaptation, where you just keep getting closer and closer to the limit."*

Two pilots from separate continents, sharing nearly identical characteristics of poor flight discipline, dispel the notion of either as total anomalies. There was a bit more to the story of the JF, which also supports my theory that these aviators serve as negative role models and may have a domino effect. Several months after JF was killed, there was another loss of life in an aircraft when a pilot attempted to do a barrel roll without sufficient altitude. It was JF's younger brother.

A final perspective

The costs of poor flight discipline are immeasurably high and the potential rewards minuscule. Why then do aviators continue to accept—and even produce—unnecessary risk out of convenience or complacency? Perhaps they believe that the unthinkable mishap cannot happen to them, or that with all other factors under control, a small infraction is no big deal. Hopefully this chapter has shattered that illusion. Murphy preys on the unprepared and the unsuspecting. No one is immune.

We have also seen that discipline must be absolute. The case study of the B-52 pilot clearly shows the danger of a single departure, as well as the contagious nature of poor discipline within the close confines of a flying organization.

Another often unconsidered cost of poor flight discipline is the un-fulfilled potential of the pilot. Poor discipline inhibits, if not totally prevents, an aviator's airmanship growth. Each bad decision is a lost opportunity for positive development, although occasionally a bad decision teaches us a valuable lesson. One can only wonder how good of a role model Lt Col Hammond or the Australian flyer JF might have been had they stuck to the high road of compliance. How many young aviators might have modeled positive future ac-tions after their lead? Controlled flight into terrain—what a humbling epitaph for such distinguished aviators.

To understand how one can fall prey to the siren song of temptation, we need a closer look at the types of discipline violations, beginning with regulatory deviations, the subject of our next chapter.

Chapter review questions

1 Why did Captain Jeppesen seek out expert mathematicians to calculate altimeter readings and then still add 500 feet to the top of the mountaintop he measured next to Salt Lake City?

2 What are the financial costs of poor flight discipline to the individual pilot and the public at large?

3 What are the training costs associated with poor discipline?

4 Why do young pilots often see rogues who practice poor flight discipline as role models?

5 What are the moral implications of choosing a path of poor flight discipline?

References

AeroKnowledge ASRS CD-ROM 1995. Accession Number: 81432.

André, Babette. 1996. "Captain Jepp, the gentile air-mapping pioneer, passes away." Internet. General Aviation News and Flyer at http://www1.drive.net/evird.acgi$pass*222.../ganflyer/dec20-1996/capt._jepp_dies.html.

Aviation Monthly Safety Summary and Report. 1996. "Accident Statistics," April.

Aviation Safety Journal. 1996. "Low altitude turn causes Baron to stall, crash," November 15.

General Accounting Office (GAO). 1996. Report to Committee on National Security (Honorable Ike Skelton). 27 pages. 1 February 1996.

Hughes Training Inc. 1995. *Aircrew Coordination Workbook.* Abilene, Texas. CS-54.

Kern, Anthony T. 1995. A historical analysis of U.S. Air Force Tactical Aircrew Error in Operations Desert Shield/Storm. U.S. Army Command and General Staff College monograph. Fort Leavenworth, KS.

McConnell, Michael G. 1994. AFR 110-14 USAF Accident Investigation Report, vol. 1, June 1994.

NTSB ASRS Directline. 1996. *Statistics.* Issue 6, p. 33.

3

The letter of the law: Regulatory discipline

Laws are a pledge that citizens...will do justice to one another.
Aristotle

Not every violation of a regulation results in an accident or incident. Thank God, because if they did, airplanes would be falling from the sky like leaves in a Fall Kansas wind. According to ASRS data, in the five-year period between 1988 and 1994, there were 128,225 self-reports of nonadherence to rules. In 1994 alone, there were 20,889 reported instances of broken regulations reported—79% of all reported deviations, and these were just the tip of the iceberg.

Regulations are—as Aristotle suggests—a pledge to support one another through compliance. As a group of aviators, we have decided that some aspects of our operations are so critical to safety and good order that we have codified them into mandatory requirements for the common good. Yet many still feel overregulated or just too busy to take the time to learn or apply the myriad of requirements that have been developed to safely deconflict and guide our ground and airborne operations. If it were only a matter of aviators ignoring the rules to their own peril, this discussion would not warrant a chapter in this book. But compliance protects not only the operator of an aircraft, but others as well, a point made abundantly clear by the passenger in the following case study.

Regulations are made to be broken, right?

by Bill Kelly, Captain, USAF

Reprinted from *USAF Flying Safety Magazine*, November 1994, pp. 17–18.

"It all began on a cool day in January, 1987. The reason I say it was cool—not cold—is because I was stationed at Eglin AFB, Florida, flying the C-21 (Learjet). I received a phone call at home asking if I wanted to fly a local training sortie up to Maxwell AFB, Alabama, to pick up some parts from another C-21 unit there. Naturally, I said yes and rushed over to the detachment where I met up with "Tom," the instructor pilot (IP) and "John," the other copilot. Tom was a major with over a million hours (at least it seemed that way to me). He was one year away from retirement and had been flying his whole career. He had flown gunships in Vietnam and other airplanes to include the C-5. Prior to the C-21 assignment, he was at the C-5 schoolhouse, and was a stan-eval pilot in every aircraft he ever flew. The point is, he was one of the most experienced pilots I have ever flown with.

John, the other copilot, was the newest member to our detachment. He had been there for only a short while and had about 100 hours in the C-21. As for me, I was a highly experienced First Pilot. I had managed to achieve a stunning 330 hours in the C- 21.

We left the detachment and went over to the hangar where we keep the aircraft. As this was such a short mission, Tom went over to fill out the flight plan and get a weather brief. John and I stayed to preflight. Have you noticed that I have not yet mentioned the mission planning? The reason I haven't mentioned it was because we didn't do any. Yes, even the C-21 had regulations covering mission planning. But what could possibly happen?

After Tom finished at Base Ops (about 30 minutes later), he met us at the airplane, and we were off. I flew the first half of the mission. We left Eglin and went up to Dannelly Field in Montgomery, Alabama, to fly some practice approaches. Naturally, being the Steve Canyon I am, the first half of the mission went flawlessly. After we finished the pattern work an Dannelly, we flew over to Maxwell to pick up the spare parts.

John was flying the second half of the mission, and we departed Maxwell for Dannelly so John could get some pattern work. He flew three perfect approaches and Tom seemed surprised at how well the new guy was flying. He even made a comment to John that he must have taken some flying pills before he went to fly.

John's next approach was a TACAN. He got completely configured at the MDA (minimum descent altitude) when Tom—unbriefed—pulled the right engine back to idle. He stated that we had just had a "simulated" bird strike and had lost the right engine. No problem. What could possibly go wrong? After all, we practice single-engine work all the time.

Let me tell you what could possibly go wrong. Apparently, John had been caught completely off guard. He immediately disengaged the yaw damper and either came in with the wrong rudder or did not come in with any rudder at all. The C-21 instantly rolled to about 135 degrees of bank. The IP took control of the aircraft and made a rudder input to correct the attitude. The aircraft then rolled to about 135 degrees of bank in the other direction. This went on for about 20 seconds. During the entire time the airplane was doing this, I kept thinking to myself, "Push up the engine!" But guess what I said—NOTHING. After all, who was I to tell Tom what to do? He was the one with all the experience.

Initially, I was sitting in the jump seat eating my box lunch, and although I realized that this was not a standard maneuver, and we would get critiqued for it when we got back to Eglin, I wasn't overly anxious. But as the dutch roll progressed, the stall warning system started to sound, and I knew we were in serious trouble. It was then I realized that we were going to hit the ground. But no problem. The gear was down and we would hit the ground and slide to a stop. After all, that's how it works in the movies, right? As we approached the ground, I sat back as far as I could in the jump seat and heard the other engine, the one that had been pulled back to idle, spool up.

That is all I can remember until I regained consciousness in the back of the plane. When I came to, the back of the plane was on fire. My right leg was broken at the femur and my left leg was badly cut at the knee. The airplane was full of smoke and I could not see Tom or John.

I heard what sounded like John's voice, and I assumed he and Tom were outside. I immediately went into survival mode and tried to get out of the plane. The door was pinned shut and the aft hatch was engulfed in fire. Luckily for me when we hit the ground, Tom had managed to put the plane into a small pond, and the pond absorbed the brunt of the impact. The plane hit the water in a slightly right-wing-low attitude, causing the wing to break away somewhat from the fuselage, leaving a hole about three feet around. I managed to pull myself up through the hole and escape from the plane.

I began to look for John and Tom but could not find them. It was then I realized that they were pinned in the cockpit and could not escape. As I crawled back to the hole to let them know that there was a way out, a low-order explosion occurred. The plane became completely engulfed in fire and Tom and John did not make it out.

There were a lot of lessons learned from this mishap. If I had only told them to push up the good engine when I thought of it, we would have immediately recovered. Just before we hit the ground, the flight data recorder revealed that when Tom did push the good engine up, we went from 100 degrees of bank to wings level. If we had another 100 feet, we would have flown away. But who was I to tell Tom what to do? The bottom line: NEVER sit through something you don't feel comfortable with.

What does this have to do with following rules and regulations? The whole mishap could have been avoided before we left the ground by following a very basic rule—the requirement to brief emergency procedures. Had we spent even the shortest amount of time on this topic, John would not have been caught off guard, he would have reacted properly and the accident would have been avoided. So the next time you come across a regulation you decide to ignore, remember my story and think about the possible outcome."

Although the passenger in this case study clearly bears some responsibility for the tragic outcome, the root cause of the mishap rests with Tom, the instructor pilot. He was responsible for the two fledgling crewmembers, and his failure to conduct adequate briefings was more than just a regulatory violation—it was a broken trust.

Three types of regulatory deviations

There are three basic types of regulatory deviations, all of which involve disciplined decision making to some extent. An aviator can be ignorant of a regulation, can be boxed in by circumstances which require an eventual deviation, or can just choose to ignore a known regulation out of convenience. Each represents a different discipline challenge. The first can be easily solved by better education and training. The second can be addressed through more detailed planning and briefing procedures and techniques. But the final category of violation—willing and arbitrary noncompliance—has at its roots the individual personality of the pilot, and is therefore the most difficult to address. Specific processes for addressing each of these areas are detailed in the discussions that follow, but first let's look at each type of regulatory deviation in turn.

Regulatory deviation #1: "I didn't know."

The first type of flight discipline violation is simply a lame excuse for failure to educate and prepare oneself for the regulatory demands of flight. It is typified by the statement, "I didn't know." To be sure, there are occasionally circumstances which can lead an aviator into the realm of the regulatory unknown, but the majority of these "close encounters of the worst kind" are born of laziness, complacency, or unfamiliarity with a new environment. Most deviations of this type go unreported, as the perpetrator is as unaware of the violation as he or she is of the regulation itself. A common example is the rural flyer who decides to fly into an urban environment and violates airspace requirements in terminal control areas unknowingly. This demonstrates that, while the rationale for the deviation may be innocent, the potential outcome can be just as disastrous as an intentional deviation, as the following example illustrates.

The pilot of a single-engine Piper Archer, callsign 91 Foxtrot, and his two passengers took off from Torrance airport on a flight to Big Bear, a resort area in the mountains west of Los Angeles. The pilot was new to the area and had only 231 total hours. He was a careful pilot who had been described as sometimes "too careful" with the rules. Flight instructors who had flown with him described him as a diligent, attentive student, but a "VFR pilot who liked to look out," more inclined to navigate by visual reference to the ground than radio navaids. Before he left for Big Bear, he discussed the route with a fellow pilot who was familiar with the course. He was advised on how to avoid the Los Angeles Terminal Control Area, some of the most heavily congested airspace in the world, by using freeways as geographical boundaries. However, having only recently moved to L.A., the pilot had difficulty sorting out the myriad of freeways intertwined in the L.A. basin transit system. On this morning, his VFR navigation techniques failed him, and he entered the terminal control area without clearance.

Aero Mexico 498, a DC9 inbound for LAX had been airborne only twenty-seven minutes since its departure from Tijuana. As they descended into the Los Angeles TCA, the controller called out unknown traffic at "eleven o'clock, one mile, altitude unknown." Flight 498 was then cleared to slow to 190 knots and descend to 6,000. During the descent, the two aircraft collided and the leading edge of Aero Mexico Flight 498's horizontal stabilizer hit 91 Foxtrot's cabin area broadside. The occupants of the smaller plane were decapitated. Both aircraft fell 6,500 feet onto the Los Angeles suburb of Cerritos, killing all on board.

Sometimes the most innocent of errors can have devastating results. Here, the failure of discipline occurred in the Piper pilot not understanding how to apply a new set of rules, ones with which he was barely familiar. To steal a line from Clint Eastwood, "A man's got to know his limitations." In the case of aviators, this means that you at least have to know how much you don't know. Furthermore, you must be willing to admit to those regulations you might have an understanding of, but may still be incapable of applying.

Self-assessment: Finding out what you need to know

The first step in getting a handle on regulatory knowledge is finding out what you need to know. Ask a senior instructor from your local flying organization or FAA representative for a complete rundown on all applicable regulations, procedures, and policies for your type of flying. After that, double-check his or her list against a couple of other "old heads" to make sure nothing was left out. If in doubt, visit your local flight examiner office and get it straight from them.

Once you know what regulations are applicable to your flight environment, you must obtain current copies for your personal use. Your training organization should be able to procure these for you, but the FAA regulations can all be obtained through direct mail to the FAA or off of the Internet, using the keywords "FAA" or "FARs." The Internet versions come with a disclaimer as to their currency. You might not get a completely up-to-date set. Make certain that you have a way of finding out when these regulations change, and develop a system for updating your publications. McGraw-Hill publishes an excellent annual guide to the AIM and FARs that is one of many commercial products on the market to accomplish the same objectives.

The second step is to briefly overview all of the regulations to familiarize yourself with their content and then set up a prioritized plan of study—and stick to it. Ask questions as they come up, and resolve any discrepancies that you might encounter between regulations and policy or procedure. The FAA has many handy study guides to their regulations that you can get free of charge from the local FAA office or off the Internet. Most other organizations and companies have similar aids to help new crewmembers learn the required knowledge. Take advantage of these. Finally, you need to assess your knowledge, hopefully before the flight examiner or mishap board does. Again, several practice examinations are available that can identify any weak areas that you might have.

Don't be discouraged by the multitude and layers of regulatory guidance. Start small. It is better to understand a little than to misunderstand a lot. But don't fly into conditions you have not adequately prepared for. You will feel more confident in yourself and others will see and admire your mature and professional approach to airmanship. Also, as a part of the airspace system, you have a right and responsibility to provide input to changing aviation regulations. Although many aviators are unaware of this opportunity, nearly all new regulations considered by the FAA have a period of time set aside for public comment, and these comments often greatly affect the final outcome of the rule or regulation. Occasionally, Congress mandates a change or a new regulation is needed as soon as possible for safety reasons. In these cases there is often no opportunity for public comment.

Regulatory competence and expertise are perhaps the least enjoyable or glamorous aspects of airmanship. As such, they are the easiest discriminator of professionalism. Serious flyers take the time and make the effort to fully comprehend the regulatory environment. They wear their regulatory expertise like a badge of honor, and rightfully so. Disciplined study and adherence to regulations is an indispensable part of airmanship.

Regulatory deviation #2: "It's not my fault, I was a victim of circumstances."

The second type of regulatory deviation is "I knew the rule, but circumstances were such that I could not avoid the violation." These regulatory deviations are often caused by poor planning or rapidly changing conditions, which can force the pilot to make a decision based on what seems like the best course of action at the time. Consider the following example from my own personal experience.

Case study: I shouldn't have trusted those weather forecasters

So there I was—a young First "Luey" (lieutenant) with about 500 total hours and 150 time in type in the Cessna T-37 Tweet. Back in those days, Strategic Air Command (SAC) crews were required to sit "alert" one week out of every three, meaning that while we were tethered to our aircraft waiting for the Russians to come, we weren't out building flying time. To compensate for that fact, the Air Force brass had seen fit to provide copilots with a "companion trainer" to fly when we weren't on the schedule to aviate the big SAC jets. In our case, this was the venerable T-37 "tweet", which we often referred to as a "6,000-pound dog whistle with wings" (for its characteristic shrill engine sound), a "flying physiological incident" (for its unpressurized cabin), and an assortment of other fond names, which I cannot print here. All in all it was still a great deal for us. We got a jet and a gas credit card and could go pretty much where we wanted to, as long as we logged a few training requirements along the way. It was perhaps not surprising that many training requirements were logged en route to old acquaintances, warmer climates, pro football games, and the like. Like I said, the Accelerated Copilot Enrichment (ACE) program was a great deal. But if you get complacent, even good deals can bite hard. In this instance, I was—and it did.

John, who was another squadron copilot, and I had decided on a weekend trip to Minneapolis, only a few hundred miles from our home base of K.I. Sawyer in northern Michigan. It was early May and the mailboxes were once again beginning to appear out of the tops of the snow banks. The forecast was marginal for the trip, but we were SAC-trained killers and had *beaucoup* experience flying into bad weather in the U.P. of Michigan, as well as recent KC-135 trips to Iceland and the British Isles. Our instrument proficiency was peaked, and our minimums were 200/¼ (200-foot ceiling and ¼ mile visibility). A few clouds weren't going to ruin our weekend. Even with the minimum navigation equipment on the "dog whistle," we should have no worries other than what restaurant to choose for dinner. Besides, we had enough fuel to fly a round-robin profile if required. We got our weather brief, briefed the mission between ourselves, and stepped to the jet around ten in the morning.

As we climbed out westbound, we broke out on top at around 13,500 feet and got a weather update on Minneapolis. The weatherman told us that things had changed significantly since we spoke to him last (only twenty minutes before). The twin cities were getting a little freezing rain, but it was a small cell and should be well past by our arrival time. Of greater concern were some building cumulus clouds at our twelve o'clock. We requested and received clearance from ARTCC to deviate north of our planned track for weather avoidance. A warning horn went off in the back of my head, and I told John we needed to keep a close eye on the fuel.

We contacted approach control about thirty miles out and were told that the runway condition reading (RCR) at our point of intended landing was currently 4, with braking action reported "poor to nil" by a recently arriving commuter. This wouldn't do. We needed a minimum RCR of "9" even with the long runway. We entered holding and contacted Minneapolis Flight Service for a weather update. The voice on the other end was soothing and said that the temperature was rising rapidly, and the "trace of ice" that had fallen would be melted off "in just a few minutes." We were in the clear at 8,000 feet, and with the airport clearly in view a scant 15 miles out, we decided to wait it out—at least until our holding fuel would require us to return to Sawyer, our designated alternate. The second warning horn sounded in my head, and I took out the "whiz-wheel" and carefully calculated our fuel burn rate, winds, mileage and threw in an extra 10% pad for Curtis Lemay. We had enough fuel to hold for fifteen minutes, so John and I traded off making 30-degree-bank left-hand turns and practicing our double-drift holding procedures. No worries, I thought, we're pros.

At the end of the fifteen minutes, the RCR was only up to a "6" and we had to go somewhere else. The only suitable "somewhere else" was back home to Sawyer. So we amended our flight plan with Minneapolis Center and were informed that our return route, the entire western third of the upper peninsula, was now "socked in with thunderstorms." Those building cumulus clouds we had passed 30 minutes earlier had fired up along a 150-mile line and were moving at a snail's pace to the east. Don't I feel stupid now.

We were now faced with a regulatory deviation no matter which way we turned, and we were running out of options faster than we were running out of jet fuel. Deteriorating fuel conditions, icy runway, thunderstorms—and two stupid pilots were all we had to work with. We swallowed our pride, declared a fuel emergency, and accepted the visual approach to the longest piece of concrete in Minneapolis. But the story does not end here. We were to be punished for our foolishness. After a perfect, on-speed landing and rollout, we slid off the edge of the taxiway while trying to taxi to the ramp, making any possibility of hiding our ineptitude an impossibility.

After multiple phone calls to our commanders and safety reps, we spent a glum weekend in the twin cities and returned home two humbled but wiser aviators. I think that's what the Air Force had in mind with the companion trainer all along.

In this situation, an actual low fuel divert through thunderstorms was clearly more risky than our eventual decision to land below RCR minimums. But what happened here? In effect, a failure to adequately plan and anticipate resulted in a self-induced emergency situation. Poor self-discipline early in the flight created a tight box we almost didn't get out of.

Situation awareness (SA): The key to "unavoidable" violations

The violation in the case study above simply did not need to happen. My intuition (those warning horns in my head) tried to tell me, the clouds tried to tell me, the weather forecasters tried to tell me, and the fuel gauge tried to tell me—that all was not well. All of the pieces of the puzzle were given to John and me, but we lacked the SA required to put the pieces together and see the deteriorating situation. There are many inputs to good SA (for a more thorough discussion on building and keeping SA, as well as recovering from episodes of lost SA, see Chapter 9 of *Redefining Airmanship* (Kern, McGraw-Hill 1997). One valuable technique to avoid rapidly shrinking boxes of options is to do what-if future scenarios based upon new information. For example, when the first thunderstorm began to form in front of us to the west, we should have speculated on the possibility of more thunderstorms closing the back door to our alternate. The knowledge of the frontal boundary was apparent from the difference in temperatures between our point of origin and our destination. This should also have keyed us to the possibility of adverse weather conditions at both locations.

The bottom line here is to see each in-flight change as having multiple implications on your near-term future and plan accordingly so as not to be forced into a regulatory deviation.

Regulatory deviation #3: Willful noncompliance

The third type of regulatory deviation is the most difficult to understand and therefore to deal with. For a variety of reasons, some pilots *routinely*—and I don't use that word casually—intentionally violate known rules and regulations. Often they get away with it, sometimes they do not, as with the pilot of the ill-fated mission below.

Case study: Unqualified for the task at hand

On November 17, 1996, at 1509 mountain standard time, a Piper Aerostar 601P, N251B, was destroyed when it collided with terrain while maneuvering near Eagle, Colorado. The private pilot and four passengers were fatally injured. Instrument meteorological conditions prevailed, and an IFR flight plan had been filed for the personal flight conducted under Title 14 CFR Part 91. The flight originated from Eagle, Colorado, and at 1455, the airplane took off on runway 7. Shortly thereafter, according to the control tower tape recording, the pilot reported he was returning to land on runway 25. The controller said he would turn the runway lights up to full brightness. When queried by the controller, the pilot advised he did not have the airport in sight, and further, he did not wish for emergency equipment to be standing by. A few minutes later, the pilot advised that he had resolved the "problem" and would continue to his destination. This last radio communication was recorded at 1509. The on-scene evidence indicates that the airplane impacted a wooded ridge at the 7,500-foot level in a left-wing-and-nose-slightly-low attitude. The airplane then skipped across a draw and collided with another ridge and burned. Examination of the propellers indicate both were rotating at impact. All major components of the airplane were located and identified. There was no evidence of an inflight fire.

At the time of this writing, no final report on the cause of this accident was available. However, within days of the accident, disturbing news was forthcoming in the local media. The Colorado Springs *Gazette-Telegraph* reported less than a week after the crash that the pilot did not possess a valid instrument rating. The Sunday, 20 November 1996 edition reported that "an investigator-intern for the NTSB in Denver said it appeared (the pilot) had taken a flight instrument test last week in Eagle and failed it" (p. B-1). The report went on to say the pilot most likely did not have the minimum training hours (15) or hours in type (5) to qualify for instrument flight without an instructor.

Further insights into the mindset of the pilot came from the man who sold him the Aerostar only two weeks prior to the accident. He stated that the aircraft wasn't equipped with deicing equipment, which would have been critical for flight into the 28-degree-F, 400-foot ceiling reported at takeoff time. "I told Dave all this, and he said 'All my flying is personal. If the weather is bad, I won't go.'" The former owner of the aircraft went on to say, "I've got 14,000 hours and I wouldn't have taken off in those conditions unless maybe I were in a jet. I also asked him if he had his instrument rating, but he was noncommittal about it" (*Gazette-Telegraph* 1996 p. B-5).

Had the pilot only been risking his own life, some measure of understanding might be possible. But also killed on the flight were the pilot's 37-year-old female companion and her three daughters, ages 15, 10, and 8.

A second case study of willful noncompliance also shows the impact of peer pressure (Chapter 8), external stressors, and the situation on flight discipline. In this example it is clear that every level of an organization can—and often does—have a direct or indirect impact on regulatory discipline. Although this is a military fighter case study, see how many of the stresses and personality factors might apply to your own brand of flying.

Case Study: A situation that got out of hand (Hughes 1995)

The mishap sortie was a scheduled two-ship military fighter training formation mission. The flight lead, an instructor pilot (IP), had briefed the mission in accordance with directives and the syllabus requirements, which specified tactical intercept training, followed by neutral basic fighter maneuvers (BFM). The IP briefed the student—mission pilot number two (MP 2)—on aircraft handling extensively and included a thorough discussion about aircraft handling during a hazardous condition known as a "scissors." The IP then mentioned to the student that he would see a scissors on the mission as a part of his training profile.

Start, taxi, takeoff, en route, and intercept training, including neutral BFM, were uneventful. MP 2 then called the IP, saying he was running low on fuel. The instructor directed the student to hold 350 KIAS and start a climb so that they could do a scissors exercise from this point. The student complied and at the "fights-on" call, the IP abruptly pulled high to the student's six o'clock position. The scissors exercise progressed rapidly to a nose-high slow-speed fight with both aircraft falling uncontrollably towards each other. A midair collision resulted. The impact killed the instructor instantly and rendered the student's aircraft uncontrollable, from which he ejected, receiving minor ejection injuries.

The IP was a 32-year-old F-15 pilot with 2,769 total flying hours, 987 F-15 hours with 389 hours as an instructor. In short, he should have known better. The student was a 38-year-old former F-15 pilot undergoing recurrency training with over 1,000 F-15 flying hours, 479 as a former instructor. MP 2 was described as one of the "best...students seen to date." What could cause two highly experienced pilots to lose control of a situation so negligently? A look at other human factors provides some insights.

The instructor pilot was heavily tasked in his daily duties. He was a flight commander, standardization and evaluation flight examiner (SEFE), squadron phase briefing monitor and project officer for an upcoming higher headquarters evaluation visit. During his off time he was pursuing a masters degree, was an instructor for a bible study group, and participated in a church basketball league. In addition, his wife and two children all had the flu the week of the mishap. The instructor had flu symptoms the day before and had taken himself off the flying schedule. The investigation board determined the IP had less than 13 hours sleep during the previous 48 hours.

Both pilots violated known directives. The instructor exceeded guidance in the syllabus by using more power than called for in the maneuvers and flying a scissors maneuver in violation of the syllabus and without first briefing the maneuver. This caused great confusion on the student's part, who did not know what to expect from the instructor, and consequently what he himself was expected to do. Finally, both pilots failed to comply with numerous rules of engagement by not stopping the scissors when the maneuver became dangerous, and when the 1,000-foot "safety bubble" was broken.

A total breakdown of discipline occurred when the IP abandoned his role as an instructor, lost self-control, and fought the student as an equal pilot. This was most likely compounded by fatigue and an underlying competitive personality.

Accident board findings indicated that organizational supervision was faulted for loading down the instructor with duties that detracted from his duties as a flight commander and flight examiner. The instructor's judgment was felt to be impaired due to excessive personal commitment, fatigue, and illness. The IP was faulted for breech of discipline due to the numerous violations of the syllabus and regulations. The student was also faulted for judgment in that he also had a responsibility for stopping the final engagement but allowed the scissors to progress below controllable airspeed.

In summary, all members of an organization, regardless of position, are responsible for the safe conduct of every mission flown. Pilots are not superhumans and need to recognize the limits of their capabilities. Once again, the rules are in the books for a reason. There are usually a lot more reasons to follow the rules than to break them.

What possesses an otherwise successful aviator to completely throw common sense to the wind and imperil the lives of others? Research shows that there are five common excuses for intentional violations of regulatory requirements.

Combating noncompliance

The very nature of aviation means that there is usually little oversight of an aviator's actions in flight. Aviators are simply trusted to do the right thing. Unwary aviators find many rationalizations for their undisciplined and often foolish actions. Five common rationalizations for poor flight discipline follow.

1 If no one knows about the infraction and nobody gets hurt, what's the problem?

2 Everyone knows that there are safety margins built into all the regulations.

3 Rules are simply to protect inept flyers from themselves.

4 This business is overregulated. Pilots did this for decades before the government stepped in.

5 I can't push the envelope and really improve if I stay within all these rules.

It is easy to see how one could buy into these rationalizations if they were looking for a reason to do so. Here are the reasons why it is unprofessional—and unsafe—to do so.

Policies, procedures, and regulations exist for a variety of reasons, and the implications of poor discipline are often unseen by the violators themselves. Small breeches of discipline that seem safe or acceptable in the sterile training environment of day-to-day operations around the "home drome," can have catastrophic implications in other more complex environments. With today's congested airspace and jumbo jet airliners which carry hundreds of passengers, the casualties from even a single accident can be so severe that margins of safety must be built into the regulations to protect innocent lives.

In general aviation, the less structured environment must rely heavily on self-regulation. This environment, coupled with lower experience levels, mandate a conservative approach. But before we move on, let's analyze each one of the "five excuses" and see if they hold water in light of the consequences of failed discipline just discussed.

If no one knows about the infraction and nobody gets hurt, what's the problem? The real problem is that you can never have an intentional infraction in which "no one knows." This is because the most significant "one" is the perpetrator him or herself. Psychologists tell us that getting away with something once is very likely to lead to subsequent attempts at the same or similar activities. Noncompliance can be a slippery downhill path. In short, you are never really alone, and an intentional deviation will likely preprogram you to try it again.

Everyone knows that there are safety margins built into all the regulations. Yes, there usually are, and for good reasons. Safety margins are designed to account for a combination of error tolerances in instruments, navigation equipment, and some human error. To disregard a regulation based on an assumed built-in margin of safety presupposes that everything else is working perfectly. When flyers make this decision, they are betting their life, and as seen in the previous case studies, the lives of others, on a flawed assumption.

Rules are simply to protect inept flyers from themselves. The logic here is that regulations are designed for the lowest common denominator, so a superior airman can violate them and still remain safe. This rationalization is partially true, in that regulations must be designed to protect all flyers, the skilled as well as the marginally proficient. That makes this excuse one of the most persuasive. The flaw in the logic is that aviators must operate under a mindset of assumed compliance if the system is to work. Just like our highway traffic system, if someone consistently rolls through stop signs or dashes under yellow lights at high speed, it is only a matter of time until someone who is following the rules gets hurt by the violator. Because no single aviator owns the sky, we share a moral responsibility to follow established guidelines, regardless of our skill or proficiency. It may be your discipline precision that provides that small safety margin between you and a less-skilled or undisciplined aviator.

This business is overregulated. Pilots did this for decades before the government stepped in.

The first half of this statement is sure to incite heated debate in many sectors of the aviation industry, but the performance capabilities, complexity, and sheer number of the aircraft operating today mandate a certain level of regulation. The depth and rationale for all of the existing regulations may appear unnecessary to individual operators from the point of view of their own aircraft and flying patterns. However, when you step back and look at the big picture of aviation where crop dusters, general aviation, experimental aircraft, commercial, and military supersonic fighters and bombers all share the same sky, the need for regulatory oversight—and disciplined compliance—becomes readily apparent.

I can't push the envelope and really improve if I stay within the rules.

There are many flyers who feel that they are developmentally constrained by the existing regulatory or organizational environment. Somehow they feel that they must practice outside of existing guidelines to become their best. Most high-risk maneuvers require instructor supervision or special-use airspace to perform. Nearly all safety regulations are based upon lessons which have been paid for in blood by those who attempted what you may be contemplating. Aviators who feel the need to "push the envelope" to improve are well advised to adhere to regulatory guidelines for doing so. Pushing a personal or aircraft envelope can rapidly lead to exiting it, which is frightening at best—and potentially fatal if the dice roll the wrong way.

Good pilots follow the rules

Excellence in aviation demands regulatory discipline, while at the same time providing a multitude of opportunities to ignore it. As a historian, I am fond of using the quotations of our forefathers to speak to current events. Joseph Addison said in 1704 that "self-discipline is that which, next to virtue, truly and essentially raises one man above another."

Although old Joe wasn't a pilot, he had the makings of one. He understood that true discipline means doing the right thing even when no one is watching. Regarding aviation regulations, doing the right thing has three parts:

1 Knowing what rules apply to your flight environment.

2 Keeping yourself out of tight boxes that might require an emergency deviation.

3 Mastering temptation to deviate unnecessarily.

When combined, these elements define regulatory discipline. Keep in mind that when we deviate from the rules of flight unnecessarily and get away with it, we are setting ourselves up for greater problems in the future. The Roman spokesman Publilius Syrus had it right in 42 B.C. when he said, "Pardon one offense—and you encourage the commission of many."

Chapter review questions

1 Can you list from memory 50% of the regulations that govern your personal flying operations?

2 Where would you go to find an all-inclusive list of current governing directives?

3 What are the three types of regulatory deviations?

4 What are the five common excuses used by crew members who rationalize deviations?

5 Explain the fallacy with each of the five excuses listed for regulatory deviations.

References

ASRS Directline No. 8. 1995.
   Internet: http://www-afo.arc.nasa.gov/ASRS/dl8.
Colorado Springs Gazette-Telegraph. 1996. "Pilot qualifications questioned in crash," November 20, pp. B-1, B-5.
Hughes Training. 1995. *Aircrew Coordination Workbook*. Hughes Training Division, Abilene, Texas, p. 80–81.
Kelly, Bill. 1994. *USAF Flying Safety Magazine*, "Regulations are made to be broken, right?" November, pp. 17–18.
National Transportation Safety Board—November 1996 Aviation Accidents, 1996. NTSB Identification: FTW97FA042, Internet. Found at http://www.ntsb.gov/Aviation/9611.html.

4

The problem with shortcuts: Procedural discipline

In the moment of action remember the value of silence and order.

Phormio of Athens to Athenian Marines before the Battle in the Crisaean Gulf, 431 B.C.

Procedural perfection is a realistic and obtainable goal for aviators. The dynamic and often chaotic environment of flight makes this goal difficult but certainly not impossible to obtain. In fact, many career fields have already reached this level of performance, albeit under more controlled conditions. In some high-risk environments, errors of procedure are simply not tolerated. How many procedural errors do you suppose are allowed in the handling of nuclear fissionable materials? Or how about in working with highly contagious viruses or biological pathogens? How many errors of procedure would you allow a neurological surgeon about to perform delicate brain surgery on yourself or a member of your family? The point is simply that in some fields, procedural error is simply unacceptable—and aviation should be one of them.

Aviators fail to accomplish procedures for four primary reasons. They either:

- Don't know that the procedures exist.
- Don't know how to accomplish them.
- Forget to accomplish them.
- Willfully decide to ignore them.

Hopefully, we dealt with the willful negligence issue in Chapter 2, where we demonstrated that the cost of poor discipline clearly outweighs any potential benefits, so we will address the other three challenges here. Appropriately, there are four key areas to address these challenges.

What is procedural error?

As earlier stated, we have ample evidence to indicate the human capability to eliminate procedural errors. But before I get myself at odds with the entire psychological community, which is fond of saying "People will always make mistakes," let me be clear about what a procedural error is—and is not. A procedural error is an error which involves a directed action from an authoritative written source, such as a flight manual or Federal Aviation Regulation. Procedural error is emphatically not a judgment or decision-making error, in the purest sense. It is simply the failure to perform a directed action properly.

Of course, there are many judgment errors which lead to procedural errors. For example, pilots who allow themselves to be rushed into approaches for which they are not ready have clearly made a judgment error, but the fact that the onrush of events causes pilots to overlook accomplishing a "Before Landing" checklist—a procedural error—is more a description of an *error chain* and should not be allowed to confuse our discussion on procedural error. Once again, procedures are directed events. The source of the mandate is usually the flight manual or governing regulations, such as the FARs. Now that we understand what a procedural error is, the next questions is, "How does one achieve procedural discipline?"

Four keys to procedural excellence

There are four keys to procedural perfection: knowledge of procedures, proficiency, attention management, and habit pattern development. Knowledge involves the understanding of what procedures apply to you, your type of flight operation, your aircraft, and knowing the difference between procedure and technique. Proficiency refers to the ability of the pilot to apply what is known and required. The third key is attention management, the ability to maintain the proper focus and the avoidance of distraction. The final element of procedural discipline is the development of systematic habit patterns which will allow you to recognize deviations in time to correct them. Let's look at each in turn.

Procedural knowledge

Procedural knowledge must be approached in much the same way as regulatory knowledge, but with one key difference—we must learn how to *apply* the procedures systematically in all situations on every flight. This can be a difficult assignment, simply because while our procedures are straightforward, the situations we face in flight often are not. For example, a complex emergency might require accomplishing one or more emergency procedure checklists, replanning a route and descent profile to a nearby emergency airfield, and communicating with the controlling agency to describe "the nature of the emergency, amount of fuel in hours and minutes, and number of souls on board"—the big three for ATC. With all that going on, who can remember to lower the landing gear? The point is simply that normal procedures must be so firmly ingrained in our habits that we are able to fight through the chaos and confusion of an emergency or other in-flight event to keep our procedural discipline.

In training U.S. Air Force Academy cadets, I find that the most difficult task for new flyers to learn is the juggling act necessary to pull off an in-flight change, be that a simulated emergency, mission change for training, or whatever. Almost invariably, the first thing to disappear is the normal procedures checklist discipline. I suspect it is much the same in other training environments.

The first task is to identify all of the procedures which apply to our personal level of flying. Subsequently we must develop habit patterns in such a way as to make procedural perfection the baseline of our flight activities. Another definition is required at this point to distinguish between procedure and technique.

Procedure versus technique

A procedure is written in some form of directive guidance. This is typically found either in your flight manual or tech order (often consolidated into a checklist) or outside governing directives such as the FARs and local-area flying procedures. A procedure is therefore *mandatory*—a word that some pilots have trouble with.

A technique, on the other hand, is just one way of accomplishing an objective. This may or may not come from written guidance, but a written technique will never carry mandatory compliance language such as will, shall, or must. Techniques are often very helpful but must not be confused with mandatory procedures. Techniques are always optional.

Confusion often comes in when techniques include procedures. For example, in pilot training I was told that at the final approach fix on an instrument approach, I should "time-turn-twist-throttle-talk-track." This meant that I was to:

1. Start timing for the missed approach point (MAP), which was not mandatory procedure because we had DME to determine the MAP.

2. Turn to the inbound course heading, which was procedure.

3. Twist in the course heading in the VOR course window, again a procedure.

4. Pull back the throttle to slow and descend, which was a technique.

5. Report that I was leaving the final approach fix—a mandatory procedure from the FARs.

6. Carefully fly the final approach course inbound, once again a mandatory FAR requirement.

A second illustration from my VFR instructor's kitbag is designed to help new students get everything done as they turn base leg in the VFR traffic pattern. It goes "flaps-pitch-power-radio call-trim-trim-trim." Extending the flaps is a mandatory checklist procedure. Likewise, the radio call is required to comply with VFR traffic procedures. The rest are techniques designed to help the new flyer get stabilized on final approach as soon as possible. While these types of memory tricks are handy for consolidating and simplifying mission crunch points, they also can be very confusing to young aviators who see the lines between procedure and technique becoming blurred. As they become more experienced, the gimmicks are no longer needed. Occasionally, flyers at this stage of development will sacrifice procedural discipline as well. For this reason, it is critically important that flight instructors clearly distinguish between technique and procedure and explain to their students why the differentiation is so vital.

It is also important to understand that an otherwise insignificant error of procedural discipline can be magnified by inappropriate techniques. Consider the various roles of individual errors and oversights that led to the death of a pilot in the following case study, which is taken from *The MAC Flyer* (Military Airlift Command 1991).

Case study: The lights were out, but someone was home (Military Airlift Command, adapted from NTSB AAR-91/03)

"On 18 January 1990, at 1904 EST, Eastern Airlines Flight 111 (EA 111), a Boeing 727-225A, collided with an Epps Air Service Beechcraft King Air 100, as EA 111 landed on runway 26R at the William B. Hartsfield International Airport, Atlanta, Georgia. The King Air was preparing to turn off the runway after landing ahead of EA 111. Both airplanes were in radio contact with the tower controller at the time. The King Air pilot was fatally injured and copilot seriously injured. There were no passengers on board, and no reported injuries on the B-727.

Eastern 111 was cleared to land about 1902, and the crew reported they flew the glide slope and localizer to the runway, landing approximately 1,200 feet from the threshold. The flightcrew said that after touchdown, they manually deployed the spoilers and lowered the nose of the airplane. As the Captain reached for the thrust reversers, the crew saw another airplane on the right side of the runway. The EA 111 flightcrew stated they were not aware of the other airplane until their landing lights illuminated it.

The First Officer stated that after he saw the other aircraft on the runway, the Captain steered the airplane to the left in an evasive maneuver; however, the B-727 continued through the landing rollout and its right wing struck the King Air. The Captain then steered EA 111 back toward the center of the runway, completing the landing rollout. The collision between the two airplanes was described by some passengers as a slight jolt, however, most of the passengers said the Captain's maneuvers to avoid the other airplane were more apparent than the actual collision.

The right wing of the B-727 struck the King Air on the aft fuselage just below the horizontal stabilizer. The empennage and left wing of the King Air were severed, and the nose gear collapsed. The cabin roof was severed from the point of impact forward to the frame of the cockpit windshield. There was no fire.

The B-727 received damage to the leading edge devices, flaps, upper surfaces, lower surfaces, and forward spar surface of the outboard portion of the right wing. The damage area included several prop slashes, surface scratches, and blue and white paint smears similar to the King Air color scheme."

Analysis

The Eastern Airlines flight crew was certified, trained, and qualified. There were no physiological factors or unusual cockpit distractions to prevent the flight crew from seeing the King Air either on final approach or on the runway.

The King Air pilot-in-command was certified, trained, and qualified for the charter flight. The presence of a copilot-in-training did not contribute to the accident.

All FAA ATC personnel involved were certified, trained, and qualified for their duties. There were no apparent physiological disabilities detracting from their ability to perform at an acceptable level on the evening of the accident. The air traffic volume in the Atlanta area at the time of the accident was average. There were no "flow control" or "holding" procedures in effect.

Both the Eastern Airlines flightcrew and King Air crew were familiar with the airport arrival procedures, runway layout, and taxi routes to their respective gates. Likewise, Atlanta ATC personnel were familiar with the King Air call sign. From previous experience, the controllers expected the airplane to exit runway 26R at taxiway Delta and proceed to the general-aviation ramp. The Safety Board could not find evidence to suggest that the King Air flightcrew delayed their exit from the landing runway.

Weather conditions were at or above criteria for visual flight rules. However, there were scattered clouds at 500 feet and fog was present about the time of the accident. In post-accident interviews, neither the Eastern flightcrew, the King Air copilot, nor the air traffic controllers identified environmental factors as a constraint to the normal performance of their duties.

The King Air 100 was found not to be in compliance with airworthiness requirements because of deficiencies in the anticollision lighting system. An Epps Air Service flight line maintenance technician observed the King Air during its initial departure and noted that the red anticollision light atop the vertical stabilizer and the white strobe light installed in the fuselage tailcone were not on. Further, the red anticollision light on the lower fuselage was not an approved type and was not functional. The "Beacon" switch controlling the red anticollision lights on both the top and bottom of the airplane was reported by ground personnel to be inoperative prior to the flight. The lighting discrepancies were not entered in the aircraft log book and not repaired before flight as required by FAA regulations. The Safety Board was not able to establish that the pilot-in-command was aware of these lighting deficiencies, although lighting is part of the preflight checklist.

The Safety Board's examination of available light bulbs from the King Air confirms that some lights were not illuminated at impact. The filaments of bulbs removed from the navigation position and logo light systems were stretched, which is common to light bulb illumination at time of impact. In the lighting configuration of "NAV" lights only, the aft portion of the King Air would present only a single rear white position light with an intensity of 20 candle power. Thus, only limited, low-power lighting would be afforded in a field of view that included a variety of runway, taxiway, and other lights.

The Safety Board also noted that EA 111 was given a clearance to land and advised by tower that they were in sight. No indication was given that they were number two for landing behind another airplane.

Had the controller provided traffic information to EA 111, the flightcrew's sense of situational awareness and motivation to search for a preceding airplane might have increased. Lacking such information, it appears the crew proceeded through their normal task of completing a routine night landing on a runway to which they had been cleared, unaware there was another airplane on that runway.

The fact that EA 111 had received a landing clearance did not relieve the flight crew of responsibility to "see and avoid" other aircraft in their vicinity. However, in the absence of conspicuous lighting on the King Air, and without prompting from ATC to direct their attention to traffic ahead, it was extremely difficult, if not impossible, for the EA 111 flightcrew to detect the other aircraft on the runway.

Role of air traffic control

The final controller was responsible for maintaining separation of succeeding airplanes on the approach to the outer marker. The monitor controller was responsible for maintaining separation of succeeding airplanes on the approach from the outer marker to within one mile of the runway. It is evident by the airspeed reductions issued by the monitor controller to the King Air that he was attempting to achieve additional separation between it and Continental flight 9687 (CO 9687), the DC-9 immediately preceding the two mishap airplanes. However, that same controller failed to compensate for the added closure rate that occurred between the King Air and the following EA 111.

The required separation between EA 111 and the King Air was two-and-one-half miles once inside the final approach fix. It appears that the monitor controller was late recognizing the potential conflict of decreasing separation between the two aircraft. About six miles from the runway, he assigned an airspeed change to EA 111. "Reduce to your final approach speed." This speed assignment did not conform with the *Air Traffic Control Handbook*, which states that a controller shall advise an aircraft to increase or decrease to a specified speed in knots. In addition, the monitor controller did not receive an acknowledgment from the flight crew of EA 111 and thus should not have assumed the instruction had been received and complied with. Therefore, the monitor controller initiated a sequence of events that caused the final approach interval spacing to quickly approach the minimum of two-and-one-half miles. Although he was relieved of direct responsibility for the ensuing loss of separation when the local controller transmitted "EA 111, you are in sight, cleared to land 26R," the Safety Board believes the monitor controller's action contributed to the speed differential and to the overtake that ultimately became a factor in the accident. At the time of the local controller's transmission, EA 111 was almost six miles from the runway, and the King Air was about three miles out. However, the distance between the two aircraft was decreasing at an unacceptable rate and was less than the required two-and-one-half miles of separation as the King Air arrived over the runway threshold.

EA 111 had about a 45-knot closure rate on the King Air, and ATC radio transcripts indicate no action was taken to reduce the rate. In addition, radio transcripts indicate, and a personal interview confirmed, that the local controller became distracted by radio communication difficulties with the flightcrew of another aircraft. The local controller stated he did not observe the touchdown or rollout of the King Air.

At 1902:38, the local controller transmitted, "Continental ninety six eighty seven, half right, stay on bravo, hold short of two six left at the end." There was no response. At 1902:49, the local controller again attempted to call the flightcrew of CO 9687 without success. At 1903:03, the CO 9687 flightcrew transmitted, "Tower Continental ninety six eighty seven bravo two holding short." The local controller responded, "Ninety six eighty seven continue straight ahead to the end and hold short of two six left." This transmission was repeated at 1903:13. There were five more attempts made to establish communication with CO 9687. The last attempt was at 1904:22. Interspersed with these efforts, at 1904:18, the controller cleared ASE 301 for takeoff on runway 26L.

At 1904:48, the flightcrew of EA 111 advised the local controller, "Tower Eastern one eleven, we just hit an aircraft on the runway."

Applications for flight discipline

In this mishap a series of mistakes of both procedural discipline and technique resulted in an accident that could and should have been avoided. The pilot of the King Air was flying an aircraft not properly illuminated, and three different controllers had the opportunity to take action to prevent this mishap. While none of the people involved should take sole responsibility, the individual who paid the highest price was the pilot who accepted an aircraft that was improperly prepared for flight. If the King Air pilot knew of the inadequate lighting, he made a willful act against the regulations; if he did not know, he failed in his procedural discipline during the preflight checklist. In either case, the failed discipline of the pilot set the stage for failed techniques by the air traffic controllers and less-than-perfect vigilance by the Eastern crew.

Skill and proficiency

Once procedures are known, a pilot must be able to execute them. Most training programs provide for a level of training which prepares an aviator with the skills required to accomplish all mandated procedures. However, skill and proficiency deteriorate over time, and unless aviators keep current, they may find themselves unable to accomplish required actions in flight. Personal training should build towards higher skill levels and smaller tolerances for error, but one eye must always be kept on the capability to execute the fundamental procedural requirements of the aircraft and flight operations.

This is not a book on skill building, but the relationship between procedural discipline and proficiency must be understood. [For a more thorough discussion on building personal skill and proficiency in all aspects of airmanship, see Chapter 3 of *Redefining Airmanship* (Kern, McGraw-Hill 1997)]. In most cases, such as accomplishing a normal procedures checklist, there is very little proficiency required. The basic requirements are outlined in the flight manual and we accomplish them. However, other procedures required for operations in special environments are not so automatic. If you haven't had to use the procedures for flying a back course localizer in a while, or leading a four-ship tactical formation to the tanker in the weather, you may want to brush up with an instructional ride before betting your life—or the life of others—on your ability to apply these procedures. Let's look at one case study where a highly experienced professional flight crew paid the ultimate price for not following procedures that they had known for years.

Case study: Final flight

A large, multiengine military jet aircraft (KC-135) disappeared from radar during a practice approach in IMC conditions. The pilot, who was also the squadron operations officer, failed to adequately plan the mission and follow prescribed instrument procedures. The mission consisted, in part, of a series of practice instrument approaches at a field which was below published minimums. Following the second missed approach and while proceeding inbound for a third practice approach, the aircraft descended below radar coverage. The aircraft subsequently crashed in a remote area $3\frac{1}{2}$ miles north of the final approach course and 8 miles northwest of the field. All occupants of the aircraft were killed. Let's back up a bit and see how this highly experienced crew could have made such a basic instrument error as descending without course guidance.

The crew departed at 0900 that morning and flew at 13,000 feet to an airport for practice instrument approaches. Upon arrival at the VORTAC, the crew did not request ceiling and visibility information prior to the approach, the first signal of procedural complacency. After holding at the VORTAC, the crew was cleared for an approach. They were instructed to maintain 9,000 feet until on a published segment of the approach and were given the local altimeter setting. The crew departed the initial approach fix (IAF) and flew without specific clearance to another IAF. They flew a DME arc until intercepting the final approach course. The crew told the air traffic control facility that they planned to fly a total of four approaches.

The TACAN portion of the VORTAC was off the air while the crew was flying the approach. The 0955 weather observation, taken shortly after the initial approach clearance, was a 1,500-foot ceiling and 1-mile visibility. This placed the field below approach minimums. The field remained below minimums for the duration of the flight.

At 1010 the crew was given missed approach instructions for its first low approach. They were told to execute the published missed approach procedure, climb to and maintain 9,000 feet, and contact ARTCC. The crew then tracked inbound on the 054-degree course on the approach centerline but did not call when departing 9,000 feet—another procedural error. They descended below radar coverage, passing 7,000 feet. The aircraft passed over the airport and made a right climbing turn until disappearing into the weather. The crew contacted the controller and requested a second approach. Radar plots indicate that the crew flew an improper missed approach, yet another indicator of procedural problems. ARTCC then cleared the crew to 9,000 feet for the second approach.

The crew called inbound for the approach, and ARTCC replied: "Radar services terminated, and you can contact (Flight Service) radio (FSS) now, and again on your missed approach, climb out to niner thousand, and vacating five, give me a call on one three three six." The crew contacted FSS, calling inbound for the approach. They were given and replied with the altimeter setting. At 1029 the aircraft was tracking inbound on the centerline at 14 DME. The aircraft remained on centerline and descended below 7,000 feet and out of radar coverage. The crew asked FSS to turn on the high-intensity runway lights and VASIs. The reply was, "Negative, I have no control on it. You have to do it with your own radio there," yet another indicator of inadequate understanding of the procedures for this field. Witnesses at the airport confirm that the runway lights were not on, and the aircraft passed directly over the runway at 500 to 1,000 feet. Once past the departure end of the runway, the aircraft made another right-climbing turn into the weather. At 1037, the crew contacted ARTCC, passing 7,000 for 9,000 feet on the second missed approach. The crew was given clearance for a third approach and instructed to maintain 9,000 feet until established on a segment of the final approach course.

At this point the aircraft was at $10\frac{1}{2}$ DME on the localizer course tracking 254 degrees and climbing through 7,000 feet. One minute later, the aircraft had crossed the course and was approximately $\frac{1}{2}$ mile north of the course at 14 DME and at 9,100 feet. The aircraft made a right turn and rolled out at 9,000 feet, tracking 052 degrees, a heading which paralleled the final approach course. At this point they were out of the radar coverage fan for the approach and had no DME information, a fact verified by subsequent FAA flight checks. Furthermore, the glide slope, navigation runway, and DME flags would have been in view, yet the crew continued inbound.

At 1040 the crew was given missed approach instructions. Radar contact was terminated, and they were told again to contact FSS. At 1041 the crew contacted FSS, "With you, we, ah, descending seven thousand for approach." At 1041:41 the aircraft descended below radar coverage. The last known position of the aircraft was at 7,100 feet tracking 053 degrees, approximately $3\frac{1}{2}$ miles north of the course and about $8\frac{3}{4}$ miles northwest of the field.

The pilot was nearing the end of this tour. This mission presented him with the opportunity to fly a memorable end-of-tour flight. The single significant medical finding relates to this end of tour or "finis-flight" attitude. His mission planning completely ignored weather, and he consistently exercised poor judgment during mission planning and throughout the flight. As the operations officer, most senior individual, and aircraft commander, he was the prime mover in this flight. He was responsible for maintaining crew discipline and ensuring that established procedures were followed. This disregard for discipline and procedures is typical of one who decides to fly his "finis-flight" in a grand or daring manner. In the military, "finis-flights" are common, and subsequent mishaps not at all rare.

All crewmembers were current and qualified at the time of the mishap; however, the pilot had flown only once in the 30-day period prior to the mishap flight, completing one approach and landing. The copilot had not flown at all in the same period. The navigator had flown only one sortie in 60 days. The lack of recent flying impaired crew proficiency and may have been one reason the remainder of the crew did not catch or point out the multiple procedural errors.

The actual cause of this accident was that the crew positioned the aircraft outside the reception envelope of the approach navigational aids and descended without course guidance in IMC conditions. The crew violated basic instrument procedures and paid the ultimate price.

This case study points out that improved proficiency promotes better situational awareness, which in turn allows for more attention to be utilized to accomplish procedures, in an ever-improving cycle of performance. The more familiar and proficient you are at flying the procedures, the more attention is available to perceive and handle unforeseen challenges. The result is a safer and more efficient aviator. There is no substitute for sound procedural proficiency, and this can be maximized through a special set of skills known as attention management.

Attention management

Sometimes in flight it seems that there is just too much to do and too little time to do it. Often when this feeling hits us, our procedures are the first things that suffer. When a situation gets so hectic that we begin to forget to accomplish mandatory procedures, big trouble is not far behind. Chapter 12 is dedicated to discussing techniques for achieving attention discipline, but it is important to reinforce the tie-in between attention management and procedural discipline here as well. If procedural discipline stood alone as a skill to master (like, for example, a crosswind landing), we could master it quite easily with practice. But practice alone will not produce procedural perfection, and an in-depth understanding of the factors which tend to pull aviators away from their checklists and habit patterns are critical to further improvement in this area.

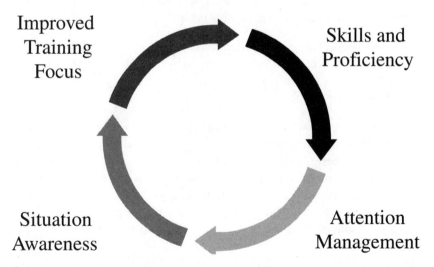

Improved Training Focus

Skills and Proficiency

Situation Awareness

Attention Management

4-1 *The procedure performance cycle. Proficiency and attention management go hand in hand. The greater the proficiency, the more time becomes available to manage the dynamic environment of flight.*

Attention management is a very complex phenomenon involving both the conscious and subconscious. It keys off of pattern recognition, or the ability of the brain to make sense out of multiple inputs by arranging them to fit patterns it has seen before. Often in aviation, there is no pattern established in your memory banks for a new situation, and this can lead to severe task saturation and channelized attention, two of the grim reaper's favorite tools for use on aviators. In order to make sure that we have the necessary attention available to complete mandatory procedures, we must learn to manage our attention.

Task saturation and channelized attention (USAF 1994)

Task saturation occurs when there is too much to attend to at one time, setting up an aviator for the possibility of missing important inputs or cues in a dynamic situation. There are basically two causes for task saturation. The first is information overload, where the sheer mass and number of sensory inputs overwhelms the brain's ability to sort and comprehend. The second type is slightly more complicated and occurs when we fail to adequately prioritize inputs, resulting in a situation where we are unwisely—and unnecessarily—time-sharing between important and unimportant tasks. When this happens, the mind often tends to find a single focus in an attempt to stabilize the situation. This singular focus is referred to as channelized attention. It is the number one human performance factor associated with lost situational awareness and can result in a breakdown of procedural discipline.

So, to deal with the dual challenges of task saturation and channelized attention, we need to be able to reduce the number and/or frequency of inputs and prioritize information. Easy enough to say, but how do we do it?

One key is to stay ahead of the aircraft and to use times of relatively low workload to accomplish future tasks. Workload management begins by determining the "crunch" times in your flight profile. Those are the times when a pilot will be most busy taking care of procedural, navigation, and communications tasks. Typically, these involve navigational action points, where we might be required to change headings, frequencies, and altitudes in a relatively short period of time. Start descent points are also busy times. Some flyers choose to use these action points for other tasks, such as fuel computations and weather updates. This might not be the best use of your attention, as it unnecessarily adds workload to an already busy segment of the flight. The risk here is that an overloaded pilot might not only delay or forget a critical procedural task, but might also cut back on the time available to clear for traffic outside the aircraft or scan cockpit instruments.

A second indispensable survival tool for pilots when dealing with task saturation is a system for prioritization when the stuff hits the fan. There are several catchy phrases that can help a pilot remember what to do first, second, last, and perhaps never—in an emergency situation. One of the simplest, and I believe most useful, are the three words "aviate—navigate—communicate." When we become overloaded in flight, the first priority must be to fly the aircraft—aviate. Check the aircraft attitude, airspeed, altitude, and climb or descent rate—and then correct them if required. Until this life-saving step is taken, you should not waste a moment of time or a fraction of your attention on any other tasks. Clear your flight path and buy yourself time to deal with the next most important issue. Sometimes this might be handling a critical action emergency procedure, which is also part of the aviate step.

Next you need to navigate. First, double-check that you are going where you intend to, or adjust the destination as required. This may mean the nearest suitable airfield or the nearest piece of flat terrain for emergency landing or bailout. Once the destination is established, verify that you are on the correct altitude, and then put the aircraft on the desired course. This step stabilizes the situation further and puts you closer to help.

Finally, communicate as required. Let the controlling agency know the nature of your problem, your intentions, and where you are going. If necessary, don't hesitate to ask for assistance or use that magic word "emergency." Once you have established these three priorities, get back to your checklist and clean up any emergency procedure items as quickly as possible so that you have time to accomplish the normal procedures for the approach and landing. This is procedural discipline, and it can be a key to survival.

The following case study indicates what can happen when the brain must deal with an unfamiliar and high-risk situation and becomes overloaded or channelized. It drives home an important point about checklist discipline.

Case study: The single-focus pilot (Hughes 1995)

The mishap aircraft was scheduled as number four on a low-level/range mission. During the rejoin after range work, the range control officer reported to the mishap crew that their aircraft was trailing vapor. The crew determined that they had a rapid fuel depletion and turned immediately for home base, 25 miles northeast. Approximately four minutes later, the pilot shut off the affected engine and on short final, the back-seater finally pushed the fire push button, an emergency procedures step that should have been accomplished long before. Obsessed with the possibility of losing the aircraft to fuel starvation, the pilot configured the aircraft late and crossed the runway threshold excessively fast in idle power. Due to excessive demands on the hydraulic system, the aircraft pitched down, striking the nosewheel first and porpoising into a nose-high attitude. The back-seat weapons system officer (WSO), recognizing the aircraft was out of control, initiated ejection.

The mishap pilot was a 25-year-old first lieutenant (1Lt) with 353 hours total time and 137 hours in the F-111. The mishap WSO was a 44-year-old major with 2,830 hours total time and 680 hours in the F-111. The mishap crew perceived that they were losing fuel at a rate that required an immediate landing. This was an important fact to the rest of the crew's decisions. The pilot was so intent on getting back and landing before running out of fuel that he failed to accomplish important emergency and landing checklists. The WSO concentrated exclusively on the fuel depletion checklist to the exclusion of monitoring the pilot's navigation and aircraft control. The result was ending up on short final excessively fast and without gear and flaps. The pilot, still believing they had only one chance to land, pulled the power to idle and planned to land midfield. The aircraft hydraulic system could not keep up and pitched the aircraft down. The WSO finally looked up from his checklist and pulled the ejection handle. They ejected with sufficient fuel for at least one more approach. Had the crew accomplished the fuel depletion checklist in a more timely manner, this recovery would have been a routine emergency recovery with an aircraft malfunction.

This mishap is a good example of why checklist discipline is so important during an emergency. It allows the crew to methodically solve an aircraft malfunction during a high-stress situation. Unfortunately, the crew took too long to do one checklist (fuel depletion) to the detriment of other important checklists (land/single engine). Rushing the approach, not monitoring aircraft performance, and forcing the landing even though it should have been taken around put this flyable aircraft in an out-of-control situation.

Habits: The key to consistency

If a key element in attention management is pattern recognition, then developing habit patterns should be an effective step towards improving our procedural discipline. The key is to clearly understand what is required by procedure and to practice these procedures the same way every time out. As your habits become ingrained in your mind, you will catch deviations far more easily. In addition, when an emergency does occur, your normal habits will make it much easier to integrate the emergency procedures with the required normal procedures, which will have become systematic and routine. But habits can sometimes work against you.

There are some inherent dangers with habits. The first is that they do not always transfer well from aircraft to aircraft. The second is that you must have a plan when your habit pattern gets broken.

Negative transfer of training

Sometimes aviators need to "unlearn" a procedure, or at least develop a process which prevents a procedure which applies to one aircraft from bleeding over into another. In addition, there are often times when emergency procedure checklists—which are (hopefully) not a part of the daily routine—can create confusion and break up a habit pattern with normal procedures. Consider the cases with the following mishaps, and ask yourself how many sets of procedures you could comfortably manage.

Case study: One too many sets of procedures to remember (NTSB 1996)

The private pilot student was receiving dual instruction in preparation for his multiengine check ride from a very experienced certified flight instructor at an airfield in Palestine, Texas. After a normal take-off and some typical pattern work, it was time for some emergency-procedures training. While attempting to set up a single-engine approach for his student, the CFI made a critical error. He shut down the left engine on the aircraft, which powered the hydraulic system.

A witness observed the airplane approaching runway 17 flying "a lot slower than they normally do and...a lot lower than usual." Another witness observed the airplane flying on an easterly heading near the midpoint of the runway; he reported that it was "not much higher than our hangar is tall." The witness also observed that the airplane's left propeller was not turning and the airplane began to "yaw to the left and roll to the left." The aircraft was not brought back under control, and it impacted the ground about 1,500 feet east of the runway.

As the accident investigation board began to piece together the nature of this mishap, it began to appear that the instructor may have made a critical error by shutting down the left engine, apparently for a practice approach. The landing gear on this airplane was extended and retracted by hydraulic pressure generated from one hydraulic pump mounted on the left engine. At the time of the accident, the left engine was shut down, the left propeller was feathered, and the gear was retracted. Examination of the engines revealed no anomalies that could have affected their performance. What might have caused a competent flight instructor to make such a critical error?

A closer look into the qualifications of the instructor may shed some light on this question, as well as to the nature of habit pattern interference and negative transfer of training. Although experience is usually a good thing for a pilot—and especially an instructor—to have, in this case it might just be too much of a good thing. The CFI was 76 years old and had flown 73 different makes/models of airplanes in the last four years. You read it right and this is not a misprint; he flew 73 different aircraft makes and models in four years prior to the accident.

The NTSB concluded that the probable cause of this accident was the loss of aircraft control during a go-around, due to failure of the flight instructor (CFI) to assure that the minimum control speed (VMC) was maintained. Factors relating to the accident were the shutdown of the left engine, which would have prevented normal gear extension, and habit interference by the CFI due to the number of aircraft that he had recently flown.

A final caution on habits

Habit pattern interference can be caused by many things: an unexpected clearance, an internal distraction in the cockpit, or in some cases, an emergency situation which draws all of your effort and concentration into a single effort. Habits are useful, but they must not be allowed to take the place of the checklist itself. Overreliance on habits may set you up for failure, especially if your mind gets overtaxed by outside distractors. It's impossible to distract the checklist. Even though some situations require accomplishing emergency procedure checklists, they must be accomplished in conjunction with the normal process and procedures for getting the aircraft safely back on the ground. Consider this point as we analyze the actions—or lack of actions—of a military fighter pilot. How well did the pilot in the next case study manage his time and workload, and what effect did this have on his procedural actions?

Case study: Checklist delay and confusion (Hughes 1995)

The mishap pilot was a 26-year-old Air Force captain with 930 total flying hours and 677 hours in the F-16. While completing checklist procedures for a possible gear-up landing, the mishap pilot inadvertently moved the fuel master switch to the closed position, closing the main fuel shutoff valve. The engine flamed out, the mishap pilot was unable to restart it, and he was forced to eject. It is likely that the gear malfunction indication was erroneous and the aircraft should have been recoverable.

The pilot first became aware of the problem when he attempted to lower his landing gear and got an unsafe gear indication—a red light in the gear handle. He was aware of the mishap aircraft's recent history of gear indication problems. He combined this information with three distinct "klunks" he heard during gear extension and his wingman's observance of what appeared to be all three gear extended normally to conclude that the red light in the gear handle and lack of a green nose gear light were false indications of unsafe gear. Assuming that it was not an actual unsafe gear condition, the pilot reviewed, but did not accomplish, applicable emergency checklists. He delayed completing the landing-gear-up landing procedure because he considered the problem to be only an indicator problem and began a VFR straight-in approach.

Supervisory personnel, who had been informed of the potential problem and the pilot's actions—a routine action in military operations—directed the Viper pilot to open the air refueling door (depressurizing external fuel tanks) and activate tank inerting (reducing internal fuel tank pressures) in accordance with the emergency procedures checklist as a precaution in case the nose gear collapsed on landing. The pilot—who was now busy flying a straight-in approach—attempted to activate both switches without visually confirming that he had the correct ones. He properly opened the air refueling door but reverted to a ground checklist habit pattern from daily main fuel shutoff valve ground checks, and he inadvertently closed the fuel master switch instead of activating the tank-inerting switch next to it. This shut off all fuel to the single engine, a bad thing indeed.

The cause of the mishap was the mishap pilot's inadvertent moving of the fuel master switch to OFF due to habit pattern interference, but the breakdown of the habit pattern can be traced further back into the flight. The pilot's decision to activate the required switches without visual confirmation was probably a technique to avoid task saturation while continuing the approach, a situation that could have been avoided by accomplishing the emergency procedure checklist when he had the time to do so on the descent. Even after supervisory personnel had directed the pilot to accomplish the checklist, the improved solution would have been to execute a go-around and accomplish the procedures in a holding pattern or at higher altitude, where visual confirmation of the correct switches could have been accomplished without diverting attention from maintaining aircraft control on a visual approach.

Techniques (not procedures!) for improving procedural discipline with habit patterns

There are many techniques for flyers to use to help prevent the breakdown of habit patterns. The first and most important rule for developing procedural perfection is simply to use your checklist unerringly. For some odd reason, many pilots take a perverse pride in being able to accomplish the flight without ever referencing the checklist. Over time this generally leads to a deterioration of procedural knowledge and eventually catches up with you when other distractions arise. Many experienced flyers still point to or "check off" every step of the normal procedures checklist on every flight. They do this even when flying solo. Perhaps it is especially important then. I have been called "anal-retentive," and an "old maid" by other flyers for my meticulous checklist discipline. Those are hits I can take, and they are far less painful than the hit delivered by the ground on a gear-up landing or a similar result of poor procedural discipline. By the way, none of these well-meaning critics has ever called me "unprofessional" or "unsafe." Those two words would really hurt.

Next, try to accomplish checklists at the same point on every flight. By developing these checkpoints in your standard mission profile, you are building a methodical database in your memory, so that the day you forget to accomplish it, a sixth sense of "something's wrong" will step in to remind you. For example, if you have a "level off/cruise" checklist in your aircraft, get used to accomplishing it immediately after you get the cruise speed set and the aircraft trimmed up or the autopilot engaged. For the descent check, many aircrews use a distance criteria for getting a weather update, followed immediately by the completion of the descent check. The exact locations don't matter nearly as much as doing it the same way each and every time.

In addition, try to accomplish certain steps of the checklist at appropriate reference points. For example, the flight manual may give you the option of extending the flaps at any point prior to beginning descent on final approach. But if you develop a personal habit pattern of accomplishing this step after rolling out on base leg every time, you are more likely to catch any deviations earlier and prevent the type of task saturation that bit the F-16 pilot in the earlier case study.

Some final words on procedure

We began this chapter saying that procedural perfection is an achievable and necessary goal for aviators. But the discipline required for this level of performance will not come easy for many aviators. It must begin with a complete understanding of the procedural requirements, and an iron will to pursue a mastery of them. Self-assessment and honest appraisal of personal performance will pave the way, but here and there we may come up against areas that will require some additional outside instruction. Recurrent training is not remedial training, and mature aviators will voluntarily request an instructional ride or two when the need exists. These refresher rides offer a unique opportunity to share your goal of procedural perfection with an instructor who is trained to see and analyze mistakes but may be hesitant to do so for an experienced aviator unless requested. Ask for and demand a thorough critique of your performance, and don't be offended when you get it.

One more disclaimer. All of the personal willpower and discipline in the world will not prevent all aviators from making procedural errors; that is too much to hope for. Rather, the goal here is that as an individual, you might begin the process of improvement towards a time when errors of procedure become a thing of the past.

Chapter review questions

1. What differentiates a procedure from a technique?
2. Why is it so critical to know the difference between the two?

3. What are the four keys to procedural perfection?

4. What does the author say is the single most important factor in developing procedural discipline?

5. What caution does the author give regarding overreliance on habit patterns for accomplishing procedures?

6. Where do procedures fit into the overall model of airmanship shown in Chapter 1?

References

Military Airlift Command, 1991. "A fatal distraction." *The MAC Flyer.* November. Scott Air Force Base, IL., pp. 20–22.

National Transportation Safety Board—Nov. 1996, "Aviation Accidents, 1996." NTSB Identification: FTW95FA226, Internet. Found at http://www.ntsb.gov/Aviation/9611.html.

Hughes Training Inc. 1995. Aircrew Coordination Workbook. Abilene, Texas. CS-80.

U.S. Air Force. 1994. Aircrew Awareness and Attention Management Workbook. Langley Air Force Base, VA.

U.S. Air Force. 1994. Human Performance Enhancement Workbook. Langley Air Force Base, VA.

5

Organizational issues for flight discipline

by Ron Westrum and Tony Kern

There are no bad regiments—only bad colonels.
Napoleon

Ninety-five percent of all aviators fly as part of some organization. In many cases, this is as part of a professional relationship—where crewmembers fly for pay as in the military or commercial airlines. In other cases, it is part of an enthusiasts' or social organization, such as the Aircraft Owners and Pilots Association (AOPA) or a local chapter of a glider club. The relationships between the organization and the individual are unique in many ways depending on the nature of the flying or the organization. But there are also many constants, and these constants can and do have a dramatic effect on the discipline of the individuals within them.

Organizational factors impacting upon discipline have been the focus of a great deal of research in recent years. Ron Westrum, a professor of sociology and interdisciplinary technology at Eastern Michigan University discusses several current concerns for organizations in the following pages.

Organizational impacts

(Reprinted by permission of *ICAO Journal.*)

Organizational factors are the macro forces that affect safety—including aircrew discipline—in an aviation organization. Every cockpit, flight deck, maintenance hangar, dispatch office, and control tower is a microcosm. In this microcosm, action is shaped by

99

the immediate decision making of those present: pilots, dispatcher, shift supervisor, etc. Yet the makeup of the microcosm is shaped by decisions made elsewhere in the organization. The decision makers—remote in time, space, or organizational linkages—set the stage. These decision makers put the actors in place and choose the props they will use for their performance.

For instance, decisions about the selection of personnel, including who sits at the controls of an aircraft, are often made in the executive suite. The type and condition of equipment in use also reflect an executive decision. Software or manuals chosen for use in operations or in training, operating rules for crew, the way errors are dealt with, and the level of communication between departments are all matters that reflect decisions made and conditions created by management.

These are organizational factors because they pertain to the organization, not just to a single crew or station in the organization. They reflect executive decisions because people at the executive level decide what trade-offs to make between safety and efficiency, investment and fluidity, buying new equipment and repairing the old, etc. These decisions, often made far from the scene of action, set the stage for what happens at the operational level. In the past, human factors studies have concentrated on the interactions between human and machine. In more recent times, studies have examined the interactions of the crewmembers. Yet the arenas in which piloting, maintenance, dispatch, or air traffic control take place are shaped by managerial decisions. The choices of people, equipment, and policies shape events. We therefore need to look at what management does, and why.

Climate versus culture

When we talk about the climate of the whole organization, we use the phrase "organizational culture." The word "culture" calls up images of a miniature social system, with its own rules and codes. We think of an organizational culture as slowly changing over time, reflecting growth, experience, and struggles. When an organization has been able to mature over a long period of time, this set way of doing things shapes all kinds of decisions. But not all aviation organizations have a culture in this sense. This model of slow growth and distillation of tradition seems poorly suited for the active world of today's flying organizations. Rapid growth, company mergers, changing leadership, military downsizing, and fierce competition have powerful effects on organizations that "culture" does not capture.

Culture reflects the established policies and values. But even new organizations have a climate—that is, each person in the system senses the corporation's overall values and norms. These values and norms are communicated by the choices it makes—the choice of people, the choice of organizational goals, the willingness to invest in training, etc. These highly visible choices communicate what is expected and valued, and what actions or accomplishments are likely to be rewarded. This can, and often does, have a large impact on pilot discipline. To a lesser extent, advertising and public relations also provide messages about a company, but for employees these messages are always compared to the day-to-day realities. The organization may carefully orchestrate its external image; within the organization itself, however, it is the key choices management makes that signal its priorities, and aviators are quick to take note.

A good example of the role of personnel actions in shaping corporate culture is a recent incident at a major U.S. carrier which decided to ground one of its captains after the pilot was involved in a dispute with another crewmember. The pilot was suspended from duty because the cockpit incident in which he was involved belonged to a larger pattern of aggressive behavior both at work and at home. He was described by the chief psychiatrist of the U.S. Federal Aviation Administration (FAA) as "stubborn, pompous, self-centered, domineering, belligerent, and aggressively intimidating," but not unfit to fly. On the advice of a medical consultant, the airline suspended the pilot. No doubt in an earlier era the captain, who sued the air carrier, would have received his certificate back without delay. But in this age of crew resource management (CRM) the airline would have put its own training programs at risk if it had not fought to have him removed. Actions taken in relation to such behavior send strong messages about what is acceptable and what is not. Organizations that permit or encourage the kind of aggressive behavior that eventually got this pilot's license suspended would have no credibility, when trying to foster crew resource management or enforce strict standards of flight discipline.

Tolerating poor flight discipline or unsafe behavior can result in serious organizational problems. An extreme example of such problems led to tragedy at Fairchild Air Force Base in the United States when a B-52 crashed while its pilot attempted unauthorized maneuvers during a practice session for an air show (see Chapter 2). The pilot, who was much liked by his superiors, had a long record of taking dangerous risks. Eight commanders had failed to take action against him. The loss of the aircraft and its crew was only the last, fatal step in a long series of extremely poor discipline by this pilot; the real failure was that he had been allowed, through his risky actions, to do things that should not have been tried and to create a culture that was not conducive to good airmanship.

Shaping a climate

Selection of personnel for key positions is only one of the ways in which top management shapes an organizational climate. By a constant succession of actions, management creates the atmosphere in which members of the organization operate. One useful indicator of the overall climate is the way that information is handled in the organization. It might be useful to suggest a range of climates, using information flow as the indicator. One such range is the pathological, bureaucratic, and generative scheme, shown in Figure 5-1.

Professor James Reason has suggested that latent pathogens—unseen, unsafe conditions—tend to build up within an organization before an accident. Obviously information flow is the means by which such conditions are spotted and acted upon. When information flow is brisk (as in a generative organization) the latent pathogen is quickly spotted and remedied. In a pathological climate, however, it is the person who spots the pathogen who is suppressed, and the problem is not resolved. In fact, we might go further and characterize organizations according to how they respond to anomalous conditions. In Figure 5-2 we see a spectrum of responses to anomalies.

On the left, we see denial responses. Those who spot latent pathogens are suppressed or isolated, unable to do anything (e.g., a high-level safety investigator who completes a confidential report showing many safety lapses is suddenly transferred to another, less politically sensitive post). In the middle of the spectrum, we see the more typical repair responses: the immediate problem is explained away or remedied, but no deeper inquiry is undertaken. This kind of "quick fix" often seems to take place when problems are first spotted, but where there has been no disaster to provide public awareness and compel action. Unhappily, a serious accident may be required to get some problems addressed. Only then are the more thorough reform solutions on the right brought into play.

"Global fixes" are often used in the aviation community to correct a problem that is common to all units of a certain aircraft type. These often take the form of airworthiness directives or in the military, new "flight crew information files" (FCIF) or Operating Instructions (OI).

Pathological	Bureaucratic	Generative
Information is hidden	Information may be ignored	Information is actively sought
Messengers are "shot"	Messengers are tolerated	Messengers are trained
Responsibilities are shirked	Responsibility is compartmentalized	Responsibilities are shared
Bridging is discouraged	Bridging is allowed but discouraged	Bridging is rewarded
Failure is covered up	Organization is just and merciful	Failure causes inquiry
New ideas are crushed	New ideas create problems	New ideas are welcomed

5-1 *How organizations respond to information concerning safety (Westrum 1996).*

5-2 *Responses to an anomaly. Organizations can respond with a wide variety of tactics to news of an anomaly, from pathological actions to generative responses which actively seek information.*

"Inquiry" takes place in those rare organizational climates where a leader decides to correct not only the immediate problem, but to attack underlying conditions as well. This happened at United Airlines following an accident in Portland in 1978, in which a Douglas DC8 ran out of fuel while the pilots circled the airport after an indicator light malfunctioned, attempting to determine whether the landing gear was down and locked. Investigation found that poor training and crew coordination was at fault, and a follow-up survey of pilots showed that there was a widespread problem that needed correction. United Airlines proceeded to develop one of the first CRM programs, its "command, leadership, and resource management" program. This thoughtful response to an underlying problem is typical of generative organizations.

In his investigation of the Dryden, Ontario, accident of 1989, Canadian Chief Justice Virgil P. Moshansky helped bring about an understanding of the accident's organizational causes.

In spite of pressure to limit his inquiry, Chief Justice Moshansky insisted on examining the background of the accident. He was able to show that the pilots and dispatchers were only the last links in a chain of unsafe practices.

Just as the corporate climate is shaped by day-to-day actions, it is also shaped by major organizational changes. As the organization grows and changes—or downsizes and changes, as is often the case in the government sector—the rapid change stresses the organizational leadership. In this situation problems can occur even when management tries to foster a positive climate. Rapid change causes managers to be overloaded with responsibilities, and these overloads can have fatal results.

Chaos and confusion

A changing organization is an organization at risk. Change increases mental workload. Psychologists talk about the mental workload of pilots as critical to situational awareness and flight discipline, but the mental workload of managers is just as important. Rapid change overloads the mind and clogs the desk of the busy commander or executive. As an aviation operation grows or shrinks, several things happen. New operations may be opened up or merged with the existing operations. Rules change as one organization comes under the control of another. Managers change positions and are forced to learn a large number of new things. Some of this workload—never enough—is passed on to their subordinates, many of whom are also new. Keeping track of events becomes more difficult. A manager can have dual responsibilities, for instance, when in transition to a new job, and often may tend to apply old skills to the new position. This overload provides for highly stressed and tired managers and the consequences may be serious, even deadly. Key issues may be deferred, forgotten, or misplaced. Some tasks may be placed in limbo as others force immediate attention. But the tasks put on hold may need immediate attention, too.

Mergers and rapid growth often cause such overloading. As two groups of personnel—often with different training—are merged, conflicts of tradition take place. Which set of standards become the new ones for the merged organization? Whose aircraft manuals take precedence? These factors may have played a major part in the Dryden accident mentioned above. The airline involved had just gone through a merger that had had several consequences—a labor strike, management turnover, and many staff departures—and it left the company with too many responsibilities covered by too few people. The organization was stretching its capabilities to the limit, and events pushed it over the limit. With ice on its wings, a Fokker F28 attempted takeoff, crashing into the trees just beyond the runway.

Pilot error? Certainly, but behind that error were management conditions that had set the stage for the fatal events.

Another example worth studying is what happened after U.S.Air acquired Pacific Southwest Airlines and Piedmont Airlines, and merged them into a single carrier. According to figures published by *The New York Times* (see Figure 5-3), pilot deviations (actions which might violate regulations) for the merged airline were high, and only gradually decreased as integration proceeded over the next five years.

Even without a merger, rapid growth can cause severe stress. Consider U.S. based Valujet. When new aircraft were added to its fleet, often of different types, it placed a major stress on the system. This temporary stress was an opportunity for incidents to occur. Many problems went unfixed while the organization worked to absorb the additional equipment and personnel. Aircraft flew with unrectified engine problems, badly rigged safety equipment, and hydraulic system leaks. Undertrained mechanics failed to complete repairs or to do them properly. Only after Flight 592 crashed into the Florida Everglades was the airline compelled to correct many of its serious problems.

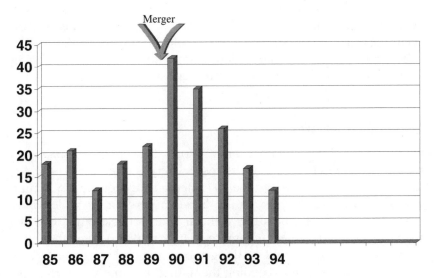

5-3 *Pilot deviations at U.S. Air. Organizational upheaval has a dramatic impact on individual performance, as illustrated here. Pilot deviations rose sharply after a corporate merger but recovered to premerger levels after pilots acclimated themselves to the new order.*

The pressures of the executive life can also be seen very clearly in the crash of an Airbus A330 in mid-1994 in Toulouse—a crash that killed the Airbus chief test pilot and six others. After a hectic day with several different kinds of activities, the Airbus crew and their guests boarded the airliner in late afternoon. Shortly after takeoff, the aircraft crashed. The crew had made a series of errors, and when things started going wrong, the captain failed to go to manual control quickly enough. It is evident in retrospect that the flight took place without sufficient planning. This accident is a graphic example of the dangers of overburdening executives when they have to make decisions about safety—including their own safety.

Labor disputes also produce stress, and not just for airlines. The 1981 strike by U.S. air traffic controllers created a very serious situation. Managers and newly recruited personnel replaced experienced air traffic controllers. Stress on the system was high. Many people predicted dire consequences. Fortunately, the FAA responded by slowing down operations to a manageable level, and the crashes that many experts expected did not occur. Nonetheless, many observers later described the extreme stress levels of air traffic controllers trying to operate with too little training and too few staff.

Similar problems occurred in Australia in 1989 when Australian Airlines' domestic pilots were on strike. Even after the strike ended, bitter feelings remained between the two sides, and hostility was directed toward pilots who stayed at work during the strike or took others' jobs after the strike ended. Even with special team-building workshops, it is clear that many problems remained after the strike was settled in 1990.

Economic pressures

Safety costs money. When economic pressures increase, managers try to decide where they can cut costs. Safety is often one of the targets for cost-cutting. Before the 1979 deregulation of the U.S. airline market, for instance, most of the large carriers had a wide safety margin. They were operating with far more checks and balances than they needed to meet federal regulations. With deregulation, this large safety margin could no longer be sustained. Meeting federal standards, rather than exceeding them, then became the norm. Fortunately, since there had previously been a wide margin of safety, moving closer to the edge did not produce catastrophe. However, many of the small commuter airlines had never had such wide safety margins, so economic pressures may have moved many to the limit or even over it. Similar conditions may exist today in the military, where steadily shrinking budgets and severe manpower downsizing have placed more and more mission demands on fewer and fewer personnel. The challenge to do "more with less" has stretched some organizations past the limit, as evidenced by the issues raised by the CT-43 crash in Croatia (see Chapter 1).

But size is no guarantee that safety will not be compromised. One U.S. carrier came under intense government scrutiny after five major accidents over a two-year period. The airline had eliminated two preflight checks in an attempt to achieve more on-time departures. A study by *The New York Times* showed that its aircraft were often leaving the gate without the kind of careful checking they needed: in one case an aircraft left with too little fuel to complete its trip. Checks revealed that there had been nine instances in which aircraft had left without enough fuel to reach the scheduled destination. It is likely that economic pressure played a significant role in these lapses. In 1994, the carrier was $2 billion in debt and losing money every day.

Not all safety problems are caused by rapid change and work overload. An accident can reveal embedded problems even in an airline with a strong safety tradition. Complacency can hide problems that need to be addressed. This was evidently the case with the Portland accident in 1978, which led United Airlines to rethink how it was training its pilots. At airlines around the world, CRM was needed long before the need for it was recognized.

As flying organizations continue to stress sound judgment and discipline in the cockpit, the maintenance hangar, and the dispatch office, we should not neglect the larger organization itself. Just as the culture of the cockpit often shapes pilot behavior, the culture of the organization affects the culture of the cockpit. All too often, this organizational culture is a neglected problem and sometimes becomes visible only through the investigation of a pilot-discipline accident. In fact it is the in-depth reporting on major accidents that has drawn the attention of the aviation community to the importance of organizational culture. It is time for the leadership within aviation organizations to begin asking some tough questions. One of the first concerns management workload. When change takes place, is mental overload an accepted fact of life, or is it something that needs to be carefully watched? How many managers pay attention to their own situational awareness? Another question concerns the role of managers in designing the work environment.

How many managers are sensitive to creating trick situations in the cockpit? How many pay careful attention to crew pairing? How often does the maintenance hangar reflect real care to make sure that working conditions encourage thorough preparation and careful handling of parts?

And finally what about the organization itself? With all the care devoted to the aircraft, how much care is devoted to the organization, its design, and its functioning? Should there be a "corporate resource management" course for managers? Who should teach it and what should its content be? The study of organizational factors is only beginning. What the future holds will be interesting to see.

One organizational challenge that is certain to remain is the ability to identify and repair or remove rogue aviators from an organization.

Rogue practitioners

Individuals can seriously injure, and in some cases even destroy an organization. While we have seen that patterns of poor discipline occur at many levels of an organization, the active presence of a rogue at any level can have disastrous implications. Rogues are literally ticking time bombs, and the challenge for the organization—and each individual within it—has three parts. The preventative challenge is to establish a culture where rogues cannot grow. The second challenge is to identify—perhaps better said unmask—the individual for what he or she is, and then finally to develop interventions that either fix or remove the rogue practitioner from the organization. The challenge is really not even complete at this point however, as the organization also has a moral obligation to make certain that the rogue practitioner does not simply pop up in another organization and continue to jeopardize lives.

Rogues develop for a variety of reasons. Some see the built-in flexibility of aviation as a chaotic environment which may be manipulated to satisfy their own egos—often with tragic results as we saw in the case study in Chapter 2. Others are simply misguided as to the actual limits of operations or are overly competitive, a combination that can be deadly. Brigadier General Chuck Yeager—a man who understands breaking barriers better than most anyone in aviation—puts it this way.

You can't really take what a pilot says he has done very seriously, especially if it comes after a couple of beers. He might

*claim that he's done a Split-S from 4,000 feet, or flew through
a mountain pass at 50 feet. My technique—if I was going to
try one of these things at all—was to double whatever I heard
had been done and try it from there. Yet some of the kids do
(take the exaggeration seriously), and they end up making a
smokin' hole. I guess I understand why pilots tend to exagger-
ate, they all want to think they're the best (Yeager, March '94).*

This type of verbal competition is innocent enough when hoisting a
pint in the local watering hole, but when it begins to take the form
of flying competition that exceeds the limits of regulatory guidance,
flight manual specifications, or common sense, it has crossed over
the line of innocent boast and gone into the realm of poor flight discipline. When this type of behavior becomes routine, you have the
makings of a rogue.

Typical patterns: Rogues cross all disciplines

Rogue practitioners are certainly not limited to aviation. In fact they
can be found in any field. However, in aviation the exploits of the
rogue often have immediate and highly visible results. A brief look
at rogues from other fields can help aviators to realize that the danger represented by rogue aviators is not entirely unique to our profession or hobby, or even to our century. It can also help us to see
common threads which may be useful in identifying these men and
women before their aberrant behaviors take a toll in human lives.

Rogues on the rails

On April 25th, 1853, one of the worst train wrecks in history occurred due to the incompetence and ego of a well-known rogue engineer. Robert Reed, a historian who specializes in the history of
train accidents, describes what occurred that fateful night at ten
o'clock in the outskirts of Chicago at a place called Grand Crossing.

*An eastbound Michigan Central express, headed for Toledo,
rammed a Michigan Southern emigrant train broadside,
killing twenty-one German emigrants. According to an eye
witness, the wrecked cars lay piled up in a swamp which
flanked the tracks. The scene was "an immense heap of iron,
splinters, doors, and baggage with the crushed locomotive of
the express train hissing steam from its ruptured boiler...cries
assailed the ears...mingled in strong discord with the deeper
groans of the dying." (156).*

The cause of the accident was "gross carelessness and ignorant rivalry between the crews of both trains." (Reed 157). Singled out as particularly responsible was one Mr. Buckman, the engineer of the Michigan Central train, who had a reputation as a "me first" engineer with an ego to match the size of his locomotive. Reed maintains that Buckman could have easily avoided the accident, "either by stopping or going on, but as he had the right of way, he took his time in passing through the intersection. His petty attitude of "me first" took twenty-one lives." (157). Like many rogues, Buckman also apparently felt that the rules did not necessarily apply to him or his operation. On the night of this disaster, he was running at night without a headlight.

From an organizational standpoint, it is understandable how this could have occurred in the mid-1850s. Railways had been extremely safe for their first twenty years of operation. In fact, between 1829 and 1853, no more than a half a dozen people had perished in any single wreck (Reed 9). But the 1850s brought on night travel and expansion. Competition was becoming tied to profit margins, and hard driving engineers like Buckman were likely seen as an asset, rather than a liability to their company. At least until April 25th.

Our next look into the mind of a rogue and the organizational response occurs a century later and half a world away, in an even larger vehicle—a U.S. Navy ship-of-the-line.

Rogues at sea

In the mid-1960s, the U.S. Navy removed Marcus Aurelius Arnheiter from command of the U.S.S. Vance, a small escort destroyer, after a series of self-serving—and quite bizarre—actions at sea. In an effort to gain fame and perhaps the command of a larger warship, Captain Arnheiter issued false position reports so he could leave his assigned area of responsibility which was several miles off the coast of Vietnam, and get to "where the action is" (Sheehan 153), which in his mind was in the shallows along the Vietnamese coastline. On several occasions, he steamed his vessel in front of firing U.S. Navy destroyers, who had to cease their combat fire missions when the Vance fouled their firing line. The commander of one of these destroyers, even began to tape-record the outlandish requests and radio transmissions from Captain Arnheiter, to send to higher headquarters as evidence of this unprofessional and dangerous behavior.

The Vance was a picket ship, designed with small three-inch guns to intercept any smugglers attempting to bring contraband or war materials to the Viet Cong in South Vietnam, especially along the

Mekong Delta area as part of a 1965 operation called *Market Time*. This patrol duty was routine but important to stop the resupply of Vietcong guerrillas operating against American and ARVN (Army of the Republic of Vietnam) troops in the south. However, it wasn't the kind of mission that would bring great opportunity for glory. In fact, while under command of the previous captain, the Vance had never fired a shot in anger. That was all about to change.

To fulfill his lifetime ambition as a naval combat commander, Captain Arnheiter called a "War Council" of his ship's officers almost immediately after taking command. He explained—behind a locked door with an armed sentry—that the Vance was to become a fighting ship and in short order, he forced his sailors to train as a Marine assault force, used ship recreation funds to purchase a small speedboat upon which he painted shark's teeth and mounted a machine gun to be used on a "hot Viet Cong target" or to draw fire from the coastline so the Vance could manipulate the rules of engagement and enter the close-in battle (Sheehan 278). When this "bait" did not work as expected, Arnheiter had the Vance fire into empty sand dunes and desolate cliffs following fictitious "urgent fire" requests and grand victories, and then wrote bogus press releases which he mailed to senior officers. To add insult to injury, he coerced junior officers by threatening their careers, to put him in for the Silver Star after apparently stalking a Chinese submarine. All of these actions were an attempt to make a name for himself as the next Admiral Nelson, his lifelong idol who had died at Trafalgar (Sheehan 3-37).

The Navy was better equipped (at least in this case) to deal with the presence of a rogue than were the railroad companies of the 1850s. Through communication with a visiting chaplain and the testimony of other ship captains, the Navy decided enough was enough, and removed Captain Arnheiter from command after 99 days at sea. A long and bitter public battle followed, and only Neil Sheehan's well-researched book, *The Arnheiter Affair*, finally convinced the public and the media that the Navy was not on a witch hunt.

The military is certainly not the only home for rogues. Sometimes they can be found as close as your nearest doctor's office or local hospital.

Medical rogues

The medical field, and likely other professions where individual contractors can move from place to place where their services are in demand, is ideally suited for rogues. Additionally, the information flow and oversight capability is limited by state and federal bureaucracy and jurisdictions. In spite of these restrictions, the information age has allowed some citizen action groups to tackle this problem, and the results are startling. In a reference text called *Questionable Doctors: Disciplined by States or the Federal Government*, the Public Citizen Health Research Group has compiled over 13,000 entries of doctors who have been disciplined for various infractions or incompetence. This is not to say that all of these entries constitute rogue doctors, certainly most do not. But you don't have to turn too many pages to find multiple entries under the same name, indicating a prolonged period of recurring infractions.

Dr. Larry Kompus, M.D. was first disciplined in September, 1980 by the state of Michigan. The forty-two year old doctor had his medical license revoked for causes which are not stated in the short report (Wolfe 19). We cannot be certain when he began to practice medicine again, but in August, 1990, he had a restriction placed on his controlled substance license and his medical license suspended in Michigan "due to an administrative complaint involving sexual misconduct...Medicaid fraud...and contributing to the delinquency of a minor." (Wolfe 19).

The trail is next found in Kansas, where in 1992 Dr. Kompus was forced to surrender his license again. The reason stemmed from a Michigan conviction for Medicaid fraud and "allegations that he traded drugs, alcohol, and other gifts for sexual favors with a minor male patient." (Wolfe 19). Although the state of Kansas was clear that Kompus "shall not apply for reinstatement of licensure in Kansas" (Wolfe 19), he simply moved west to Colorado and began again. The final entry on Dr. Kompus is listed on May 21, 1993, where the Colorado authorities ordered that he surrender his license again. There is no indication provided as to where Dr. Kompus might be practicing today.

The medical community faces difficult organizational challenges when dealing with rogue practitioners. Although there is the loosely aligned American Medical Association which is attempting to deal with the issue of negligence and incompetence among its own ranks, the job of rooting out rogues really falls into the hands of the licensing authorities at the state and federal governments—two groups which are notorious for bureaucratic constrictions to information flow. But information flow does seem to be key in identifying and dealing with rogue practitioners from all walks of life, including the business community. Consider the case of Barings bank in identifying the reckless activities of the most well-known rogue of our time, futures trader Nick Leeson.

Rogues in business: Nick Leeson

Barings bank was over two centuries old when a single individual— Nick Leeson—destroyed it. It had been used to finance the Louisiana Purchase for the fledgling United States, and funded the Napoleonic Wars. On the cover flap of Judith Rawnsley's *Total Risk: Nick Leeson and the Fall of Barings Bank*, there is the following quote from Duc Le Richelieu in 1818, "There are six great powers in Europe: England, France, Prussia, Austria, Russia, and Barings Brothers." (Rawnsley cover). The notion that one man could manipulate the system and bring down such a durable and powerful financial institution is reason enough to learn more about the nature of rogues, regardless of your particular profession.

Nick Leeson worked for Barings' most profitable subsidiary in Singapore, a freewheeling, entrepreneurial Securities operation (Rawnsley back cover). To make an exceedingly complex story short and to the point—Leeson overextended Barings by several hundred million pounds through the invention of a fictitious client and by defrauding bank officials with bogus reports. He had been operating in this manner for over two years. He had been a corporate hero, Barings' finest, called "the Michael Jordan of the trading floor" by others and "Superman" by himself. But it was all a lie. By the time anyone found out about the scam, the powerful blue-blooded bank was financial history. In relatively short order, the bank was placed into "administration" the British equivalent to U.S. Chapter Eleven proceedings.

Back in London, Chairman Peter Baring apologized to his fellow bankers and blamed the entire mess on a "rogue trader" (Rawnsley 24). At the corporate headquarters research department, panic broke out when it was announced that a bailout of the bank had fallen through. After one employee made a joke, saying "grab what you can and run," many employees took him to heart and began to copy files and load boxes into the backs of their cars. For Barings, things had hit rock bottom. On March 6, 1995, a Dutch financial corporation bought out all interests and liabilities from Barings Bank for one pound.

In the aftermath of the Leeson affair, financial analysts, management and business gurus, and CEOs from around the world all asked themselves the same question. How could this have happened? By their own admission, officials at Barings were aware that the "systems and control culture are distinctly flaky," a memo to Barings CEO Peter Norris stated fully three months before the collapse (Rawnsley 131). There was more fundamental issue at hand. The organization had been unable to identify or control an organizational rogue.

Common characteristics

Rogues appear to share similar characteristics and patterns of behavior. If these patterns can be recognized by an organization before a problem arises, the possibility for effective intervention is increased. This is not to say that all who fall into these behavior patterns will become a rogue or hurt an organization. On the contrary, many who exhibit these same qualities do extremely well. The point is simply that organizations—just like pilots—need to have situational awareness, and the following characteristics can help to focus some attention on potentially hazardous patterns of behavior.

Rogues are socially adept. They communicate effectively and often are accomplished at corporate politics. They seem to have the ability to play off of hot-button issues to impress superiors, while simultaneously sensing the strengths and weaknesses of peers and subordinates, and utilizing them for personal leverage. This typically results in social strata where the individual is perceived much differently by superiors than by peers or subordinates, and where there is very little ambivalence. People tend to take sides when a rogue is around and are either groupies who believe the rogue can do no wrong or enemies who sense the impending disaster.

Rogues are often untruthful. Information is power, and rogues seem to be able to use fact and fiction for personal gain. By rationalizing that their cause is worth any means of accomplishment, lying becomes simply another tool in the rogue's bag of tricks.

Rogues and potential rogues often feel that they are trapped in a system that was designed for "lesser people." They see themselves as superior to others in the organization, even those above them on the career ladder. They often seem to believe that circumstances, and not talent, loyalty, or hard work, is what has made them subordinate to others. As such, the rules of the organization should not apply to them, because they were designed to protect against the lowest common denominator, not superior beings such as themselves.

Rogues are driven, but often by different motivations. Some seek fame and glory, others only to gratify their egos, which are typically quite large. Many seek to fulfill a childhood dream or quest to be the best...something. For others it is simply money. But whatever the motivation, the ambition is strong enough to overcome the normal and healthy inhibitions that govern behavior. These characteristics are often bolstered by some early success, which the rogues see as validating their personal superiority.

Even when identified, rogues can be difficult to deal with. Keep in mind that they are often quite popular in the organization, and in some cases their exploits may have been used as a positive example by senior leadership or management earlier in their careers. They can, and often do become organizational "cult heroes." Rogues have learned what rules they can break, when, and with whom. Because of their perceived skill and expertise, some younger members of the organization see rogues as role models and begin to copy their style and actions. When this occurs, the negative impact can expand exponentially.

Don't kill creativity

Care must be taken not to overreact to behaviors which may identify a potential organizational rogue. Indeed many of these same behaviors are demonstrated by highly motivated team players, and an organization certainly does not wish to curb creativity or innovation, not if it wants to flourish or perhaps even survive in today's competitive and rapidly changing world. To the enlightened managerial eye, these personality characteristics and behavior patterns may mean either good or bad things for the organization, but they should never be ignored. For good or evil, high energy people who display the profile of the rogue bear close watching.

How does an organization protect itself against rogue behavior? Primarily in three ways.

- Select carefully and establish a culture which is non-receptive to disloyalty or rule-bending.
- Know your people by staying tapped into the flow of official and unofficial information.
- Take action when warranted—bad news doesn't get better with age.

Establishasing a nonreceptive culture: Hero or villain?

Rogues flourish in chaos. A well-organized system with clear standards and values—what many organizational management types call "cultural norms"—is very difficult to infiltrate and corrupt. Like bacteria in a petrie dish, a rogue needs nutrients to grow. These nutrients can be thought of as poor vertical communication, unwritten or ill-defined standards, malaise, and unresolved conflict. The first proactive step in this preventative process is knowing your people

Good organizational information flow is built upon trust, which begins the moment an individual enters an organization. In flying organizations, a pilots' reputations sometimes precede their arrival at a company or unit. At the first opportunity, organizational leadership should clearly lay out the standards of conduct in the organization and clear up any doubts or misconceptions that might be present as a result of rumors or reputations. If the previously noted patterns of behavior arise, resolution must be reached immediately, because to a rogue, silence is often construed as acceptance

Taking action: Bad news doesn't improve with age

At the first hint of deviant behavior, leadership must act. The specific form of the intervention is of course dependent on the situation and the individual, but one bit of advice is universal. Document the counseling, reprimand, etc. Make certain the subordinate is crystal clear about who is in charge, and what the ramifications of further deviance will be. If appropriate, discuss the situation with the rest of the organization as a whole, to preclude any chance that deviant behavior might be misconstrued as acceptable and cut any hero worship of the rogue off at the knees.

What action to take?

The particular type of intervention to take is often very difficult to determine, and is almost always extremely uncomfortable for both the leadership and the individual in question. Initially, two questions must be asked. "Is the individual salvageable?" and "What were the motivating factors behind the incident?" Once you have answered these two questions, you are prepared to act. Once you understand the motive and intent, a course of action can be determined. It is also advisable to look into the personal history of the individual by talking to former employers, peers, and subordinates. In each of the cases listed above, there was a clear trail of deviant behavior long before any official action took place.

In many cases, removing the stage on which the rogue operates will be sufficient to curb deviant behavior and may well cause a true rogue to self-select out of the organization. If this does not curb the behavior, dismissal and follow-up communication with the rest of the organization will be required. Explain in sufficient detail the reason and necessity for the action, and use the opportunity to reestablish organizational and cultural standards.

Seven principles for organizational flight discipline

Flight discipline ends in the cockpit, but often begins with the organization in which an individual flies. In order to provide a healthy environment for uncompromising flight discipline, conscious and deliberate attention should be paid to the issue. The following guidelines are offered to help organizations establish conditions conducive to uncompromising flight discipline.

1. Talk the talk by speaking often on the issue of standards. Encourage two-way communication both vertically and horizontally within the organization on all issues that concern regulations, procedures, and company policy.

2. Walk the walk, and set the example by following all directives to the letter. Discipline is fostered by congruence between word and deed. If you slip up as a leader, admit it and correct it. Leadership must not take advantage of positions to cut corners in any areas, on the ground or in the air. If you follow the rules, you can expect compliance. If you act like a part of a team, you can ask for teamwork.

3. Select personnel carefully. This applies to initial hires as well as internal moves within the organization. Consideration must be given not only to who is best qualified for the job, but to what signals might be sent to others in the organization by the action being contemplated. Do not leave an individual in a position of authority who practices poor discipline.

4. Resolve conflicts quickly and openly. Often organization policy can conflict with required procedures or even regulatory guidance—especially if the procedures or regulations change. Consider the case study in Chapter 1, where the CT-43 Airlift Wing was unable to change lower-order organizational policies to comply with new regulatory guidance. The unintended—but obvious—signal sent to the aircrew of IFO 21 was that

compliance was optional. It should not have come as a great surprise when the crew did not follow existing regulations—they were following the lead of their commanders.

5. Identify the "latent pathogens" or unsafe conditions and attitudes. Keep open lines of communication, and actively look for problem areas. When an unsafe condition or attitude is identified, attack it resolutely. The longer a pathogen exists, the deeper it becomes ingrained in the culture of the organization. This can lead to an undisciplined mindset that perceives an organization willing to tolerate unsafe conditions.

6. Fix the problem, not the person—unless the person is the problem. Don't shoot the messenger or the whistle blower when bad news is uncovered. One of the most effective leadership tools available for instilling discipline is to shift focus and resources to resolutely address a problem that has been identified by lower echelons. However, if rogue practitioners are suspected in your organization, they should be dealt with in a forthright, documented, face-to-face manner. This will preclude any notion of managerial complacency or malaise.

7. Trust your intuition. If open channels of communication exist, some hint of poor discipline will manifest itself in one form or another. Perry Smith, in *Taking Charge: A Practical Guide for Leaders*, states that any hint that something may be amiss should result in "raising additional questions, and postponing important decisions until you are reasonably comfortable you are choosing the right course of action." Quoting Emerson, he goes on to say, "The essence of genius is spontaneity and instinct. Trust thyself." (Smith 7.)

Chapter review questions

1. In what ways do organizational policies and decisions affect personal flight discipline?

2. What trade-offs are decision makers required to make that might affect flight discipline?

3. What is the difference between organizational climate and culture?

4. What are "latent pathogens" and how might they affect discipline?

5. How might organizational change affect flight discipline?

6. What are the characteristics of a rogue practitioner?

7. What are the seven organizational principles listed to foster flight discipline? Are there others?

References

Rawnsley, Judith H. 1995. *Total Risk: Nick Leeson and the Fall of Barings Bank.* New York: HarperCollins.

Reed, Robert C. 1968. *Train Wrecks: A Pictorial History of Accidents on the Main Line.* New York: Bonanza Books.

Sheehan, Neil. 1971. *The Arnheiter Affair.* New York: Random House.

Smith, Perry M. 1986. *Taking Charge: A Practical Guide for Leaders.* National Defense University Press: Washington, D.C.

Westrum, Ron. 1996. "Human Factors Experts Beginning to Focus on Organizational Factors in Safety." *ICAO Journal.* October. pp 6–8, 26–27.

Wolfe, Sidney, and Mary Gabay, et al. 1996. *Questionable Doctors: Disciplined by the States or the Federal Government.* 1996 edition. Public Citizen Health Research Group: Washington, D.C.

Yeager, Chuck. 1994. Personal Interview. Edwards Air Force Base, CA. March.

Part Two

The anatomy of flight discipline

6

Personality factors
and flight discipline

People have one thing in common—they're all different.
Robert Zend

Since the earliest days of human flight, pilots have held that they are a different lot. Research findings bear this out, but not always in the way we might think. A quote from Roderic Beaukieu's "The pilot and the thinking machine" provides us with the personality profile of the daring pilot of myth.

> *"...the aircraft pilot is tough, courageous under fire, decisive, strong in leadership skills, and independent. The pilot is always a male, kissing his girlfriend or wife (or both) good-bye as he goes off cheerfully to face his destiny. A pilot has a confident, reassuring voice, generally with a southern drawl like Chuck Yeager. He chases the stewardesses, can stay alert in six time zones, can foil a hijacker, and at the same time he has the skills to float a hundred-ton piece of machinery onto a runway with barely a tremor." (Dietz and Thorns 152.)*

While this personality profile is certainly a myth (at least we hope so), there are many aspects of pilot personality that we now know as fact. From across the spectrum of aviation, we see unique similarities in aviators. For example, studies of Navy pilots show that they differ from "normal" people in many ways. They are more achievement oriented, exhibitionist, dominant, heterosexual, and aggressive than the population at large (Fry and Reinhart 1969). A study of male and female general-aviation pilots in 1974 demonstrated that many pilots of both genders have a "hero" or "heroine" personality type, often seeking thrills and adventure (Novello and Youssef 1974). Additional studies show that personality types can be

123

used as predictors of success in flight operations, as well as determining who is likely to have the next accident. Other research has shown pilots to be more likely to ignore stress and other physiological concerns, including illness and the effects of alcohol. Do any of these traits sound familiar? Does our personality impact upon our discipline? You bet it does, and in this chapter we will see how critical it is to understand our own personality and how it fits into the flight discipline equation.

The early picture of pilot personality

The stereotype of the pilot as a lone eagle, a risk-taking daredevil may well have been true in the beginning. In a 1918 edition of *Lancet*, a British medical journal, the first-documented scientific analysis of pilot personalities was done by one pilot and a medical doctor trying to determine the general makeup of a "good" pilot. The article is at once entertaining and enlightening, and the excerpts below illuminate that, in many ways, pilots have not changed much from the earliest days of human flight.

The essential characteristics of the successful and unsuccessful aviators: with special reference to temperament
(Rippon and Manual, 1918)

"The enormous number of pilots who have qualified lately is proof that the aviator is not a "super-man." It is true that we see certain men who perform miraculous stunts, but when we come to talk to them and examine them we find that they are quite ordinary people.

Flying is now confined to the public school boy, the cavalry officer, or the athlete (early preselection criteria). We take many of our pilots from the lower middle classes and some from the artisan class. The most useful method for discovering whether or not the pupil is likely to become successful is to study the life history of the pilots who we know to be efficient. This demonstrates the there are many characteristics common to the successful pilot which are absent from the pupil who has been withdrawn from instruction in flying on account of lack of aptitude (washouts).

Sport

The successful aviator has always had the attributes of a sportsman...He joins the Air Force because he is keen on flying and it appeals to his sporting proclivities.

Character

He possesses resolution, initiative, presence of mind (early SA?), sense of humour, judgment, is alert, cheerful, optimistic, happy go lucky, generally a good fellow, and frequently lacking in imagination.

Age

The majority of successful pilots are under 25 years of age, the explanation obviously being that the resiliency of youth enables them to accustom themselves more rapidly to a new occupation and to recover quickly...from strain and stress...

High-spirited

Anyone who has lived with pilots for any length of time cannot fail to notice that they possess a very high degree of...animal spirits and excessive vitality.

Amusements

When they have finished flying for the day, their favourite amusements are theatres, music—chiefly ragtime, cards, and dancing, and it appears necessary for the well being of the average pilot that he should indulge in a really riotous evening at least once or twice a month.

Alcohol

Alcohol is taken freely by the older men, but the young, fit pilot hardly ever touches it...pilots are well aware of the danger in taking too much before flying ...the fit pilot needs no stimulant.

Occupation

We found that the best type of pilot was seldom drawn from a sedentary occupation, that those who had lived a sheltered life were not as good as those who had 'roughed it.'

Hands

One of the most important characteristics we have noticed in successful aviators is "hands." This characteristic is difficult to define (and it still is today), but may be described as follows. A horse rider with good hands is able to sense the mentality of the horse by the feel of the reins and also to convey his desires accurately to his mount. We have never known of a man who has consistently been in the first flight making anything but a good pilot. In the same way, a pilot with good hands senses unconsciously the movement of the aeroplane, and rectifies any unusual or abnormal evolutions almost before they occur. Skillful pilots appear to anticipate 'bumps.' He is invariably a graceful flyer, never unconsciously throws undue strain on the machine, just as a good riding man will never make a horse's mouth bleed.

Mechanical knowledge

The question whether this type should possess a knowledge of mechanism, and of the ways and wherefores of flying is a very debatable point. The authors, however, desire to express their definite conviction that the less the fighting scout knows about his machine from a mechanical point of view, the better. From the very nature of his work he must be prepared to throw the machine about, and at times subject it to such strains, that did he realise how near he was to the breaking point, his nerve would go very quickly...

Characteristics of unsuccessful pupils

Early life

...Did not play games...Amusements usually indoors...

Object in joining the Flying Service

Usually on account of pay or to avoid the infantry. Do not show any enthusiasm for flying.

Mentally

They are sluggish, dreamy, emotional, self-conscious, lacking in sense of humour, and devoid of that elusive quality commonly known as "guts."

> Whilst under instruction the unsuccessful pilot flies erratically, grips the control lever much too rigidly, makes unnecessary movements jerkily, and to onlookers gives one the same impression of a badly ridden horse. In short, he does not possess 'hands.'"

The contents of this study are illuminating. The discussion of the temperament of both successful and unsuccessful pupils suggests much the same as more recent rigorous studies, basically that extroverts and good communicators make good flyers. There were other interesting aspects of the study as well which have been replicated over the past 80 years.

Successful pilots were characterized as "good on the hunt," meaning that they were usually excellent with horses and had "good hands" and the ability to manage changing situations while in motion. They were also described as competitive and typically good at games on the school yard. Even from this early study, we can see several elements of a modern pilot's personality profile. Good pilots are still competitive, although not to a fault. They still manage developing situations well. We now call this trait situational awareness. Good pilots still require excellent hand-eye coordination, which is more physiological in nature, but the confidence born of this physical coordination is typical in the successful pilot's personality profile. We have come a long way since 1918, in aviation as well as aviation psychology. We now know much more about what makes the pilot mind tick.

New demands on the pilot

Modern airmanship demands more than individual achievement, and certain personality characteristics are seen as beneficial to the modern team-oriented environment. Communication and team decision-making skills are far more important today than at any time in the history of human flight. Personality traits are now seen as the unchangeable foundation upon which a pilot makes her or his airmanship develop. For many who possess the traits necessary for shared decision making and good communication, this is good news. For those who do not, the key is self-awareness and conscious effort in these key areas. For all of us, regardless of personality type, an awareness of our strengths and frailties—as well as being sensitive to those of others—is critical to safe and effective airborne operations.

It is also worth noting that the 1918 *Lancet* study said that pilots did not need to be too smart or mechanically inclined because if they understood what held the aircraft together and the aerodynamic principles that held them aloft, they may become too frightened to continue flying at all. I sincerely hope we have overcome these tendencies, although one can certainly understand the fear factor involved in flying the Wright, Blériot, Nieuport, Fokker, and Curtiss aircraft of the first two decades of powered flight.

From the dawn of human flight, personality characteristics have shaped pilot performance, and for good or evil, we are typically stuck with what we have.

Personalities are permanent

Personality traits must be viewed as "enduring components" (Gregorich et. al. 1989) or permanent fixtures of ourselves. Unlike attitudes, technical skills, or knowledge acquisition—all of which can be improved with education and training—you must play the hand you were dealt regarding personality traits. Most psychologists agree that personality traits are developed very early in childhood, and some researchers even postulate that our personality makeup is determined in the womb or at the moment of conception.

How personality shapes performance

Researchers have found that personality traits impact pilot behaviors and performance in a variety of ways (Foushee 1982; Kanki 1991). If we are to strive for maximum performance, it becomes incumbent upon us to understand, even if we cannot change, what effects our personality may have on our flight behaviors and discipline. Our personality affects three primary areas related to flight discipline: how we communicate, how we handle stress, and how we make decisions.

Personality factors and communications

Your personality type can either enhance or severely inhibit the flow of information both inside and outside of your cockpit. This is especially true if you are in a leadership position, such as a pilot in command of a crewed aircraft, or if you are acting as an instructor.

Communications research has become very technical, so in order to discuss this phenomenon further, a few definitions are in order. Personality traits are typically discussed in terms of instrumentality—or goal orientation, and expressiveness—or people/communication orientation. If we break this down a bit more into everyday language, pilots with a high level of instrumentality will focus on achievement. They will seek mastery and will express an interest in improving current skills and learning new ones. Instrumental personalities also see the value of hard work and are not afraid to roll up their sleeves and get busy. Finally, instrumental pilots are competitive and will attempt to outdo others if the opportunity arises to do so (Gregorich et. al., 1989).

On the flip side, expressive attributes "reflect an interpersonal orientation and the notions of warmth and sensitivity to others. Verbal aggression is a form of negative expressivity, and refers to a type of nagging hostility directed toward others." (Gregorich et al., 1989, 2)

People—especially pilots—are far too complex for a single-category approach. For example, you may be primarily instrumental and yet occasionally practice verbal aggression to achieve your objectives. To get a better handle on what these interactions of personality types mean to aviators, researchers at the University of Texas and NASA-Ames Research Center developed "clusters" of personality traits which more accurately portray the complexity of individuals found in cockpits around the world. Even better, they communicated their findings in a language pilots can easily understand, labeling their clusters as "right stuff," "wrong stuff," and "no stuff." Let's take a look at each.

Pilots with the right stuff are identified as individuals who are strong in both goal orientation and communication. Also, pilots with the right stuff lack any tendency towards verbal aggression or an autocratic or dictatorial interpersonal style. This makes sense, as goal orientation and the ability to communicate should make for better performance in both leaders and followers in the aviation environment.

Pilots with the wrong stuff were identified as possessing high motivation for goal achievement, but improper means of obtaining it, including verbal aggression. They are often seen as "rugged individualists"—more intent on forcing their own objectives and solutions down the throats of others than in being a team player. It is easy to see why this label applies.

An example from the cockpit of a major U.S. airline provides a classic illustration of this personality style, as well as how internal and external factors can contribute to magnifying latent personality traits.

According to an article published in *The Wall Street Journal,* Captain Wayne O. Witter, who liked to be called Captain WOW (for his initials) in the cockpit, was "a rigid and domineering personality" (Brannigan A10). On one European trip it all came to a head. As Brannigan writes:

> *"The mood was set from the first flight. The engineer refused to address Mr. Witter as "Captain WOW," saying in a later report to superiors that he didn't want to feed an 'over-inflated ego.' More importantly, the engineer charged that Capt. Witter had reacted hostilely when his crew tried to point out errors...The conflict culminated at the gate in Frankfurt, Germany 10 days into their assignment; Second Officer Sweeny's report to superiors said that Capt. Witter had 'a screaming fit in the cockpit.' (A10)*

The flight engineer summed up the difficulties involved with working in an environment dominated by a wrong-stuff pilot in command. "Capt. Witter's terribly bitter attitude, along with his violent, aggressive personality make it extremely difficult for other crew members to perform their duties." (Brannigan A10). As difficult as this may seem, pilots with no stuff can be just as hazardous.

Pilots with no stuff score low in both goal orientation and communication capabilities. These pilots are often drawn to aviation by the desire to interact with the technology rather than people. As pilots in command, no-stuff aviators are not overly concerned with either the mission or the rest of the crew, are typically content to let a situation develop without intervention, and often must rely on others to speak up on their own in an assertive fashion to prompt action that is not specifically directed by policy or procedure.

So what does this mean for us? Many studies have shown that aircrew performance is linked to communication patterns, which deal directly with personality traits. Each of us should conduct an honest self-appraisal and determine what this means for our flying activities. As for me, I have come to realize that I tend to be overly competitive and occasionally feel that my solutions to problems are inherently better than others. In short, I have a bit of the wrong stuff in me. I have consciously tried to counteract these tendencies by first recognizing scenarios where my competitiveness could lead to poor flight discipline, and secondly by actively seeking more input to my decision-making process before acting on my own internal gut instincts. I may not be able to change my personality, but I can be aware of the dangers it poses and change my behaviors to account for it.

Personality factors and stress

Different personality types handle stress in different ways, some better than others. The CRM program of Transport Canada provides us with a simple and yet useful way in which to view these differences at a macro level. They suggest that people are either "hot reactors," who are almost immediately susceptible to stress in most situations or "cool reactors," who are less likely to respond immediately to stress inputs.

Hot reactors are often very achievement oriented and see stressors as having an immediate impact on their ability to accomplish the mission. If you are in this category, as I am, you should realize your potential for worsening a problem through self-induced stress. This can often come about as a direct result of worrying inordinately about mission accomplishment. Additionally, hot reactors can become more easily distracted and run an increased risk of losing situational awareness. Finally, reacting too strongly to outside stressors can impair normal decision making and hence have a negative impact on flight discipline.

A closer look into the case of Captain WOW clearly reveals the hazards associated with being a "hot reactor." On the European trip, the flight engineer reported that the captain busted several altitude restrictions as a result of his conflict with the crew. "Calling the crew insubordinate, he (Witter) said, 'I felt like I was flying solo.'" (Brannigan 1996, A10) Perhaps due to his hot reaction to the stress of the moment, he was.

Cool reactors deal more effectively with mounting stress, a situation which can often mask the buildup, both to themselves and others on the cockpit team. Make no mistake about it; stress impacts us all, and regardless of personality type, we can be overwhelmed. The next chapter on external factors and flight discipline will discuss the stress phenomenon in greater detail.

Personality factors and decision making

In a landmark study by noted aviation safety specialist John Lauber, it was discovered that seven common elements contribute to most accidents and incidents in the flight regime. Based on the discussions above on the relationship between personality characteristics and communication and personality and stress, see if you can identify links between the personality profiles we have discussed and the contributory elements of aviation accidents listed below.

1. Preoccupation with minor mechanical problems.
2. Inadequate leadership.
3. Failure to set priorities.
4. Inadequate monitoring.
5. Failure to delegate tasks and assign responsibilities.
6. Failure to utilize available data.
7. Failure to communicate intent and plans.

According to Robert Helmreich, "It does not take a large leap of faith to posit that the personalities of crew members should be related to their style of management." (Helmreich 1982, 9). Armed with this list, and a subjective analysis of your own personality style, you should be able to make some connections to help you avoid some of the most prevalent human errors. For example, if you share some wrong stuff with me, and tend towards an autocratic decision-making style in the cockpit, then perhaps we should focus on errors #5 and #6. By seeking more inputs to our decision making and then using the data that becomes available, theoretically we should become more disciplined and better decision makers. If you tend towards a *laissez faire*, no-stuff style, you might want to focus on the first three errors, as this should force you to take a greater role in the command of your aircraft.

Perhaps nowhere in aviation are the demands placed upon a pilot greater than in naval aviation. Yet a great deal of negative press has clouded the picture of the naval aviator. Let's shift gears for a moment and take a holistic look at the personality type associated with this most demanding of flight regimes.

The "top-gun mentality"

Somewhere in the nether regions between attitudes and personality exists a condition I will choose to call a mindset. I'm certain that psychologists have identified the specific domain of what I refer to in more sophisticated terms, but for our purposes I would like you to think of it as a state of mind which predisposes individuals towards certain types of actions when confronted with appropriate stimulus or opportunity. Often, these predispositions end tragically if they are not known and consciously countered.

In early 1996, Navy Lieutenant Commander Stacy Bates took off from Nashville, Tennessee, and accomplished an extremely nose-high afterburner climb into the weather with his F-14 Tomcat, apparently for the benefit of family and friends who had come to the airport to see him and his back-seater off. Moments later, the aircraft was seen descending from the clouds in a nearly vertical attitude and impacted the ground, killing three civilians on the ground. No ejection was attempted. Perhaps due to the loss of civilian life, or perhaps because this was the second F-14 that Lt. Commander Bates had crashed, the event was quickly picked up by the wire services and made headline news nationwide. A human-factors debate was played out in the public eye, with many asking difficult questions about the personalities who fly in defense of our nation.

The discussion that followed this incident centered around the term "top-gun mentality." The secretary of defense, obviously suffering from the negative media coverage of the accident, and still stinging from the aftermath of the Tailhook incident, used the term negatively. He cited the need to bring the "top-gun mentality" under control in the military. He was obviously referring to the tendency for some aviators to "hot dog" with their aircraft, much like Maverick, the renegade naval aviator in the hit movie. There is a lesson here for military fighter jocks, as well as all the untold thousands of wanna-bes.

Certainly, combat aircraft are not toys for exhibitionism. However, the layman's understanding of the term "top gun" is based on fiction—the movie—which is loosely based on fact—the military's fighter weapon's schools. Let's cut through these confusing layers and get back to the original concept of the term, the one based on fact. The lessons from the real fighter weapon school approach is far more valuable than the Hollywood version.

Military operations in combat are truly the crucible of expertise and flight discipline. Donald Bringle, a career Naval aviator and former

Navy Fighter Weapons School instructor, comments on his view of the real top-gun mentality in excerpts from a *Naval Proceedings* commentary in April 1996.

The real top-gun mentality is a positive force...Never have I been associated with a more professional, highly talented, intensely dedicated group of aviators. The standards expected of every top-gun instructor, and consequently of every student, remain higher than those of any other organization in aviation.

The standards that Commander Bringle speaks of are a necessary prerequisite to carrier aviation, due in large part to the demanding nature of the mission. Few of us will ever approach the airmanship and discipline demanded of carrier-based aviators, which is a tough admission for me to make as an Air Force pilot, but true nonetheless. The commander explains.

> *"Every day of the year, on every carrier deployed overseas, there are nearly 300 young aviators on six-month deployments, braving extreme conditions, night arrested landings with no divert field in the middle of the Indian Ocean or North Atlantic, where the deck can pitch 30–40 feet in a cycle. Yet the leadership still launches the aircraft, and the young men and women launch without question because that is what they are expected to do. In fact, they revel in the challenge of it all."*

The external demands placed on carrier aviators have molded these aviators into a disciplined lot—at least at sea—in spite of Hollywood characterizations to the contrary. There is much to learn from the positive personality characteristics displayed by these professional aviators, and Commander Bringle leaves us with a few valuable insights as to the internal makeup of these gifted flyers.

> *"This lifestyle demands an individual who must have supreme confidence in his own abilities, yet maintain a high respect for the aircraft which he flies, and the conditions through which he has to operate the aircraft...there has to be a sense of purpose for an individual to want to make a career of this. He does it for the love of flying, the camaraderie of his associates, and for the love of his country. This is the true top-gun mentality."*

From all of the research into the psychological makeup of successful aviators, two aspects of pilot personality are especially crucial. First, with regards to personality profiles, we play with what we bring to the game. Secondly, some personality traits are more conducive to success in modern aviation than others, and in fact, some are downright destructive. So how does a knowledge of personality styles—our own as well as others'—benefit flight discipline?

Self-assessment is the key

The key to better flight discipline is honest appraisal and self-assessment. While there are many excellent diagnostic tools for determining what personality traits you possess, you probably don't need one to tell you what type of personality pattern you fall into. Intrinsically, you know whether you are goal oriented or people oriented, whether you initiate conflict or avoid it. You already recognize whether you are a "hot reactor" or a "cool reactor" to stressful situations. Based on these uncomplicated initial assessments, you can prepare yourself for the possible consequences.

If these common-sense assessments are not enough, there are many psychological instruments available, some of which can be taken online from your home computer.

A short list of references is provided at the end of this chapter for further study and self-analysis with regards to your individual personality traits.

Chapter review questions

1. What are the four typical characteristics of the instrumental personality type?
2. What are the hazards associated with a "hot reactor" to stress?
3. What are the seven major human error contributors responsible for most aviation accidents?
4. What does Donald Bringle argue is the real top-gun mentality?
5. What can be done to compensate for personality characteristics that might be conducive to poor flight discipline?

Additional resources on personality types for self-assessment and understanding

Books

Briggs-Myers, Isabel (with Peter Myers). *Gifts Differing*. Consulting Psychologists Press, 1980. ISBN 0-89106-011-1 (pb) 0-89106-015-4 (hb).

Briggs-Myers, Isabel and McCaulley, Mary H. 1985. *Manual: A Guide to the Development and Use of the Myers Briggs Type Indicator*. Consulting Psychologists Press.

Hergenhahn, B. R. 1990. *An Introduction to Theories of Personality*. Prentice-Hall, New Jersey.

Jung, C. G., Baynes, H. G. (translator). *Psychological Types*, Bollingen Series, Princeton U.P., 1971. ISBN 0-691-01813-8 (pb) 0-691-09770-4 (hb).

Keirsey, David and Bates, Marilyn. *Please Understand Me*, An Essay on Temperament Styles. Prometheus Nemesis Book Company, P.O. Box 2748, Del Mar, CA 92014. One of the more widely known books describing the Myers-Briggs Type Indicator. It includes a self-test.

Keirsey, David. 1987. *Portraits of Temperament*. Prometheus Nemesis Book Company, P.O. Box 2748, Del Mar, CA 92014 (619-632-1575).

Kroeger, Otto and Thuesen, Janet M. *Type Talk*. Bantam Doubleday Dell Publishing Group, Inc. (Tilden Press also mentioned.) ISBN 0-385-29828-59. An easy-to-read book that gives profiles for all sixteen personality types.

Lawrence, Gordon. *People Types and Tiger Stripes*. Available from Center for Application of Psychological Type, Gainesville, Florida. ISBN 0-935652-08-6.

Lowen, Walter (with Lawrence Miike). 1982. *Dichotomies of the Mind: A System Science Model of the Mind and Personality*, John Wiley. ISBN 0-471-08331-3.

Schemel, George J. and Borbely, James A. *Facing Your Type*, Published by Typofile Press, Church Road, Box 223, Wernersville, PA 19565.

Internet sites

http://www.teleport.com/~perceive/p_self.htm#Self-Help

This page has links to multiple resources and self-tests with full scoring capabilities and explanations.

http://sunsite.unc.edu/jembin/mb.pl

This site allows you to take the The Keirsey Temperament Sorter, a personality test which scores results according to the Myers-Briggs system.

http://user.cs.tu-berlin.de/~doering/test.htm#personality

A potpourri of interesting and fun personality tests and explanations.

References

Beaty, David. 1995. *The Naked Pilot: The Human Factor in Aircraft Accidents*. Airlife Publishing Ltd.: Shrewsbury, UK.

Brannigan, M. 1996. "Captain WOW: When is the Mental State of a Pilot Grounds for Grounding Him?" *The Wall Street Journal*. March 7. New York.

Bringle, Donald. 1996. "The Top-gun Mentality." *Proceedings*. April. p. 8–9.

Deitz, Shelia R., and William E Thoms, eds. 1991. *Pilots, Personality, and Performance: Human Behavior and Stress in the Skies*. Quorum Books: New York.

Edens, E. 1991. "Individual Differences Underlying Cockpit Error." Unpublished doctoral dissertation. National Technical Information Service, Springfield, VA.

Foushee, H. C. 1982. "The Role of Communications, Socio-psychological, and Personality Factors in the Maintenance of Crew Coordination." *Aviation Psychology II*. November. pp. 1062–1066.

Fry, G. E. and R. F. Reinhardt. 1969. Personality Characteristics of Jet Pilots As Measured by the Edwards Personal Preference Schedule. *Aerospace Medicine*, 40, pp 484–486.

Gregorich, S., Robert L. Helmreich, John A. Wilhelm, and Thomas Chidester. 1989. "Personality Based Clusters as Predictors of Aviator Attitudes and Performance. In: R. S. Jensen's (ed) *Proceedings of the 5th International Symposium on Aviation Psychology*, vol. II. Columbus, Ohio. pp 686–691.

Helmreich, R. 1982. "Pilot Selection and Training." Paper presented to the annual meeting of the American Psychological Association, Washington D.C. August.

Kanki, B. G., M. T. Palmer, and E. Veinott. 1991. "Communication Variations Related to Leader Personality." NASA-Ames Research Center, Moffett Field, CA. Publication #231.

Novello, J. R., and Z. I. Youssef. 1974. Psycho-social Studies in General Aviation: Personality Profiles of Male Pilots. *Aerospace Medicine*, 45, 185–188.

Novello, J. R., and Z. I. Youssef. 1974b. Psycho-social Studies in General Aviation: Personality Profiles of Female Pilots. *Aerospace Medicine*, 45, 630–633.

Rippon, T. S., and E. G. Manuel. 1918. "The Characteristics of Successful and Unsuccessful Aviators, with Special Reference to Temperament. *The Lancet*. September 28, 1918. pp. 411–415. London.

Transport Canada. 1997. "Crew Resource Management: Stress Management." Internet: www.caar.db.erau.edu/crm/resources/misc/transcan/transcan6.html.

7

Hazardous attitudes

When you see a snake, never mind where it came from.
W. G. Benham

Good news and bad news, the good first: Psychologists and operators have been able to categorize certain hazardous attitudes into
nicely defined packages, along with "antidotes" for pilots to prevent
them. The bad news? The word is not getting out, evidenced by
dozens—if not hundreds—of accidents and incidents that are caused
by pilots who fall victim to classic hazardous attitudes each year. Attitudes, unlike personality characteristics, can be modified. Awareness and assertive action by individuals can purge hazardous
attitudes. That is the goal of this chapter. Before we get into listing
and discussing these demons, let's look at a case study and see how
many we can identify for ourselves. The following case study illustrates a number of both internal and external factors which negatively affected an otherwise excellent pilot. The result was tragic, but
predictable.

Case study: Pushing the limits (Hughes 1995)

Until recently, the U.S. Air Force practiced a unique and rather
challenging mission to off-load cargo from an extremely low flying
(5–7 feet above the ground) aircraft. These missions were called
the Low Altitude Parachute Extraction System (LAPES). The aircraft
flying just barely above the ground with the rear cargo door open
executed a LAPES drop. As the crew approached the extraction
zone (EZ), a drogue parachute, attached to the cargo, was inflated
out the rear of the aircraft and the cargo was "skidded" into the
landing zone. The idea was to deliver cargo into a hostile area that
needed precise airdrop but was too "hot" to land at. The perfect
example of an effective LAPES operation was at the battle of Khe
Sanh in Vietnam, where C-130 deliveries were the lifeblood for

trapped U.S. Army forces surrounded by North Vietnamese regulars. LAPES deliveries require exact discipline and effective crew coordination. Occasionally, however, a pilot sees the LAPES mission as an opportunity to strut his stuff.

During a practice LAPES mission for a demonstration the following day, the mishap pilot executed a steep approach, two-step flare, and a very aggressive climb out that was outside established regulatory standards (MACR 55-130). The entire practice sequence was observed by wing and squadron supervisors. The aggressiveness of the practice maneuver frightened the copilot and the flight engineer, and after the mission, the copilot confronted the pilot about the lack of discipline on the practice LAPES delivery, and who agreed to "shallow out" the maneuver the next day.

Although the copilot's assertiveness was a step in the right direction towards restoring discipline, the squadron commander later praised the aircrews for an outstanding performance on the practice flight, thereby inappropriately reinforcing the mishap pilot's undisciplined behavior. Prior to the actual LAPES demonstration flight the following day, a copilot change was made, thereby negating the previous agreement to "shallow it out" and significantly reducing the pilot crew experience level. The accident chain was building.

On the day of the demonstration, the pilot was intent on making a steep approach to enhance the entertainment value of the event. He established an excessive nose low attitude and a vertical velocity of 2900 feet per minute within 100 feet of the ground, a flight condition from which recovery was not possible. The aircraft impacted the ground with sufficient force to fracture the aircraft structure.

As the aircraft skidded out of the extraction zone, it passed over a crest, and at approximately 1,680 feet from touchdown the engines and propellers departed the aircraft. At approximately 1,800 feet, the right wing tip contacted a parked vehicle, which was propelled down track about 35 feet and exploded. At approximately 1,850 feet, a jeep was struck by the left external fuel tank and an Army soldier in the passenger seat of the jeep was killed instantly. At approximately 1,900 feet, the aircraft flipped over and fell to the ground 2,040 feet from the initial impact. During the rotation the empennage separated and came to rest vertically 160 feet beyond the center wing section. There were many sources of ignition: hot engines and components, hot electrical components and possible sparking. All four main fuel tanks, oil, and fuel vapors ignited resulting in a fireball and flash fire.

Both loadmasters sustained fatal injuries during the mishap sequence. The pilot and navigator survived the crash, but received horrific facial burn injuries in the post-crash fire. The copilot and flight engineer were found outside the flight deck area but within a broken perimeter of fire and received serious burn injuries. The aircraft was destroyed by fire.

Air Force regulations stated, "It is permissible to use an approach altitude higher than 200 feet if operational requirements dictate; however, in all cases, base the descent point on gross weight, ground speed, and pilot judgment. The descent rate should normally be 1000 to 1500 feet per minute." This pilot had nearly doubled the recommended descent rate to put on a good show.

The mishap chain of events began with the mission briefing for the Monday practice. Significant emphasis to provide a "big show" was directed at all aircrews from the airborne division commanding general. Although the wing commander stressed it was just another training exercise and to perform the mission safely, the mishap pilot intended to put on a good show. After Monday's practice mission, the mishap pilot made certain conscious decisions regarding the manner in which he would perform the next extraction (hidden agenda). At the onset of the mishap the pilot intentionally delayed drogue chute deployment and descent in order to steepen the approach to the EZ. His confidence level played an important role in his decision-making process. The mishap pilot's confidence level can best be exemplified by the comments he made to the originally scheduled copilot. After the copilot expressed his concerns about the aggressive nose high climb that exceeded 20 degrees, accompanied by a significant loss of airspeed, the mishap pilot explained that he had a lot of experience flying LAPES that way, was very comfortable with the maneuver, and that it made for a good show. However, he relented, stating that if the maneuver really bothered the copilot that he would "shallow all that stuff out." Later Monday afternoon, the mishap pilot was overheard to say to another crewmember that he would make the next run-in a little *steeper.*

Other squadron members confirmed the mishap pilot to be compulsive, devoted, and a perfectionist. These traits were important in determining how the pilot became internally motivated. The underlying factor was self-motivation, a desire to magnify his self-image, reputation, or career. The role of outside motivating factors can be illustrated by the premission briefing. The commanding general underscored the need to "fill the sky with parachutes" and to put on a "big show." Although the mishap pilot's aberrant behavior on the previous day's demonstration caused concern for the wing director of operations (DO) and the squadron commander, they applauded his effort at the postmission debrief.

The skill and experience of the pilot was well known by all crewmembers on the aircraft, and this could have resulted in overconfidence in his abilities. A no drop/go-around was not suggested by the mishap copilot or any other crewmembers, in spite of the fact that they certainly recognized the abnormality of the situation. The mishap pilot was perceived by the other crewmembers as being highly experienced and skilled (excessive deference). The copilot had only 200.1 hours of C-130 flying time. He was a dedicated, highly motivated individual, but lacked knowledge and experience in not only LAPES, but in the C-130 in general. During the mishap sequence, the crew relied entirely on the mishap pilot's ability to conduct the approach. What developed was a copilot syndrome whereby the copilot and possibly other crewmembers were satisfied to have the pilot in complete control of the final approach. Their trust was poorly placed.

This scenario depicted a number of hazardous attitudes, including the airshow syndrome, pressing, macho attitude on the part of the pilot, and complacency on the part of the copilot and the rest of the crew. Any one of these in isolation can be deadly, and in combination can be too much to overcome, even for so-called "golden hands" aviators. Let's look a bit closer at each of the hazardous attitudes.

Hazardous attitudes impacting flight discipline

Aviators understand the need for control, but we typically think of it in terms of airspeed, altitude, and heading. Equally important from a flight discipline perspective is an inner control. Pilots come to aviation from across the spectrum of society, and we fly for a wide variety of reasons. Some motivations lend themselves to a solid base of professionalism, and others are potentially hazardous. It is both appropriate and necessary that we can identify hazardous attitudes and possess tools to modify them when we encounter them. Flight discipline demands no less. These descriptions are likely not complete or conclusive, and you may know of many others, but the important point is to recognize any attitude that may impact upon airmanship and have a plan to counteract it if it has a negative impact and enhance it if it is positive. The following are descriptions of some of the more common hazardous attitudes.

Pressing (also known as get-home-itis)

More critical judgment errors are made in the name of "getting the job done" than perhaps all the rest of the hazardous attitudes combined. In the military, these attitudes are often described by other, less harsh-sounding terms. Aircrews are often rewarded for "hacking the mission"—meaning bending or possibly breaking the rules in the sacred name of the mission. During Desert Storm, several aircraft were either lost or damaged when overaggressive pilots descended below a standard operating procedure (SOP) altitude to attack relatively low priority targets. In one case, more casualties resulted during the rescue attempt to get the pilot out of enemy territory.

In commercial flying, pilots occasionally press lateral distance limits from thunderstorms, to get through to their destinations. In general aviation, "scud-running," the VFR pilot's attempt to stay out of the weather by hanging just under the ragged edge of a bad weather system, is a common manifestation of pressing. In one incident, taken from the ASRS database, a helicopter pilot with only a VFR rating was lucky to survive to share his lesson.

Case study: Pressing for the customer (ASRS 1995)

The weather in the Los Angeles basin was marginal VFR, though for helicopters, acceptable. The mission was to pick up the owner of a race track at the track, fly direct (VFR) to a country club near Palm Springs, and return approximately 1.5–2 hours later. The weather east of Los Angeles was extremely marginal VFR; Riverside at measured 700 ft overcast, visibility 1.5 and fog. March Air Force Base was about the same. Beaumont was carrying an indefinite ceiling, visibility ⅛ mile. Obviously, the bad part was from March to Beaumont. I got both an FSS and DUAT weather forecast briefing, both of which were poor to say the least.

The flight was relatively routine (SVFR at Riverside and March) until just past March, when the ground began to rise into low jagged hills.

I began flying toward the hills, still following the freeway. The ceiling got lower as the terrain rose and soon I was not only avoiding hills but dodging clouds as well. Events happened rapidly after that, as I was still stupidly flying at about 100 knots. Poof!—into the clouds—ground contact lost! I knew "down" meant death, so I, a lowly VFR only pilot, pulled aft cyclic and climbed into the clouds. Thank God I had an awesome 12 hours instrument training and knew enough to go to the attitude indicator immediately. I stayed on it until breaking out in what seemed an eternity later, though probably less than a minute in reality. I had flown the same route a number of times before, so I knew that the really tall mountains were still many miles away. My immediate concern, upon entering the clouds, was power lines hidden somewhere ahead.

I managed to keep everything on an even keel all the way, but I've read the article stating that my life span as a VFR pilot upon entering the clouds was about 178 seconds. So, I suppose that if I'm ever stupid enough to do this again, I now have only about 108 seconds left. Figuring up all the possible FAR violations I made, I came up with a total of a whopping 62!

What caused me to do such a stupid, deadly thing? Hazardous attitudes, lots of them! The "let's take a look" syndrome, the accomplish the mission, get there-itis, pressing, thing; the macho, it can't happen to me attitude, and probably just poor pilot judgment. After all, as a helicopter pilot, the world is my airport. I can land virtually anywhere if I need to—as long as I can see the ground!

If this flight had resulted in a crash and the death of myself and my passenger, I think a case could be made for suicide and first-degree murder. I sincerely thank God and my instrument instructor for the ability to pull this one out of the hat.

What will I do when (not if) this situation comes again? I will be more assertive about the realities of flying in poor weather and tell my customer that the weather is bad and the flight will be delayed/canceled. After all, he who chickens out and runs away, lives to fly another day.

Avoiding the "pressing" trap

The key to avoiding this trap is personal honesty. Be honest with yourself. Just how bad do you want or need to get to the given destination or accomplish the mission objectives? If you find yourself taking risks that you normally would never consider, ask yourself why. When you find yourself saying that you MUST get somewhere, ask yourself what price you would be willing to pay for your arrival. Don't let your desires write checks that your aircraft and skills can't cash. Remember, this is the number one hazardous attitude. Fine-tune your flight discipline antenna to recognize it, and then react with common sense. Stay inside your limits, always leave yourself an out.

Let's "take a look" syndrome

The pilot in the previous case study got himself in trouble by thinking that he could alter his plan if the weather was really as bad as it was forecast. He decided that there was no harm in taking a look. The reality of the sudden IFR encounter left him with only one option, and it was a bad one he was lucky to survive. Pilots must be aware of the potential for disaster by merely "taking a look" or "sticking our nose in it to see if there is really ice in those clouds." Once into a situation, be it clouds, winds, or whatever, you have narrowed your options considerably. In effect, by "taking a look" you may well have dealt yourself a bad poker hand, and you may end up needing a very good draw to keep from going bust.

Resisting the "take a look" syndrome

Before you stick the nose of your aircraft into something you might regret, visualize the worst possible conditions you might encounter, and then analyze your options from that point. If it is something you are comfortable handling, press on. If not, sit this one out until conditions improve.

Antiauthority

Some aviators see the relatively unsupervised environment of flight as their chance to finally be their own boss. Although a pilot in command always has final authority over the conduct of his or her flight, as aviators, we are required to abide by the rules and regulations that govern our type of flight. The authorities or air traffic controllers are not trying to "fly your aircraft for you," but merely insuring that the airspace you are transiting remains accessible and safe for everyone with whom we share the skies. Noncompliance in aviation is not the act of a rebel, but rather that of a child—someone who is probably too immature to be in command of an aircraft.

Avoiding the antiauthority impulse

Although occasionally this hazardous attitude is manifested consistently by an immature aviator, more often than not, it comes upon us quickly as a result of something that happens in flight. Perhaps a controller gives us a vector we don't feel we need or want, so we cancel IFR sooner than we should. Or maybe, the suggestion to deviate around a harmless-looking weather cell seems overly intrusive, and we react impulsively and take an unnecessary risk with a developing thunderstorm.

The regulations are not put in place for someone else; they are for all of us. If you decide to aviate without following the rules, you are endangering the lives of others, putting disciplined airmen at risk and casting all of us in a bad light. Avoid this behavior by recognizing anger in flight, and realize that it can seriously impair your judgment. Simply follow the regulations, controller's instructions, or other guidance unless an emergency or safety-of-flight concern dictates otherwise.

Machismo

Many pilots see the sky as their own personal playground where they can prove their prowess to lesser mortals. They may also see their aircraft as an instrument to these ends. While justified confidence is critical to safe and effective operations, the macho pilot is often mostly boast, which can place him or her in a position of having to attempt a maneuver merely to prove—either to themselves or others—that they can actually do what they claim. Ironically, dual instructor crews are often most vulnerable to this hazardous attitude. Even with all of their accumulated experience, two instructors are often tempted to engage in a battle of egos in the cockpit, often resulting in unhealthy competition where flight discipline is the first casualty.

Although the macho hazardous attitude permeates all sectors of aviation, it is conspicuous in the military and competition environments, where hot aircraft and supercharged egos often combine like a gallon of jet fuel and a match.

Avoiding the machismo trap

Real machismo is characterized by humble confidence, not arrogance. Let your actions in flight be indicative of a confident and conservative approach to normal and emergency situations. Impress friends and aviation colleagues by your conscientious approach to planning and flying and never attempt a maneuver to prove something to yourself or someone else, unless you are legitimately trying to improve yourself in a training environment. Be particularly vigilant when you are flying as part of a two-instructor team. Regardless of the environment in which you fly, always remember the two most dangerous words in aviation—"Watch this!"

Invulnerability

Some aviators—usually pilots—feel bullet proof. They think disaster or bad luck will always happen to the other person. These wishful thinkers are prime candidates to run out of fuel, fly into thunderstorms, or ice up and crash. This attitude often develops over time, perhaps after we have had several close encounters with bad weather, mechanical problems, etc., and survived. The more experience we have in the air, the more susceptible we are likely to be to this hazardous attitude. After all, if you survive ten years of flying without a scratch you must be leading a charmed life, right? Wrong! It's better to think that the law of averages is catching up with you. A wind shear doesn't care if you are a rookie general aviation pilot or if you have 30 years experience flying the line for a major carrier. It can and will kill you if you get overly complacent.

Fighting the invulnerability syndrome

Of course, none of us actually believes that we can't be injured or killed; it's just a matter of not realizing danger at the right moment. This is manifested in many ways, one of the most obvious being the military aviator who delays ejection until he or she is out of the safe ejection envelope, and then action is futile. Delayed decision making is another manifestation of the invulnerability syndrome. I am not suggesting that we should reach for the ejection handles (literally or figuratively) every time that the Master Caution light comes on, but that we realize the danger residing in each situation, and react with appropriate promptness and respect.

The key to combating feelings of invulnerability is to realize that every year, literally hundreds of outstanding pilots—many that are more skilled and proficient than we are—get to meet Elvis, because they did not believe it could happen to them. They took all their skill and judgment into a situation, and came out on the losing end. Unless we are foolish, we will see that there, but by the grace of God, go us. We fight a constant battle against ourselves and the elements every time we take off, and if we want to keep the sum of our take-offs and landings an even number, we must maintain a healthy respect for the often-hostile environment of flight.

Although we may get lucky and survive situations we should have avoided, disciplined and smart aviators learn to avoid similar situations in the future. But others do not. They develop the undisciplined attitude that says, "I made it once, I'll probably succeed again." Yeah, maybe you will, but make certain your life insurance is up to date.

Impulsiveness

The opposite of the invulnerability syndrome is impulsiveness. Pilots can be a hasty bunch. As controllers, we want to take charge, and this often leads to situations where we act impetuously. This is especially true when we are confronted with situations that have no clear-cut answer. We want to make a decision and we want to make it NOW, and yet we may not have all of the required information to make the call.

There are many examples where pilots acted in haste, later to regret their impulsiveness. An Airbus departing from London had engine problems and the crew took prompt action to shut down the engine—the wrong engine, which is a critical mistake on a two-engine aircraft that already has one bad motor. This critical mistake was repeated on an Army Blackhawk helicopter less than two years later, once again driving home the point that high stress situations can overcome professional aviators if they are not aware of the hazards of impulsive actions.

Overcoming impulsiveness

Airborne situations seldom require immediate decisions or action. Analyze the situation, think before you act, and give your mind a chance to overcome the excitement and adrenaline of the moment. Many instructors give students a "wind-the-clock" step, when reacting to an emergency situation, just to provide those valuable few moments to settle down and rationally react to a situation. In the front of several flight instructional manuals' sections on emergency procedures are five worth repeating at the onset of any emergency, stop—think—collect your wits. This is sound advice. Whatever technique you use, whether it's winding the clock or repeating a mantra designed to slow you down in a crisis situation, be aware of the pilot's tendency towards immediate action and prepare to counter it.

Resignation

Aviators can't afford to give up in flight. We can get frustrated and even angry at our situation, but when we give up, it's usually all over but the wake. In the same way that a driver cannot afford to let go of the steering wheel, a pilot cannot give up control of his or her aircraft or the situation. Sometimes however, we put ourselves in a situation in the air where we feel like there is no way out, and resign ourselves to fate. This is often referred to as the "what's-the-use" syndrome, or resignation.

Occasionally, situations get out of control. Perhaps we have lost situational awareness, the aircraft malfunctions or a weather phenomena gets the best of us. We must then take some actions to bring the situation back under control. The nature of the situation will dictate what options are available to you, but giving up is not one of them. One of my first instructor pilots gave me a bit of advice that has served me well over the years. He said "when you start to feel control begin to slip, squeeze the situation until you are certain you are back in command." What he advocated was a two-step process of recognition and action. Let's set up a hypothetical example and see how this works.

Suppose you are planning a visual approach to a strange airfield with multiple runways. Based upon your weather briefing and the review of the airfield diagram, you are planning a straight-in to runway 17. As you switch from approach control to tower frequency you are told that you are cleared "for a visual approach to 17, circle to land past taxiway Bravo on Runway 23." As the adrenaline cuts loose and your larynx constricts, the tower controller, now annoyed by your delayed response to his clearance, asks if you copied his clearance, and inquires if you are aware of the NOTAM on the displaced threshold on runway 23. You recall seeing something about the first 500 feet of some runway being closed, but you hadn't really paid much attention to it since you were planning on landing on 17, and just where in the heck is taxiway Bravo, anyway? What do you do? The clock is ticking.

A pilot resigned to his helplessness in this situation may simply reply "affirmative" and attempt a maneuver he is not prepared to execute. An assertive pilot must take back control of the situation, and quickly. Perhaps a response like the following would do the trick, "Tower, Cessna 123, unable to execute circling approach to runway 23. Request clearance to land on 17, or a go-around to give me time to review the procedure."

Fighting resignation

The best way to avoid resignation is to always leave yourself an "out." Don't bet your life on a single course of action. Mountain flyers learn to avoid box canyons; we all train to recover from stalls and other abnormal flight conditions. We must also train ourselves to be proactive, especially when things change in flight. By looking ahead and structuring the changing situation to meet our capabilities, we can regain control. The more training you have and the more you work on polishing your skills, the more "outs" you have. The more you know about your aircraft systems and operating procedures the better your chances that you can handle any situation.

Regardless of your experience and training, you may find yourself in a novel situation and you have two options. You can give up, and not work at a solution and let fate take its course, or you can realize that you are never helpless and that you can make a difference. A classic example of a crew that avoided this hazardous attitude was United Flight 232, where a DC-10 experienced a complete loss of hydraulics and flight controls. One pilot on this remarkable flight commented that as he sought solutions to this previously unseen problem, he recalled the words of his flight instructor who told him "Never give up—try something, anything, but don't ever give up." The crew of Flight 232 never said "what's the use." Instead, they realized that they could—and eventually did—make a difference.

Complacency

According to Chuck Yeager, complacency is the number one challenge for experienced pilots. Many of us who fly similar routes and missions routinely, understand this tendency well. We find ourselves on the same route of flight for the hundredth time with nothing new on the horizon. The result is often a dangerous level of inattention. New instructor pilots are particularly vulnerable. Once you become an instructor in an aircraft, aviation becomes more routine than when you were merely struggling to master the machine and the mission. When you combine distraction and complacency, trouble awaits. Complacency and distraction are found at the root of nearly all gear-up landings. In 1983, two separate C-5 Galaxy aircraft landed gear-up (all 28 wheels), and both crews had at least one instructor pilot on board.

Combating complacency

Fight complacency by keeping your mind busy with flight mission related activities. Challenge yourself even on routine flights, better said especially on routine flights. There are many ways of doing this. Sharpen your estimation skills by trying to guess the exact time or fuel state that you will arrive at the next waypoint. During lulls in the flight play the guessing game. What if my engine quit now? Where is the nearest emergency airfield? What is the highest terrain within 50 miles of my flight path? Avoid boredom and inattention that can result in complacency and improve your knowledge and situational awareness simultaneously.

John W. Olcott, the President of the National Business Aircraft Association sums up his opinion on pilots and complacency this way (1996).

> *Fundamentally, airmanship is about acceptance of responsibility for everything that involves your aircraft—its position, its equipment, its crew and the instructions you choose to follow...from air traffic controllers. When you choose to move throttle or power levers forward, you and you alone are responsible for a safe outcome. While successful airmanship requires using knowledgeable input from all available resources, nothing allows abdication of responsibility. Maintaining such vigilance in an atmosphere that so easily nurtures complacency is an awesome challenge (Olcott 3).*

Airshow syndrome

It seems like everyone wants to be a demonstration pilot these days. Me too. I love watching the Blue Angels and the Thunderbirds, and often wish I could perform like they do in front of an adoring crowd. But I am realistic enough to know that I cannot, and furthermore that every year pilots lose their lives, and take the lives of others, by trying to perform maneuvers they are not qualified to attempt. Even though I am trained and qualified to perform the full set of aerobatic maneuvers, I would never, repeat never, attempt to perform them at low attitude or to show someone else how good I was. I have been asked to, but have always resisted the urge because I realize that my qualifications do not extend into these parameters.

The airshow syndrome occurs when a pilot decides that "it's time to make a name for myself and impress somebody." Tragically, aviators often succumb to the temptation to "show their stuff" to friends and families. Far too often, the "show" the family and friends see, is not the one intended. A Navy F-14 crash in Nashville, Tennessee was caused in part because the pilot was performing a maximum performance takeoff into the weather while his parents were watching. The results of this unnecessary maneuver were the death of two Navy crewmen, two civilians, and the destruction of an expensive national asset. On another occasion, two Air Force pilots on a cross-country flight did an unapproved airshow for the parents of one of the pilots, who had the tragic opportunity to videotape their son's death.

Although I have no firm statistics on this, I have been told recently that we have lost over 40 vintage aircraft—also known as Warbirds, at airshows over the past 20 years. This number seems astounding, but just this summer I have read of three tragic crashes of vintage aircraft performing at summer air demonstrations around the country. Most of these events are preplanned and practiced, which makes it clear that to attempt an impromptu airshow is foolish at best, and can result in a complete loss of discipline and tragedy at the other end of the fortune spectrum.

Case study: An impromptu airshow (Hughes 1995)

A military fighter pilot had to divert to an intermediate stop for maintenance. While the aircraft was being repaired, the pilot talked with the personnel there and planned his return flight. Prior to starting his engines, he spoke with the local supervisor of flying, advising him of a planned "high-speed pass" on departure. After starting, the pilot received permission from the tower for an opposite-direction low pass prior to departure from the airfield. After takeoff, the pilot performed a series of turns to align himself with the runway 180 degrees opposite to his departure. He overflew the runway at a very high rate of speed, 100–200 feet from the runway, and when approximately 1000 feet from the departure end, he performed an abrupt pull-up with afterburner engaged. The aircraft suddenly and violently broke up, burst into flames, and crashed, killing the pilot.

The pilot was 24 years old, on his first operational tour in the F-15, and was considered inexperienced with a total of 315 hours in type, 513 hours total flying time. The investigation revealed no psychological problems. He was flying regularly and was qualified for the mission. Crew rest and duty day were not factors. Weather was clear. The pilot had planned and precoordinated a fly-by with the local on-scene supervisor (SOF). This was a fairly routine occurrence and was condoned as being a morale booster for the local maintenance personnel. If this were such a routine event, why would the pilot feel the need to max perform the aircraft to the point of destruction?

The answer may simply lie in the excitement factor associated with the airshow syndrome. This hypothesis is supported by a few other facts determined by the accident board.

On the day of the mishap, the pilot failed to arm his ejection seat, fasten his parachute chest strap, or to insert the presence or weight of external fuel tanks into the armament control panel (AP). Without the proper configuration into the aircraft's computer system, a conflict would have developed within the overload warning system (OWS), most likely causing it to shut down. As the mishap pilot made his fly-by at an estimated 550–605 KIAS, he probably pulled back on the stick listening for the warning tone of an approaching overload. The wings then failed at approximately 8.5Gs due to the high speed and high fuel weight.

Defending against the airshow syndrome

Part of the problem is that many pilots choose to fly to prove to themselves and others that they have "the right stuff" to be a pilot. Once this goal is met, they look for how this new found persona can be used to prove even more. If your motivation for flying sounds similar, and mine certainly was at the beginning, be honest with yourself. Recognize the hazardous attitude that may lie within you, and recoil from any temptation to prove your worth through your aircraft like you were touched with a hot poker.

Occasionally, you may have a legitimate request and reason to do a "fly over" or similar event. Make certain that you understand and review all of the regulations and restrictions associated with the event, and develop and practice a profile that does not exceed the limitations of you or your aircraft. Build in a margin of safety. Avoid the airshow syndrome by developing, briefing, and flying a safe plan for all flights. The crowd will likely not know the difference, unless you end the show in a smoking hole, which many others have done.

Emotional "jet lag"

Pilots are often perfectionists and react strongly to mistakes, especially their own. Emotional jet lag is particularly dangerous to aviators who pride themselves on being perfectionists. When an error does occur, these perfectionists can't seem to put it out of their minds, and their brain stays at the point of the mistake, dwelling on the error or its cause. This can be even more distracting when you are flying with others, as you may tend to fret about what others are thinking. The obvious problem with this is that the aircraft keeps moving while our thinking does not.

The inability to snap back after a mistake is particularly crippling to students. Almost by definition, a student makes mistakes, usually lots of them. They are supposed to, that is part of the learning experience. For those who suffer from emotional jet lag, or cannot accept criticism in flight, the job of learning to fly becomes a difficult one indeed. In ten years of instructing military pilots, I have seen two critical differences between fast learners and slow learners. The first is preparation, and the second is the ability to shrug off a mistake and get the most out of the remaining flight.

Bouncing back: Fighting emotional jet lag

As much as we all hate mistakes, they are a part of life—and flying. We all make mistakes, but while you're in the air is not the time to dwell on it. You cannot allow your mind to get behind the aircraft. If you do, much bigger—and perhaps deadlier mistakes are likely to follow. Emotional jet lag often results in lost situational awareness, and the recovery process is similar to that for lost SA. First, admit the mistake. Second, ask yourself what impact the error has on the remainder of the mission. This will require you to project yourself into the future and begins the recovery process immediately. If you find that the error causes you to modify some part of the remainder of your flight, you are back out in front of the aircraft. If not, make a note to address the error in the postflight debrief or analysis and forget about it until then!

If you are particularly susceptible to this problem, it is sometimes helpful to "give yourself" one or two small errors before you start to get down on yourself. We all want to fly perfectly every time out, the odds of it occurring are slim. Don't let your errors snowball. Stay caught up and worry about mistakes when you are debriefing on the ground.

The hazardous attitudes discussed above can impact either aircrew members flying alone or with others. Some attitudes specifically impact flight safety when pilots—especially experienced or high-ranking pilots—are flying as a passenger or as an additional crewmember.

Excessive deference

If we are flying as a copilot or passenger with a pilot that has more experience or perceived skill than we do, or in the case of military aviation a higher rank, we may hesitate to call attention to their deficient performance. When we do call out deficient performance, it tends to be vague. Instead of the first officer on probation telling the captain she is 15 knots slow on final approach, he will often cloak the criticism with a comment like "you're a little slow" or even more an even softer approach like, "We might have a little windshear out here today." This tap dance is not particularly helpful and can in fact be dangerous because it may mask the severity of the situation.

Assertiveness and specifics: The answers to excessive deference

Assertiveness with respect is seldom taken personally. In fact, it will often mark the junior person as someone with integrity who can be counted on to provide timely, accurate, and important information. In fact, a senior pilot who catches his own mistake and corrects it may wonder why it went unnoticed by the other pilot, and may begin to form opinions about your diligence or attentiveness.

Counter the tendency to be vague and ambiguous by using specifics. Instead of, "Ma'am, you're a little slow," try "Ma'am, airspeed is 10 knots below reference and decreasing." Avoid recommendations unless absolutely necessary. In the above situation, it is seldom helpful to say, "You need to add power" unless the situation has deteriorated to a dangerous state. The experienced pilot will realize what to do after you point out precisely what is wrong.

No matter who you are flying with, if you notice something wrong speak up, be specific, and worry about the consequences later. Most pilots will appreciate the help in the cockpit and it's better to say something before it's too late.

A final perspective on hazardous attitudes

Hazardous attitudes are only hazardous if we allow them to manifest themselves in our behaviors. By taking time to understand hazardous attitudes, and by looking for any hint of them in our flying activities, we can abolish them. Also look for hazardous attitudes in those you fly with. At a minimum, it will increase your vigilance. If you are able to share what you have learned with the perpetrator, you may save a friend's life.

Secondly, watch for conditions that can set you up for a hazardous attitude: the need to get somewhere quickly, the unexpected request for a flyby, the flight with a senior or more experienced pilot who you feel can do no wrong. Simple awareness can be the most effective broad-spectrum antibiotic to cure the whole family of hazardous attitudes. Keep your eyes and ears open for cues, but most importantly, listen to your conscience.

Chapter review questions

1. Why are hazardous attitudes seen as easier to deal with than problem personality traits?

2. What factors could cause you to take unnecessary risks in order to accomplish flight objectives?

3. Have you ever exhibited any of these attitudes in your flying activities? If so, when and where, and perhaps most importantly, why?

4. List all of the hazardous attitudes and possible ways to combat them.

5. Are there any other hazardous attitudes that were not covered here?

References

Hughes Training Inc., 1995. *ACT Workbook*. Abilene, TX.: Hughes Training Inc. Reprinted by permission.

Olcott, J. 1996. *National Business Pilots Association Magazine*. "Complacency: The Silent Killer." October.

AeroKnowledge ASRS CD-ROM. 1995. Accession number 136756.

8

Peer pressure

*Against a foe I can defend myself, but Heaven protect me
from a blundering friend.*
D'Arcy W. Thompson

In spite of its beauty, the airborne environment is extremely hostile.
It is filled with dozens of stressors, traps, false dilemmas, and advice
that often create box canyons in the mind. The flyer who first said
that aviating is "thousands of hours of boredom, spiced with a few
moments of stark terror" understood the dualistic nature of the beast.
Pressure from peers only compounds the challenge.

To function safely and effectively in this environment, and to apply
our airmanship skills appropriately, we must understand the multiple
external factors that impact upon our decision making, some of
which come from an unlikely source—our friends and associates—
in short, peer pressure. The great Chinese warrior-philosopher Sun-
Tzu postulated that if you know your enemy well enough, you will
achieve "a thousand victories in a thousand battles." As aviators,
those are the kind of odds we are looking for, but we need to know
our friends as well as our enemies because sometimes it's hard to tell
the difference.

Peer pressure defined

Peer pressure is a suggestive social presence that prompts one to conform to others, to be part of a group. The reason we tend to conform
to suggestions and norms formed by others lies in two fundamental
needs of all human beings—the desire to be liked and accepted, and
the desire to be right (Insko 1985; Insko et al., 1988). Peer pressure
often asks for action; at other times it suggests inactive compliance and
urges one to "not rock the boat." Peer pressure can be obvious or subtle, spoken or nonverbal. Its origin may be personal or organizational.

Sometimes peer pressure is unintentional, while other times it is carefully calculated. It has been blamed for everything from airplane crashes to teen pregnancy. But in spite of all of the nuances of peer pressure, one thing can be stated with some certainty. It is a powerful force, one that can bend judgment for better or worse. In aviation, peer pressure often manifests itself in poor decision making, but it is important to note that it can be a positive force as well, a subject we will discuss at the end of this chapter.

Sources of peer pressure

It is difficult to codify the elements that impact on our flight discipline as strictly internal or external, since all stressors are eventually internalized in the mind of the pilot. However, there are certain key sources of peer pressure that blend both internal and external pressures and can have definite influences on flight discipline. Peer pressure comes from a variety of sources, including self, family, friends, colleagues, supervisors, air traffic controllers, and other aviators we don't even know. They can distort the relative importance of the flight mission (which I term mission weight), the nature of the organizational environment with regards to cultural norms, the internal model of our own capabilities and limitations, and many other aspects of our normal decision-making process. Figure 8-1 shows the multiple sources of peer pressure on the pilot with a variety of negative outcomes.

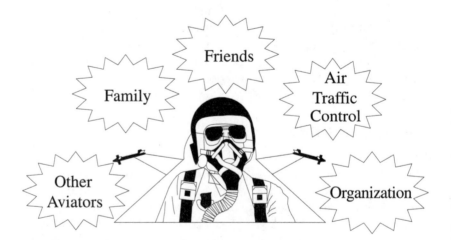

8-1 *Sources of peer pressure. Aviators face pressure from many sources, often simultaneously. The first step in counteracting these pressures is awareness.*

How peer pressure works

Peer pressure accounts for failed flight discipline in six principle ways.

1. It acts through pilots' personal pride. When aviators feel the need to show that they can accomplish something as well as other pilots, they will often do things they would not do under ordinary circumstances.

2. Peer pressure places unwarranted pressure on flyers by creating added importance to complete a mission— i.e., mission weighting, peer pressure can cause mis-prioritization of tasks, where mission accomplishment is stressed above safety.

3. Through actively advocating poor flight discipline, peers can often influence poor judgment through the power of their own personalities and suggestion.

4. By the mere presence of family or friends, pilots often forsake good judgment to show their prowess.

5. We can also have an unhealthy competition with our most critical peer—ourselves, by constantly trying to better our last performance. Although it might be difficult to visualize our ideal self as a peer, in many ways we are our most demanding critics. Unfortunately, for some of us, this can lead to unrealistic expectations and poor judgment.

6. By virtue of organizational position or rank, undue influence can be placed on a subordinate to either exercise poor judgment or withhold information in fear of retribution.

Any one of these types of peer pressure can result in a sudden loss of judgment and poor flight discipline, and when they occur simultaneously the situation can become extremely difficult to overcome unless you are aware of the pressures in advance and make conscious and deliberate steps to avoid them.

Pride of performance

The first type of peer pressure is simple pride of performance, a condition where a pilot feels the pressure to do something as well or better than a peer. Under normal circumstances, this can be healthy, as a little competition can spur improvement on both sides. But when taken to extremes, it can be destructive. In many cases, the peers may not even know that they have played a part in your poor decision, as it may be something as simple as accomplishing a successful instrument approach to minimums, as in the following example.

In the mid 1990s, a military B-1B bomber crashed short of the runway at Ellsworth AFB, South Dakota, after descending through the minimum descent altitude (MDA) on a TACAN approach, clipping a telephone poll, and crashing into a rising slope just short of the runway threshold. The aircraft was completely destroyed. One of the contributing factors in this mishap was the fact that the aircraft commander was a senior instructor pilot (IP) and flight examiner who had just heard a younger aircraft commander successfully land out of the TACAN approach. Although this was a routine training mission, and the crew had more than enough fuel to divert to a base with more favorable weather conditions, the senior officer was determined to take another shot at the TACAN approach.

After hearing that another B-1 had successfully landed out of the TACAN approach, the senior IP pressed his approach below minimums and ended up in the dirt. Fortunately, all four crewmembers survived the accident due to the quick reflexes and decision skills of the copilot, who activated the automatic ejection sequence for all four of the crew after they had impacted the pole.

The fallacy with this decision is that weather, especially bad weather, changes rapidly. The fact that a younger, less-skilled pilot was able to land out of the approach, probably had more to do with a changing overcast cloud deck that his ability to fly the approach better or worse than the mishap pilot.

A second type of peer pressure occurs by simply having someone present to observe your actions and performance. No one likes to fly badly, especially in front of a friend or colleague. The desire to "hack it" is often increased in this environment, even when things begin to spiral out of control, as in the next two case studies from the files of the ASRS. In the first, a student pilot practicing landings at an uncontrolled airport finds out that we all have our own limitations.

Case study: "If they can hack it, so can I."

I was 2–3 feet above the runway and was just about to set the airplane on the runway when I pulled the yoke too much and ballooned a few feet. Because of the crosswind, which was gusting, I started to drift sideways towards the left side of the runway. My airspeed at this moment was rather low, so directly after ballooning I started to sink again and touched down. Due to my poor experience, I think, I was concentrated on keeping the airplane on the runway instead of executing a go-around. Directly after touchdown, when the airplane was on the runway, it was rolling towards the edge of the runway so I pushed right rudder to get the airplane (towards the middle of the runway). As the airplane started to turn, it hit a runway light with the left wing. The runway light was totally damaged and the wing became partially damaged.

Supplemental information (from Accession No. 126780): My student was having a difficult time with landings that day but decided to keep going anyway. I feel the incident could have been prevented in a number of ways:

1. The student should not have continued doing landings since he was having a difficult time with the crosswind.

2. Upon ballooning back up, he could have executed a go-around. I think he continued the landings because there were other students in the pattern who did not stop because of the wind, so he felt a kind of "peer pressure" to continue the landings. (Aeroknowledge 1995. Accession Number: 127191.)

The next report sounds hauntingly familiar to the case study on the CT-43 accident in Dubrovnik, Croatia, which is detailed in Chapter 1. In this case however, the end result was much less tragic after a significant heading and altitude deviation during an NDB approach at a large urban airport on the East Coast.

Case study: Peer pressure prevents effective crew coordination
The captain was flying. After beginning the NANCI #2 arrival in the vicinity of Philadelphia, I left the captain with ATC to get the ATIS. The weather at the destination was reported as 1,400' overcast, visibility about 6 miles.

The captain commented kiddingly that flying an NDB is something he only does "once a year" at recurrent and to "keep an eye on him so that he doesn't hit anything"!

This comment was prompted by a special weather advisory from approach control after just passing the Yardley VOR. We briefed the approach, and approach turned us toward the final approach course to intercept. The problem was (our) aircraft was high and fast (clean). I called the DME from LGA VOR repeatedly, hoping that his reaction would be to start configuring. He did so very timidly, however, and we were still not properly configured at this stage of the

approach. We passed GRENE intersection at approximately 4,000'
going about 220 knots (altitude at GRENE is 2,700).

While all of this was going on, he intercepted by trying to fly to the
beacon. I was focusing on speed/alt while planning on a missed approach (I never thought he'd land out of it), so I wasn't really paying
attention to our track when approach gave us a "090-degree heading
to 're-intercept'" and cleared us for the NDB 4. The captain complied,
and as he was approaching the 1,400' level-off point, he shallowed
out the descent rate, which had been very high. He still, however, was
descending when we got to 1,400', and I repeated "altitude" and
"1,400" twice while reaching behind yoke to arrest the descent. He
saw this and said, "I've got it." "OK." Now we had broken out of the
overcast and since we could see the ground, he wanted to continue
visually to the airport. He was having trouble establishing himself on
the approach course, and he asked me if I could see the airport. I responded by saying, "We're supposed to be at 1,400" since we hadn't
passed PETHS yet. He made a half-hearted attempt to climb, but we
only went up about 50' and then we got station passage, allowing us
to continue to 680' MSL. I didn't push it too much since I could see the
ground and felt we had sufficient obstacle clearance; however, we
were definitely too low. I saw the airport and he continued flying visually to runway 4. After landing he said, "Well I wasn't too proud of
that." I didn't say anything, but he knew I was not very happy with
what had transpired. We had been flying for five days together and
this was not what I would call the first operational error in his conduct
of a flight (second altitude bust,...and at least the third drug-in-duck-
under-the-glideslope approach).

This incident is particularly disturbing since the captain serves on a
voluntary "training committee." Additionally, he is friends with
everyone in the domicile, or so it seems, and while I feel he is somewhat aloof and very unprofessional in his cockpit behavior, I am re-
luctant to say anything to the flight office because I feel his network
is so deep. There is no doubt, however, that he knows he is "weak."
He has friends at our training center and has managed to arrange to
fly in the simulator with his own private instructor for an extra period prior to reporting for his annual recurrent training. Not only is
this unfair to all of the others who are checked annually, but I tend
to think he learns all of the gouges to pass the check and then go
out on the line and fly marginally at best!

*"With regard to this incident, I should have been more vocal
and specific about our position relative to the field, and if he*

didn't start to slow down I should have used references to our speed and altitude relative to our position from the airport. From a human-factors perspective, I can say that being in a two-man crew makes it very difficult to be critical of another crewmember since there are "no witnesses." An incident like this never happened to me while flying in a 3-man crew, because I think the peer pressure to do things properly is greater, since the ramifications of not doing things SOP (standard operating procedures) can be corroborated. I think I'm going to bid a 3-man/woman airplane before they're all gone." (Aeroknowledge 1995. Accession number: 113070.)

Mission weighting

Pilots sometimes fly by different sets of rules, depending on the perceived relative importance of the mission. Although this is closely related to pressing, which we discussed in detail in Chapter 7, it bears some additional discussion here because of the origin of the motivation to perform on a given flight mission. The relative importance of a flight mission in these cases is usually determined by someone other than the pilot. For example, a military pilot understands what a busted ORI (operational readiness inspection) flight means to his boss and his peers. For the wing commander, it can mean the difference between promotion and pass over; for the peers, it means either relaxing for a few months following the inspection or frantically attempting to "get things fixed" before being reinspected. In other circumstances the pressure can originate from other sources, as in the case of our next case study, which received widespread national attention.

Case study: Freezing rain, gusty winds, broken dreams

The initial NTSB report was simply catalogued as follows.

Accident occurred APR-11-96 at CHEYENNE, WY

Aircraft: Cessna 177B, registration: N35207

Injuries: 3 Fatal.

To many Americans, however, the story of seven-year-old Jessica Dubroff's untimely death stirred deep emotions. The courage, initiative, and daring of the four-foot two-inch student pilot had been the center of a media frenzy. Jessica was trying to become the youngest person ever to fly coast to coast, and her radiant smile peeking out from under the bill of her baseball cap had captured the hearts of America during the first day of their trip. The crew was engaged in a transcontinental record attempt flying over eight consecutive days. But when she, her father Lloyd, and flight instructor Joe Reid showed up at the Cheyenne, Wyoming airport on the morning of 11 April 1996, they felt the sting of cold rain as thunderstorms approached the field and the wind gusted to near 30 knots. But the media was still there, so Jessica and crew delayed their departure, participated in media interviews, preflight, and then loaded the airplane. It would later be determined that the aircraft was 96 pounds over the certified maximum gross weight at the time of departure.

As the bad weather rolled into Cheyenne, the crew was undoubtedly in a hurry to depart. This was clearly evident when they tried to taxi without removing the wheel chocks. Ironically, this cost them even more time, as they had to shut down the engine and have the chocks removed. As they approached the runway, they received a litany of weather warnings and were advised of moderate icing conditions, turbulence, IFR flight precautions, and a cold front in the area of the departure airport. The airplane was taxied in rain to takeoff on runway 30. While taxiing, Reid—the pilot in command (PIC)—acknowledged receiving information that the wind was from 280 degrees at 20 gusting 30 knots. At the same time, "a United Express pilot facing similar conditions made a different choice. His commercial flight, with 19 passengers on board...decided to delay." (Alter 1996, p. 25.)

Dubroff's aircraft departed on runway 30 towards a nearby thunderstorm and began a gradual turn to an easterly heading. Witnesses described the airplane's climb rate and speed as slow, and they observed the airplane enter a roll and descent that airport manager Jerry Olson called "a classic stall" (Alter 1996). They crashed in a residential neighborhood below (NTSB 1996).

The formal accident report hints at peer pressure from the media as a possible cause for the failed flight discipline.

> *"Probable Cause*
>
> *The pilot-in-command's improper decision to take off into deteriorating weather conditions (including turbulence, gusty winds, and an advancing thunderstorm and associated precipitation) when the airplane was overweight and when the density altitude was higher than he was accustomed to, resulted in a stall caused by failure to maintain airspeed. Contributing to the pilot-in-command's decision to*

take off was a desire to adhere to an overly ambitious itiner-
ary, in part, because of media commitments." (NTSB 1996.)

The investigation that followed this tragic accident focused on the nature of the record attempt itself, pointing out that children should not be attempting world records in dangerous activities. But they missed the real point here. It wasn't the fact that a seven-year-old was attempting to set a new record that killed these people. It was a poor decision by the pilot-in-command, who apparently felt that this flight was important enough to warrant an overweight takeoff at extremely high-pressure altitudes (which he had little experience with), into weather so severe that a commercial pilot delayed his takeoff. The peer pressure created by the media frenzy had overwhelmed a good pilot's normally sound judgment, causing an erroneous and fatal belief that it was vitally important that they launch from Cheyenne that morning.

Overzealous instructors

Another form of mission weighting occurs when an instructor is overzealous in attempting to accomplish training objectives. When this occurs, the lesson left with the student may not be the one that is intended.

When I was a new aircraft commander in the B-1, I was having a little trouble with night-air refueling. I recognized my lack of proficiency and went to my training officer in the squadron and asked for some additional instructor rides. After completion of the training, I felt that I was proficient, but the local flight examiners (we called them the "Black Hats") wanted to make certain, so they scheduled a "no-notice" flight evaluation for my next night flight. The mission called for an air-refueling rendezvous almost immediately after takeoff, and an "onload" of 100,000 pounds of fuel, or a little over 15,000 gallons. As soon as we got airborne, our navigator informed us that there were rapidly building cumulus clouds in all four quadrants, and that we needed to leave the local area or land quickly if we wanted to maintain a safe distance from the weather.

The evaluator pilot intervened and said that we really needed to get the air refueling as scheduled so that he could be sure that I was ready for the upcoming inspection. Over protests from the rest of the crew, we completed the air refueling as scheduled, dodging thunderstorms all the way down the track. Dropping off the tanker, we were now completely surrounded by BIG weather, and since we had on-loaded so much fuel, we were far too heavy to land safely. (Due to landing-gear restrictions, the B-1 can take off much heavier than it can land. In this case, we had taken off and then increased our gross weight by nearly 100,000 pounds, further complicating our dilemma.)

After a great deal of discussion among the crew as to options (there were precious few), the evaluator once again overruled our request to dump fuel to land within limits and directed that we land at the heavier gross weights, claiming, that "the aircraft has been flight tested far beyond what the flight manual says."

Although we landed without serious incident—we did get hot brakes—the lesson I took from that flight was not about night refueling. Rather, I learned that a single-minded approach to a training issue can create serious distraction and impair good judgment and flight discipline. I believe this observation has made me a better instructor pilot, as I now realize that what you do as an instructor leaves every bit as much of an impression as what you say, and sometimes more.

Active advocacy

The third type of peer pressure is what we typically think of when we hear the term. It occurs when a peer actively advocates doing something out of the ordinary. This can take many forms, from a passenger wanting to see some aerobatics or to overfly a friend's party at the beach, to a company asking an air taxi pilot to meet a weekly quota, as in our next example from the files of the ASRS.

Case study: Meeting the quota

"Because of a hurried schedule and the pressure to fly my weekly quota, I launched into marginal VFR weather. As I started to taxi I had radio problems which distracted me and added to my poor judgment. I had passengers on board, so I attempted to maintain my...VFR status. Because of rising terrain outside the ATA, I was unable to maintain VFR and elected to enter IMC conditions rather than turn back because I was a considerable distance from the airport. Although Flight Service (FSS) had not forecast icing I also encountered ice on climbout. I knew I would (re)enter VFR eventually, but should not have compromised my passengers' safety. From now on I will not let office and peer pressure lead me into launching into questionably marginal conditions." (Aeroknowledge 1995, Accession number: 156148.)

In the case above, the pressure came from an organizational policy or "quota." In almost all organizations, supervision would much prefer you to take a conservative approach to safety than to press the limits of yourself or your aircraft in the name of some arbitrary goal or number. If this is not the case in your organization, perhaps you should start looking for another employer.

Sometimes peer pressure comes from a well-meaning source, such as an air traffic controller, whom most pilots strive to help out whenever possible. Accepting a marginal clearance, even when just trying to help out, can be extremely hazardous, as our next case study points out. Remember this when you decide to "help out" a controller by accepting a marginal clearance: a near mishap looks just as bad on the controller's record as it does on yours.

"I was captain and pilot not flying (pnf) on a commuter flight from LAX to Palmdale. We had been cleared for a visual approach to runway 22 by the Joshua approach controller. When we were switched to the Palmdale tower, he cleared us to land runway 22 (left-hand traffic) with a short approach west of the VOR and a long landing. The VOR is about 1,000 feet down runway 22, which is approximately 12,000 feet long. We acknowledged the clearance, but I knew that this would make it a very tight approach, as we were still very high (approximately 9,000-ft. MSL). We also noted that the tower had another aircraft in right-closed traffic for runway 25 and that this must be the reason for the request for the base turn west of the VOR. As the first officer started the base turn just inside the VOR at approximately 4,500 ft. MSL, I started having strong doubts that we could make the landing. We were configured with full flaps, gear down, which can really bring the airplane down fast. As we continued I knew it was not looking good. The last distance marker I saw was the 4,000-ft. marker. I was worried about an attempted go-around at this point with this configuration and the airspeed bleeding off quickly. I told the first officer to get it on the runway. I estimate we touched down at just less than 3,000 ft. remaining. We had to use full reverse and full brakes to stop on the remaining runway. We stopped with about 25 ft. of runway left. Too damn scary!

"In retrospect, there is no question it was an extremely poor decision on my part to continue the approach. I thought

about a go-around several times during the approach, but never had the first officer initialize it or call it out. I was uneasy with the approach from the time we got the clearance, but still accepted it. We were trying to help the controller, but that is no reason to be put in a bad position. Contributing factors—but not excuses for my weak decision:

1. *"We were on the last day of a four-day trip on a reduced rest of eight hours the night before. The day before we had flown 8.2 hrs with 10 legs and seven approaches. There is no question we were both tired, but of course this is all legal—and becoming more common for cost-cutting measures by the airlines.*

2. *"The mindset in the commuters to stay on tight schedules and don't get late. As much as I feel I don't have a problem being late, I still feel the pressure from me as well as the airline. From me, being late means getting home late. After 4 days, one wants to get home. Arriving late means later flow times back to LAX and it can snowball easily from there. It's the old get home-itis thing. Stupid, but still there. Also, shorter turnaround times in schedules to save money by airlines.*

3. *"Manhood versus safety issue, the concept that doing a go-around makes one feel they have erred or could not fly the airplane well enough." (Aeroknowledge 1995. Accession number: 246995.)*

Ah yes, one of the three most worthless things to a pilot—runway behind him!

Regardless of who the peer is, the coercion to do something one would not ordinarily do in flight is a risk to flight discipline. Sometimes, however, the peer is not asking you to do anything out of the ordinary with regards to flight operations, yet is still a hazard to good flight order and discipline. In the following report, the pressure comes from the captain, who becomes so engrossed in his topic of conversation that he forgets to fly the aircraft.

"Enroute to Deep Park VOR at FL (flight level) 250, we received clearance to cross Deep Park at FL180. Throughout the flight the captain had been explaining his position and views on being a union member (I am not). When we were

*cleared to FL180 I set the altitude warning to FL180 and re-
peated the clearance, yet he elected to withhold descent until
closer to the VOR. He then returned to his discussion of the
merits of unionism and we missed the (descent point). Center
called and asked what our clearance was just outside
of the VOR, and we initiated an immediate descent but missed
the assigned altitude of FL180 by several thousand feet.*

*"I had allowed peer pressure to involve me in an emotional
and distracting discussion during a fairly routine phase of
flight and it still caused problems! A simple enroute descent
to 7,000' was missed because a discussion was started during flight that involved a topic of high emotional content.
Unionism, abortion, religion—a professional crew should
keep the topics to those relating to the flight, or at least to those
issues not likely to distract the crew!" (Aeroknowledge 1995.
Accession number: 159477.)*

Competing with yourself

Perhaps the toughest peer to deal with is our perception of ourselves. Most pilots strive to continually improve their performance,
but we must realize that a personal best is not in the cards for every
flight. There is an old saying from "Grandpa Pettibone's" Naval Aviation Newsletter that says, "No pilot is any better than his last landing," meaning it is performance, and not reputation, that counts in a
flyer. But this quest for perfection can be a double-edged sword, as
the following reporter came to find out when he tried to accomplish
a personal best.

*"During approach into Green Bay, WI, I set up an approach
that would allow me to land farther down the runway than
the (normal) touchdown zone. This would reduce the taxi
time to my ramp destination (and beat my previous best
time). I believe there was a 5-knot tail wind for my runway
at the time of landing. Upon application of brakes after landing, I noticed a reduction in braking effectiveness and proceeded to apply brakes more heavily than usual since my
concern now was if there would be sufficient runway left to
stop the aircraft with this reduced braking action. Once completely stopped, I added power to taxi to the ramp. It was then
that I realized something was wrong, because near takeoff
power was required to move the airplane and I could feel the*

tire imbalance as I taxied. Once clear of the runway, I shut down the airplane and conducted a visual inspection...I discovered that both main gear tires had blown. Factors that I feel contributed to the situation include: landing with a slight tail wind; nighttime effects on judging distances; peer, not company, pressure to always try to arrive at the destination sooner than previous trips." (Aeroknowledge 1995 accession number: 286625.)

In a personal version on this theme, as a B-1 instructor pilot, I was required to demonstrate "air-refueling boom limits" on my annual flight evaluation. This meant that in addition to demonstrating proficiency in in-flight refueling, I had to show that I could fly and instruct to the upper, lower, inner, outer, right, and left limits of the air-refueling envelope. The idea was that an instructor should be able to demonstrate to a student the acceptable refueling picture, as well as recognize a dangerous condition. After some practice, most instructors could perform these maneuvers with relative ease, so we invented new games to play. We would combine these "limits" into double and triple limits, meaning that we would go from the upper-outer-right limit to the inner-lower-left limit and various combinations thereof. After a time, we even began to see how quickly we could hit all of the corners of the envelope and return to the center—30-degree elevation, 12-foot extension—on centerline. Of course, there was no regulation against what we were doing, but neither was there any logical reason for doing it. We were just competing with ourselves to see if we could get better at something.

In a less mundane example, I was told of a student and instructor pilot practicing touch and goes with a beer bet on the best landing. The student won the first exchange, but the instructor blamed it on a sudden gust in the flare, so they went double or nothing on the second set of patterns. Once again, the student greased on a roller, and the IP, now clearly feeling the pressure to perform, clunked on his second attempt. Angry at his ineptitude, the IP added full power for the go-around, held the aircraft on the runway to build airspeed, and yanked the aircraft into a tight overhead pattern. During the pull up, the IP nearly stalled the aircraft, had to perform a traffic pattern stall recovery, and flew within 50 feet of the control tower during his recovery. Luckily, both the tower controller and the airport manager were aware of this IP's tendencies, and corrective actions were administered. This instructor wasn't really competing with the student, and his anger wasn't the result of losing a six-pack. He was angry because he was not performing to his own set of high standards, and it was this self-induced peer pressure that got the best of him. His commander gave him six weeks on the ground to think about it.

Pressure from above

Although pressure from a senior-ranking aviator is not technically peer pressure in the purest sense of the word *peer*, it does occur often enough to address in this section, especially in light of the current emphasis on breaking down rank and seniority barriers in the cockpit through good CRM practices. The old joke of airline captains who tell their copilots that they understand crew resource management as "you are the resource and I am the management" is no longer funny. Far too many graves have been filled with single-seat mentalities in crewed aircraft. The following case study illustrates one such captain whose aggressive pressure tactics eventually caught up with him.

Case study: The captain is always right?

On December 1st, 1993, Captain Marvin Falitz had a decision to make. The Northwest Airlink pilot was en route from Minneapolis-St. Paul to Hibbing, Minnesota, in a Jetstream BA-3100 with 18 passengers on board. He was cruising above icing conditions and was concerned about the descent through the ice into the Hibbing airport. The Jetstream has a reputation for "poor handling during icing conditions" (Oslund 1994, p. 12A) and Captain Falitz made the call to rapidly descend through the icing conditions. Perhaps the term rapidly is an understatement, as NTSB investigators later determined that the aircraft was descending at an average of 2,225 feet per minute, more than twice the rate allowed by the flight manual.

As the captain started his descent for Hibbing, first officer Chad Erickson was busy. He had been directed to ensure that the radio-activated lighting system at Hibbing was correctly configured and was struggling with this task as the captain began his nosedive into the clouds. Perhaps because he was task saturated, or perhaps because Captain Falitz was known as very difficult to work with, the first officer failed to make required call-outs of the plane's altitude during the approach to Hibbing. The safety board found that Erickson "was distracted from his duties (to monitor the descent)...as a result of poorly planned instructions from the captain" (Osland 1994, 12A). Safety board member John Lauber went a bit further in his analysis of the breakdown in crew coordination, stating, "At critical times, the captain couldn't resist piling the workload on this guy." (Osland 1994, p. 12A.)

In addition to the peer pressure applied by the captain which prevented the first officer from accomplishing his monitoring duties, the aircraft was not equipped with the modern Ground Proximity Warning System (GPWS), which would have warned the crew that they had missed their level off altitude. Flight 5719 never did level off, and it impacted the ground in a steep descent approximately three miles short of the Hibbing airport.

A closer look into the captain's background provided evidence of an aggressive, and often abusive, crewmember who once punched a fellow crewmember and on another occasion walked out of the cockpit claiming that his colleague "made him sick" (Thomma 1994). He had been called "headstrong, argumentative, and extremely overbearing" and had been accused of intentionally "jostling the flight controls to give passengers a rough ride out of contempt for the airline" (Thomma 1994). Clearly, this troubled captain made crew coordination difficult, if not impossible.

So what is a pilot to do when confronted with pressure from above? Many aviators feel that they must simply "live with it" because of concerns for their continued employment or promotion within the system. But living may not be part of the picture if it gets too bad, as our previous and following case study points out.

Case Study: A company out of control

About 8:55 p.m., on May 30, 1979, Downeast Airlines, Inc., Flight 46 crashed into a heavily wooded area about 1.2 miles southwest of the Knox County Regional Airport, Rockland, Maine. The crash occurred during a nonprecision instrument approach to runway 3 in instrument meteorological conditions (IMC). Of the 16 passengers and two crewmembers aboard, only one passenger survived the accident. The aircraft was destroyed.

The National Transportation Safety Board determined that the probable cause of the accident was the failure of the flight crew to arrest the aircraft's descent at the minimum descent altitude for the nonprecision approach, without the runway environment in sight, for unknown reasons. Although the Safety Board was unable to determine conclusively the reason(s) for the flight crew's deviation from standard instrument approach procedures, it is believed that inordinate management pressures, the first officer's marginal instrument proficiency, the captain's inadequate supervision of the flight, inadequate crew training and procedures, and the captain's chronic fatigue were all factors in the accident.

According to John Nance, the author of *Blind Trust*, there was more to the story, much more. Downeast Airlines was a small commuter air carrier operating out of Rockland, Maine, a location notorious for its bad weather and sea fog. At first glance, it seemed as if it would be extremely difficult for a small company to make a go of it at this location, flying in an area where the weather was very often below takeoff and landing minimums for a FAR Part 135 carrier. But Downeast had beaten the odds, in no small part due to the single-minded approach of the owner, who routinely pressured pilots to break minimums, fly with overloaded airplanes, and accept mechanical defects. He "knew all about sea fog and approach minimums, and had scant respect for any pilot of his who would cancel or divert a passenger-carrying, money-making flight because the actual cloud ceiling and visibility were slightly below the legal minimums" (Nance 1986). The owner made it clear that pilots who could not or would not push (or bust) the limits did not fit in his company. In short, the owner's position was that he felt that pilots who wouldn't violate the regulations were cowards. This attitude permeated Downeast operations.

There was more than supervisory pressure impacting the decisions of the pilots at Downeast. Nance explains.

> *"There was a more insidious force that had perpetrated this attitude: peer pressure. When pilots who had no previous airline, commuter, or military flight experience found that they could meet most of the schedules despite the fog (which was notorious in the Maine area), by sneaking around the minimums and the rules, they began to develop a sort of perverse pride in their own abilities. A pilot who could fly an overweight airplane, take off in less than legal weather, ignore mechanical problems in order to bring the airplane back to Rockland with revenue passengers...was more often than not proud of himself. Any pilot who came on board and couldn't do as well was less of a pilot—less of a man." (Nance 1986.)*

There was even an informal set of "Downeast minimums," or approach altitudes that were considerably lower than the FAA allowed but routinely flown by these "intrepid" aviators. There were multiple stories told by passengers who recalled seeing trees brush by only inches below their aircraft, and the night a loud thump was heard on the third attempt to get into Rockland. When the aircraft landed at Augusta, a large dent was found on the leading edge of the aircraft's right wing, but the pilot played it off a bird strike from a sea gull, in spite of the obvious fact that sea gulls are too smart to fly in "pea-soup fog." (Nance 1986.)

In a nutshell, peer pressure played a role in nearly every facet of Downeast Airlines operations, and until the NTSB unraveled the corporate culture issues during an extensive accident investigation, no one blew the whistle.

We have seen that the six types of peer pressure can have a profound impact on an aviator's flight discipline, but there is one critical aspect of flight discipline we have not yet touched on—how to resist it.

Resisting peer pressure

Even rugged individualists find it easier to go along than to buck the system, but resisting the type of peer pressure which induces poor flight discipline is one of the marks of a mature airman. In order to resist poor advice and peer pressure, we must first have a clear picture in our own minds of what constitutes right from wrong as it relates to our particular moment in time. In aviation we call this situation awareness. This is not an argument for a sliding scale of ethics or morality, what some social psychologists refer to as situational ethics, but rather that the nature of the situation must be considered in the aviation environment. A pilot must consider dozens of factors when making a call to resist or go along with peer pressure. Once again, to resist unwanted peer pressure, we must possess a clear picture of airmanship and hold any potentially hazardous suggestion up to the light of true airmanship to determine whether or not we should accept or reject a course of action.

In addition to the desire to conform, people also have an intrinsic desire to be unique and maintain our individuality (Snyder and Fromkin, 1980). If "everybody's doin' it" (practicing one or more forms of poor flight discipline), then you can assert your unique individuality by standing alone for higher standards. Distinguish yourself by understanding what good discipline is and refusing to deviate simply to be part of the crowd. Stand for something positive.

Principled decision making

Without a clear picture of airmanship in our own minds, we will be far more likely to submit to the emotion of the moment or the

charisma of the peer advocating (either intentionally or unintention-
ally) a questionable course of action. The movements and actions of
a crowd can take on almost mystical status to affirm or excuse undisciplined behaviors. To fight this battle, we must have principles
within—the ones that truly determine what we feel about ourselves
as airmen. It is this internal picture of what we are which will serve
as a backdrop for standing up against a trend. This often requires that
we reject institutional "heroes" who have made their claim to fame by
"beating the system" or "bending the rules." Uncompromising discipline can be an unpopular path to walk in some aviation circles.

Airmanship integrity—flight discipline—is often unpopular because
it is restrictive. Most of today's aviators grew up in the "Age of
Aquarius," where the motto was "If it feels good, do it." Enforcing
anything with this lot is a challenge on a good day, and next to impossible with the hardcore hedonists, some of whom hold multiple
FAA ratings. But there is a way to make a difference, and it begins
with a positive example.

Peer pressure as a positive aspect of flight discipline

Peer pressure can cut both ways. If we hold ourselves to the highest
possible standards, we can become a conduit for good airmanship
in others. Captain R. J. Phillips, a chaplain with the United States
Navy, put it this way:

> *"Peers can hold an individual to the highest standards or
> lure them into the gutter. Peers can coax one to honesty or
> derail one to dishonesty. Peers can keep a friend morally
> sane and sober, or merrily invite a friend to lose sanity and
> sobriety along with the rest of the bubbas." (Phillips 1996.)*

In short, social psychology is critical to flight discipline. You make a
difference. You have an impact on other flyers even if you don't
want to. Every action, every inaction, contributes to a culture which
either actively promotes or decays flight discipline.

We should all realize that we are swayed by the actions and words
of others. Don't kid yourself. Many flyers who have once thought
certain acts as "unthinkable," have found themselves trying the same
undisciplined antics within a few short months. One example highlights how far this can go. For many years there was an informal

"mesa club" in the western United States, where pilots could become members only by accomplishing a touch and go off a small—and very high (1,200 feet AGL)—stone mesa in the middle of the desert. For all I know, the club still exists. Some idiots even went as far as designing a patch to publicize their stupidity to the world. These pilots weren't always undisciplined; they became that way because of peer pressure.

Realize that you have the option—better said the responsibility—to choose a different path from that of the crowd if you feel your flight discipline may be at risk. It's not just a matter of integrity. It's a matter of survival.

Chapter review questions

1. What two fundamental human needs makes peer pressure so effective?

2. What is meant by mission weight? Can you think of flight objectives that might cause you to change your normal operating procedures? Be honest.

3. What are the six ways that peer pressure can affect flight discipline? Can you think of personal examples that fit these descriptions?

4. How can peer pressure affect cockpit communication?

5. Do you think that peer pressure was a significant contributor in the tragic crash of Jessica Dubroff? From whom might it have come?

6. What can be done to resist negative peer pressure and preserve flight discipline?

7. How can peer pressure be utilized to enhance flight discipline?

References

Alter, Jonathan. 1996. "National Affairs: Maiden Flight" *Newsweek*. 22 April.

Inkso, C. A. 1985. "Balance Theory, the Jordan Paradigm, and the West Tetrahedron." In L. Berkowitz (ed.), *Advances in Experimental Social Psychology*. New York: Academic Press.

Inkso, C.A., R. H. Hoyle, R. L. Pinkley, G. Y. Hong, R. M. Slim, B. Dalton, Y. H. Lin, P. P. Ruffin, G. J. Dardis, P. R. Brenthal, and J. Schloper. 1988. "Individual-Group Discontinuity: The Role of a Consensus Rule." *Journal of Experimental Social Psychology*, 24 505–519.

Osland, John J., and J. Christensen. 1994. "Hibbing Air Crash Blamed on Captain." *Minneapolis Star-Tribune*, 25 May, 1994.

NASA ASRS. 1994. Aeroknowledge ASRS CD-ROM.

Nance, John J. 1986. *Blind Trust: How Deregulation Has Jeopardized Airline Safety and What You Can Do About It.* New York: William Morrow and Company.

National Transportation Safety Board, accident identification number SEA96FA079. Internet site at http://www.ntsb.gov.Aviation/SEA/96A079.html.

Phillips, R. J. 1996. *A Principle Within: Ethical Military Leadership.* Presented at the Joint Services Conference on Professional Ethics XVII. Washington, D.C. January 25–26, 1996.

Snyder, C. R., and H. L. Fromkin. 1980. *Uniqueness: The Human Pursuit of Difference.* New York: Plenum.

Thomma, Steven. 1994. "Pilot in Crash Near Hibbing Had Checkered Flight Record." *The Minneapolis Star Tribune.* Dec 17, 1993. B-1.

Part Three

Practical issues for flight discipline

9

Guiding lights

The critical role of instructors and mentoring to flight discipline

Let such teach others who themselves excel.
Alexander Pope: Essay on Criticism, 1711

One of the most difficult maneuvers in aviation is the critical balancing act between two forces—flight discipline and flight instruction. It may seem at first that these two forces are pulling in the same direction, and they should. However, flight instructors must not only preach and practice flight discipline, but they must also deal with a host of stressors not present in the normal flight environment. These include maintaining personal proficiency, knowing when and how to intervene, and adjusting instructional styles to the needs of the student.

It has been said that aviation is nearly equal parts of art and science. If that is true, then flight instruction tips the scale well onto the side of art. Reading the minds of students, searching for teaching opportunities, and improvising to capitalize on the personality dynamics of the student-instructor team can be beautiful art indeed.

While science is designed around disciplined cause-and-effect quantitative measures, the idea of a "disciplined artist" almost sounds like an oxymoron. Yet that is what we must be. The penalties for undisciplined actions or instruction by a flight instructor are too severe to be otherwise.

Before we look into just how this delicate balancing act is achieved, let's view the other end of the spectrum, a place where flight discipline might not have been passed along, a place where I sincerely hope you never have to be.

On losing a student—and a friend

I was in a deep sleep in one of the newly refurbished rooms of the Visiting Officer's Quarters (VOQ) at Ellsworth (AFB in South Dakota) when the phone rang. I was on temporary duty from my job as an instructor pilot/flight examiner at the B-1 bomber Combat Crew Training Squadron (CCTS) at Dyess Air Force Base in Abilene, Texas, working with a group of scientists from Armstrong Labs on a special fatigue study using the crews and simulators at Ellsworth. I'd been gone a week and I already missed my family, the camaraderie of the squadron, and the comfortable feel of the jet. I was losing proficiency, and I knew it, but this was my one chance to get the real research done that I needed for my doctoral dissertation, so I couldn't pass up the opportunity.

It was a cold night, dark and cold—November 30th, 1992. Actually, I guess it was the morning of December 1st when the phone rang. I remember seeing the silhouettes of blowing snow in the orange glow of the mercury-vapor street lamps shining in from the parking lot outside my window when I sat up abruptly, like one always does when the phone rings in the middle of the night. Where was I? Who could be calling at this hour? This can't be good news, I thought. Maybe it's a wrong number.

"Tony?" I recognized my wife's voice at once. "Hi darling, what's goin' on?" I asked quickly, not really wanting to know. "There's been an accident." My heart sank. I took a deep breath and let it out slowly. I was wide awake now. As the father of two boys, then four and two, I feared the worst. "What kind of accident?" I replied.

"A bomber crashed last night, I just heard it on the radio." I looked at the clock, it was 4:33 A.M. I remember wondering why my wife was up so early, but quickly put this out of my mind and started the rapid-fire series of mandatory questions. "Where did it happen?" "Who was on board?" "Which squadron?" "Was it a student and instructor team?" "Has anyone called you?" "Have you talked to Jay?" (my best friend at the squadron).

She didn't know many details, only that the aircraft had gone down somewhere down south, about 30 miles from one of those tiny little Texas towns with a funny name. That told me a lot of things.

They must have been low level, probably on IR-165, our "back-yard" training route. It was mountainous terrain, which meant the crew—whoever they were, was probably practicing mountainous night terrain following (TF), a currency requirement for all B-1 crews. "Something must have gone wrong with the aircraft," I said, immediately doubting the validity of my words. I had flown the "bone" long enough to know that the TF system had a fail-safe fly-up system that was reliable to a fault, often trying to "save" you when there was no danger to either life or limb, usually caused by a renegade electron burp in the system. No, it must have been the crew, I thought. They must have done something wrong down there. At 540 knots and 400 feet, there isn't much time to recover if you screw something up.

"Tony, are you still there?" she interrupted my thought string. "Yeah, I'm here, sweetheart. I'm sorry...just lost in thought. Listen, thanks for calling. I'll find out who they were and call you back. Don't call any other wives or anything. I'll let you know."

It's an unwritten Air Force rule that has been followed for decades. When someone goes down, you stay off the phones. The unfortu-nate families will find out soon enough, when the commander and chaplain appear at the door, usually in dress blues. Until then, you just wait. At least Shari didn't have to go through it this time, I thought. With me away, she didn't have the same agonizing wait as the other wives whose husbands were on the schedule that night. I didn't know why, but I was already feeling guilty.

I found out all too soon. The mishap crew—God, how I hate that term—was made up of four close friends of mine, two of them for-mer students. Zen, Paul, Tim, and Scott, a fully qualified crew from our sister squadron, had impacted a cliff at nearly 600 knots on a moonless night. I had flown with Zen, the aircraft commander when he had first gone through the initial qualification program almost two years before. He was a great guy, a superb officer and a steady aircraft commander. He had some recent difficulties in upgrading to instructor pilot, but Zen knew his stuff.

Paul, the copilot, had just qualified in the aircraft, and I knew him very well. I had flown with him on several of his early training sor-ties, instructed him in the simulator, and had given him his flight evaluation which had qualified him to fly night TF. He was a super kid, a great student, a former member of the Air Force Academy's parachute team, the "Wings of Blue." Paul had graduated at the top of his pilot training class and had picked the B-1, his dream jet. I was impressed with everything I knew about Paul. As one of the first "straight from UPT" co-pilots into the weapon system, I had taken a special interest in his training. Where had I failed him?

I immediately started second-guessing myself. What hadn't I taught him? Or worse, was there something I had taught him that caused him to misinterpret an in-flight cue? I replayed every training flight, simulator, and debrief in my mind. Had I left anything out? Had I shortchanged these guys in any way? "Wait a minute," I told myself, "You don't even know what happened yet." But somewhere inside I did know, and it hurt—badly.

The accident investigation confirmed my intuition—crew error. They also determined that it was likely that Paul had disconnected the automatic fly-up and pushed over into the ridgeline to cause the crash. There was some legalistic mumbo-jumbo about incomplete training paperwork, etc., but I knew, or at least I thought I knew, that somewhere I had missed an opportunity to provide one more critical piece of information—one life-saving technique or procedure.

Flight instruction has never been the same for me since then. No, I did not choose to hang up my wings like the mythical Daedalus. I still wear my silver Command Pilot wings proudly and flying is still fun—but instructing, the teaching and mentoring of the next generation of flyers, has forever changed for me.

Paul and Zen are buried just down the road from where I live today, at the USAF Academy cemetery. I stop by often to renew the pledge I made to them then:

I will stay, or at least do my best to stay, at the peak of my instructional game. I will debrief thoroughly, look up every student question I do not know the answer to. I will do my homework and come prepared to instruct. I will never willingly bend the rules (and I used to), and I will speak out against those who do. I will sternly critique every safety of flight issue I observe. I will "call 'em as I see 'em" and not be pressured by time lines, syllabus flow, or outside pressures to pass or fail a student, for I—and I alone—am responsible for the quality of product I place into the cockpit as a qualified pilot. I will not pass a marginal student on to the next level for someone else to deal with. I will use my God-given talents in research and education to promote positive flight discipline and airmanship for as long as I'm given the privilege to play this game—and I will never second-guess myself again.

The proficiency challenge

One of the most difficult challenges for any instructor is getting enough "stick time" to maintain personal proficiency. As instructors, our primary job is to train others, but we must—repeat must—keep our own skills honed to a sharp edge. There are three reasons for this.

Students, by definition, are lacking in some aspect of their training or performance, or presumably they would not be flying with you. Your skills must be sharp enough to recognize a deteriorating situation, which can happen very quickly, and to intervene with sufficient skill and in time to not only save the aircraft, but to preserve the training environment. Recovering an aircraft at 100 AGL will do little to enhance the learning curve of the student for the remainder of the sortie. Frightened students learn very little.

A second critical reason for maintaining proficiency is to provide "the picture that is worth a thousand words." If you as an experienced instructor are not able to provide an expert demonstration of a maneuver, you have done two negative things. First, you have lost your credibility as an IP, which you may never get back. Second, you may have built a learning barrier in the mind of the student, who sees the instructor's inability to perform a particular maneuver as an insurmountable obstacle for a mere student.

Finally, you must keep your skills sharp for the piloting challenges inherent in everyday flight. The IP must maintain the capability to handle all the tough flight tasks—the ability to shoot an approach to minimums, land the aircraft within the design crosswind limits, or successfully fly an unfamiliar circling approach, such as in our next case study.

Case study: Proficiency and pride—When do you say "this won't work?" (Hughes 1995)

The mishap occurred in an Air Force Northrup T-38 Talon during a student-training sortie, on the final leg of a three-day, six-sortie cross-country flight. The mission progressed normally through two precision approaches at the home field. During the final pattern, a circling approach, the aircraft overshot the final approach course, began correcting towards the extended runway centerline, then impacted the ground. The aircraft was destroyed. There was no attempt at ejection and both crewmembers were fatally injured.

The mishap pilot was a 24-year-old instructor pilot (IP) with 540 total flight hours. The rear cockpit crewmember was an undergraduate pilot training (UPT) student with 152 hours of flying time. There were no deficiencies found in the student's training and since the IP was flying at the time of the mishap, the student's performance in UPT was not considered a factor. The IP was relatively inexperienced, but fully qualified. According to other members of the squadron in which he flew, his flying skills were average to slightly below average. The IP was described as "very sharp" with outstanding book knowledge. His approach to flying was an attempt to capitalize on his superior mind by mentally constructing a series of preplanned responses to situations. This made him more mechanical and slowed his development in aircraft "feel" and "air sense."

A circling approach is flown infrequently and allows little time to establish a course and glide path. Higher power settings are required to maintain airspeed and the pilot is much more likely to misjudge the approach due to the different visual aspect from a lower altitude in comparison to the more familiar overhead pattern. The IP had never flown a circling approach to this runway. Additional factors that possibly affected the IP's judgment at the time of the mishap were limited rest the night prior to the mishap, poor nutrition, and low fuel.

The safety board's findings identified the IP as causing this mishap. He misjudged the base turn of the circling approach and overshot to a degree that interception of the final approach course could only take place well inside the desired roll-out point. He then failed to initiate a go-around. Possible factors include impaired judgment due to fatigue and lack of recent food intake, overconfidence in his ability to deal with situations, and/or overconcern with a low fuel state. The IP failed to recognize the high sink rate, the low airspeed, increasing airframe buffet, and high angle-of-attack until there was insufficient altitude to recover. He then failed to direct and initiate ejection.

In short, this instructor failed in his responsibilities at multiple levels and did not have the discipline to take around what must have clearly been an ugly approach. He also failed his student by not keeping proficiency in circling approaches. With so little circling experience, his recognition of a dangerous situation was slow, and his decision to continue obviously flawed. Through this lack of disciplined training and inability to abandon a poor approach, he failed in his moral responsibility to protect the life of the student he had been given care and custody of.

A second case study from the military environment also demonstrates these factors, but additionally highlights the stress that can occur between an instructor and a student, resulting in an unhealthy and dangerous mindset for an instructor who might feel the need to prove herself in a tough emergency situation.

Case study: Multiple stressors

The mission was a cross-country navigation-training sortie in a Cessna T-37 "Tweet." On the takeoff leg, the number one engine-overheat warning light came on. The instructor pilot (IP) declared an emergency and asked to return to the field. She then followed emergency procedures and shut down the engine and maneuvered the aircraft to accomplish a single engine, overhead pattern and landing.

It started poorly. During the mishap sequence, the IP misjudged the downwind spacing and angled toward the runway resulting in an overshooting turn to final. She made a timely decision to initiate what appeared to be a successful go-around then apparently decided to attempt a landing near mid-field. Nearing mid-field, the aircraft was observed to turn to align with the runway in what appeared to be a second attempt to land out of the overshoot. This one didn't work any better than the first. This decision probably limited their options to ejection or riding it in. She reestablished level flight momentarily, but the aircraft was in a thrust deficient situation, and she was headed into an area of rising terrain. The aircraft turned away from the field and hit a tree approximately one mile from the runway. The IP and student ejected out of the safe ejection envelope and both were fatally injured. She had made the decision to eject too late and too low.

A closer look into the human factors of the incident is revealing. The IP was relatively inexperienced, with only 530 total flight hours. She was considered to be below average in flying skills, but was fully qualified. In addition to the pressures associated with being an inexperienced female instructor in a male-dominated Air Force, she had recently experienced several medical problems and was scheduled to undergo surgery several weeks after the mishap date. Outwardly, she appeared to be happy and well adjusted, but was most likely under significant internal stress. These stressors were compounded by the student with whom she was assigned.

The student pilot had approximately 66 hours of flying time and was described as below average. He was dissatisfied with the instructor, had doubts about her flying skills, and believed she was inconsistent in her grading practices. In a word, he was a complainer. This undoubtedly placed additional pressure on the IP during the emergency, as she tried to prove to herself, and to the student, that she could hack it.

The findings were that an equipment malfunction and the IP caused this mishap. Although the engine overheated, the aircraft was still recoverable. The IP accomplished appropriate emergency procedures and attempted to land via a single engine pattern. To this point her actions were flawless. But for some unknown reason, she discontinued that attempt at low airspeed and altitude and was unable to accomplish a successful go-around.

Indecisiveness, and perhaps a lack of confidence in both the aircraft and her own abilities to make a second single-engine approach, created a situation from which they could not recover. Failing to recognize the unsalvageable situation in time, a late ejection command was the last error in this chain.

Both of the case studies above illustrate the need for instructors to maintain high skill levels in both normal and emergency procedures. Additional stressors will occur from time to time, both in life and with students. They cannot be allowed to be compounded by marginal skill or proficiency. The most disciplined decision an instructor can make is to stop instructing long enough to insure personal capabilities are at least maintained, if not improved.

Case study: Disciplined instruction means expecting the unexpected

A large, multiengine aircraft was making a VFR touch-and-go landing with a simulated engine-failure, takeoff-continued, emergency to be practiced during the takeoff phase. During this maneuver, the aircraft departed the left side of the runway with approximately 5300 feet of runway remaining. The aircraft broke up, caught fire and was destroyed. Five of the seven crewmembers were fatalities; two crew members were uninjured. The aircraft was airworthy for the entire duration of the flight up to the point of the mishap itself.

The aircraft launched on a normal training mission and after a cruise portion of the flight returned to the home field for transition training. They began their descent from the TACAN initial approach fix for a Hi-TACAN penetration and approach to a missed approach. The crew subsequently performed five additional instrument approaches. One was followed by a missed approach, the last four by touch-and-go landings. Following the fourth touch-and-go landing, the crew requested and received a clearance from tower for a VFR pattern. When the aircraft was on downwind the instructor pilot briefed the crew that this would be a VFR four-engine approach and touch-and-go landing. He also briefed that they would be doing a simulated engine failure; takeoff continued during the takeoff portion of the touch-and-go. During this maneuver, the instructor pilot thoroughly briefs the crew of his intention to pull one of the outboard engines to idle to simulate an engine failure on takeoff leg. The student is trained and expected to counteract the yaw with appropriate rudder and delay rotation until a safe airspeed is reached.

Training records indicate that this was the student pilot's first simulated engine-failure, takeoff-continued maneuver in 20 days and his first ever from the left seat. There is no indication that the instructor had demonstrated the maneuver in flight prior to this attempt. However, he had performed this maneuver several times in the cockpit procedural trainer (CPT), a low-fidelity simulator. The training received in the CPT teaches the student to look into the cockpit to determine which engine is failed, since there are no visual presentations in this device to cue the pilot to outside references. This response is contradictory to what is taught in the aircraft, which is to look outside. Visual cues should be the only method used to determine control inputs while on the ground.

The student pilot had a tendency to "undercontrol" the rudder during the previous engine-failure, takeoff-continued maneuvers, which he had accomplished only from the right seat. In addition, previous instructors stated that he tended to rotate before the aircraft was under control while performing this maneuver. There were enough indicators of student trouble with this maneuver to put the IP "on guard."

On the final approach, the aircraft touched down in the center of the runway and bounced two or three times. Following the touchdown, the aircraft proceeded down the runway in a three-point attitude. The flaps and trim were reset. Power was advanced and the engines accelerated. After the power was advanced and stabilized at touch and go power settings, and prior to the 6,000-feet remaining marker, the instructor pilot simulated failure of number 1 (left outboard) engine by retarding the throttle to idle. At that moment the student pilot rotated the aircraft sufficiently to lift the nose wheel off the runway and simultaneously applied full left rudder, even though the aircraft was already yawing to the left. The absence of nose-wheel skid marks indicated the nose wheel never returned to the runway surface.

One second after incorrect control input was completed the aircraft was two feet right of centerline, slightly nose up, right wing low and beginning a rapidly increasing side slip to the left. At this point, the instructor pilot has approximately 2.25 seconds to react and accomplish full rudder reversal before the vertical fin stalls. If the instructor pilot does not initiate positive counteraction by the time the rudder reaches full left, recovery is virtually impossible because the yawing moment cannot be reversed.

At this point the aircraft continued left and the right main gear crossed the centerline and lifted off the runway. At 3.25 seconds of elapsed time the left main gear began to leave the runway and the aircraft heading increased from 16 to 25 degrees left of centerline. The side slip angle increased from 8 to 16 degrees and the vertical fin stalled. The violent left yaw and roll rate continued to increase until the outboard number one engine nacelle contacted the runway and departed the aircraft. Elapsed time for incorrect control input is now six seconds. The nose section impacted the ground approximately 70 feet from the edge of the runway. The initial impact was on the left side of the fuselage just forward of the crew entry hatch. The aircraft pivoted on its nose after initial impact, as the tail continued in a counterclockwise rotation. This impact caused the cockpit compartment to separate from the main fuselage in the area of the cargo door. The aircraft slid backwards on its tail nearly 3,500 feet to its final resting place after turning 285 degrees from runway centerline. The two crewmembers in the rear of the aircraft exited uninjured by jumping out of the aft escape hatch. The cockpit, fuselage and other components of the wreckage continued to burn after coming to rest.

There were several factors that contributed to the instructor's poor reaction to the situation, including the initial shock of the student pilot using incorrect control input; the lack of caution (a.k.a. complacency) on the part of the instructor pilot since the student was older, previously experienced, and knowledgeable; pressure to get all the training required by the mission profile; and the instructor's self-image of being able to salvage the maneuver.

The bottom line here is that the IP got caught off guard. The student pilot induced a surprise factor into the cockpit environment, and the IP couldn't cope. Perhaps the instructor pilot was not aware of the full implication of the maneuver should control be used incorrectly. Perhaps he was unaware of the limited amount of time available for his action if a mistake occurred. Perhaps he was momentarily distracted by the violence of the aircraft maneuver. Perhaps, perhaps, perhaps...

How can we practice better instructional discipline during critical phases of flight? One method is to provide "human stops" to critical control inputs.

Human stops

All flight controls have mechanical stops, which limit control movement past the point where damage could be done. Instructor's hands and feet, and even words if used carefully, can work the same way. Most flight instructors have learned to "guard the controls" during certain flight regimes. Here are a few examples.

During a recent informal survey of fellow instructors, I discovered an interesting fact. Many of them do not fly with their feet on the rudders at all times with a student. I think this is a very bad technique. Improper rudder inputs are often an involuntary response when a student tenses up, and you need to feel it as it happens—not after the fact. You can almost feel some students' fear level by the pressure they apply with their feet. Fly with your feet on the rudders, and you can also prevent the type of improper control input which caused the mishap in the last case study.

I have also developed several techniques with the throttle as a result of some close calls. For example, I think most of us insure that the power is full forward during takeoff and stall recoveries, but how many of you guard the throttle on final approach during gusty wind conditions. John, a friend of mine, tells the story of landing on the tail of a T-37 when his student pulled the power too early on a no-flap touch and go in strong headwinds. Although he immediately slammed the power back up to the stops, the spool-up time was too great and the damage was done. He (and now I) flies with his hand about one-half inch behind the normal throttle position the entire way down final approach as a matter of habit. No more surprises for John.

Likewise, most instructors keep their hands prepositioned next to the stick or yoke during stalls, approach to stalls, slow flight, and recoveries. Students can push over all they want to, but I'll not have them snatching the stick and putting me in a spin unexpectedly.

Care must be taken to insure that you are not "riding" the controls, as this can have bizarre instructional implications too. I have flown with students who continuously slammed the aircraft on the runway without the slightest attempt to correct the problem, in spite of my demonstrations and best instructional techniques. Upon further inquiry I discovered that their "normal" instructor likes to "help out in the flare" with his early students to help give them "more confidence." False confidence is a bad thing. Both student and instructor must know who is flying the aircraft. This also drives home another important point: instructors are not all alike and previous instructors may have planted a seed which will sprout out when you least expect it, such as in the case that follows.

After an ugly and unstabilized final approach which I should have directed a go-around out of, the student landed very hard (5.5 Gs) and bounced the aircraft 25 feet into the air. For those of you who fly with green students—what I like to refer to as "molding virgin clay"—you realize that this is not all that uncommon an experience, but what happened next is. Without so much as a word of warning, the student let go of all the controls and started crying uncontrollably. After assuming control and recovering the aircraft to pattern altitude, I asked why he had done that. He explained that his "normal" instructor took control of the aircraft after every bad landing and flew the next approach as punishment. Interesting technique, I must say.

Flight discipline and role models: Students never forget

by Randall W. Gibb

Teachers of all types are held to a higher level of conduct. As role models for students, teachers are not only instructing knowledge and skills, they are also teaching intangibles such as integrity, character, and yes, discipline. The reason the position of a flight instructor is so important is that students never forget what teachers do, both the good and the bad. We all remember our teachers—from our seventh-grade gym teacher to our high school English teacher and especially any level of flight instructor through our years of training. As students we remember many details about these teachers—how they carried themselves, how they acted when confronted with adversity, how well they listened when we had questions. It is impossible for instructors to remember all of their students, but as students we always remember our teachers.

As a former instructor pilot for the U.S. Air Force Undergraduate Pilot Training (UPT), I can remember only a handful of students that I trained. I can remember only those with whom I flew a majority of their training flights. I would have to check my logbook to confirm the name of many of the students I have flown with. Yet, even having flown once with a student, that student recorded and absorbed my approach to professionalism in the aircraft—my flight discipline. Years later I am constantly amazed when encountering former students who remember details of a flight from yesteryear, when I can barely recognize their faces. I only hope the details they remember were positive ones. As teachers, we must remember that the specifics of what we instruct about the aircraft or specific maneuver may well be forgotten. However, what lasts in students' minds is the impression that you, as the instructor, made on them in terms of flight discipline.

For example, flying in the UPT environment, I had numerous opportunities for teaching flight discipline at the beginner's level. One illustration of a seemingly insignificant and mundane activity that communicates flight discipline was documenting specific training maneuvers on a computer "bubble sheet." In the Air Force, it is often said that the student doesn't graduate, their grade book does. Therefore, the grade book must reflect accurately what the student has accomplished. Students observe the instructor properly documenting what was accomplished, or quickly and sloppily—and sometimes inaccurately completing the "required paperwork." The student has either learned that flight discipline extends outside of the cockpit, or a more negative lesson that disciplined documentation is not important.

Another example occurs when an instructor tries to stop the earth from revolving in order to get a few more minutes of daylight. Certain specific maneuvers cannot be accomplished after official sunset, emergency landings for example. This results in some undisciplined instructors conveniently forgetting or ignoring the time of day in order to accomplish one more maneuver. What message is being sent to the student if their IP squeezes in a prohibited maneuver after official sunset? Flight discipline should not be taught on a sliding scale.

Students look up to their instructors and take mental notes of all of their actions. From the students' perspective, flight instructors are "all-knowing entities" representing a level of knowledge, skill, and all-around expertise that they could never dream of reaching. In reality, of course, given enough time, anyone can learn to fly. All maneuvers flown by instructor pilots will some day be replicated by their students. Some, and hopefully most, students will go on to exceed the skills of the IP. That is why it is imperative that instructors set the perfect example—especially in terms of flight discipline. Set the bar high and keep the standards of discipline and airmanship such that each generation of aviator will be more skilled and disciplined than those who have come before.

Truly exceptional IP's are those that can be role models for flight discipline, a skill much more difficult to master than a perfect cloverleaf or an NDB approach. While this sounds simple in theory, it can be extremely difficult in actual practice. Remember, to a student, you are not what you say you are—you are what you do. They probably won't remember all of the procedurally correct things you did in an hour's flight; however, they'll definitely recall that one short-cut you took. It may have been something very benign at the time, something that to you was maximizing the student's training or getting yourself some proficiency. But if your actions did not follow the standard operating procedures, regulations, or local guidance—it was wrong—an insidious form of condoning poor discipline.

A good instructor never consciously demonstrates poor flight discipline. Unfortunately, any unintended act of poor discipline may surface years later in a student, when he or she will preface some illegal or unsafe action/maneuver with the phrase "back in training, my IP..."

Flight discipline is a matter of character. Do you have the character to stand up for doing the right thing? I was once at an IP meeting in which the topic of flight discipline was addressed. A suggestion thrown out to the group was "before accomplishing a particular maneuver, would you do it if the wing commander was in the airplane with you?" I totally disagree with that approach to tackling flight discipline problems. True flight discipline is displayed when no one is looking. Regardless of who may or may not be there to witness—do the right thing. You follow the rules, local guidance, and regulations and log the accomplished maneuvers whether the wing commander, a new student, or no one is in the airplane with you. That is the integrity of a professional aviator and that is flight discipline. This is what needs to be modeled to young and impressionable student pilots.

Another common saying in flying training organizations is "fly what you want and log what you need." Again, this mentality is counter to proper flight discipline. Flying requirements are established to ensure proper proficiency is maintained. Logging approaches not flown is a perfect example of a breakdown in personal integrity and flight discipline. Students overhear instructors saying this and immediately the seed has been planted that it is OK to lie about flying training requirements.

Role modeling is tough work

Unfortunately, it isn't always easy to be a perfect role model. There are many internal and external pressures on the flight instructor that can hinder this "perfect role model" stature. These pressures can originate with good intentions, and yet still be counterproductive to flight discipline. Take, for example, the all-important time line, the training world's measure of success. Is the organization graduating students on the standardized timetable?

Occasionally, the time line drives instructors into attempting to accomplish training in hazardous weather conditions. You may feel career pressures or fear a loss of prestige if your student is falling behind. It may be construed that you are a less than capable flight instructor if you can't manage to accomplish the student's required profile in accordance with a predetermined timetable. All the other instructors are doing it, why not you? This then puts you in an awkward situation to push some cloud clearance limits or document training accomplished when the student never really did a particular maneuver. Internally, you may struggle to do the right thing versus comprising your integrity to succeed. Your desire is to move up the ladder of instructor progression and you realize the implications of falling behind or not meeting the quota. Resist. Take the high road.

Flight discipline is preventative medicine

There is seldom any need to push the limits, and never a need to break them. There are the unfortunate types who need to break rules for enjoyment. Poor flight discipline can then become an addiction. Once you break a simple rule or regulation, it is then much easier to break others and a habit pattern can manifest itself in ever-increasing deviations from the norm. Once you form a rule-breaking habit pattern, you lose the ability to differentiate between necessity and convenience. When this stage is reached, the wheels are in motion for an aircraft accident that may kill. Worse, as an instructor, a student may witness your addiction. Now, not only are you prone for an accident but you become an accomplice for role modeling a behavior that may be mimicked and result in the death of a naive student.

Often after an accident other pilots question, "where did they ever get that technique?" referring to some crazy stunt a pilot attempted and unfortunately became a mishap statistic. As I think back over my years of flying, I start to recall some bizarre things I have witnessed. Someone who may be more impressionable or have less self-confidence could assimilate these events as acceptable pilot behavior. But people must be sure of their beliefs and question even those of higher rank or status concerning their actions.

Confrontation

Confrontation is a necessary part of flight discipline. Once as a second lieutenant I was flying with a major. On the recovery back to the traffic pattern he did an illegal aileron roll. We were easily 7,000 AGL and an aileron roll in a T-38 is a very quick and harmless maneuver. In fact, it was a very effective way to clear as we were descending. But no matter how you try to explain it, he was wrong to have done it. Any bank angle greater than 90 degrees is not allowed except in the confines of the Military Operating Area. In fact, I was upset that he did it without asking, assuming that I wouldn't care. That insulted me. I informed him that I was very uncomfortable with his actions and I never flew with him again.

Silence condones. As aviators it is our professional responsibility to confront those lacking in flight discipline. Lack of correction is as good as praising the deviation. Over time, individuals who consistently push the rules and regulations and are not corrected by other pilots will believe that what they're doing is acceptable. It doesn't do any good to inform the safety investigation board after a crash on a low level that you had noticed Pilot X flying too low previously. Instructor pilots carry an awesome burden of teaching all pilots that they fly with. Even between fellow instructors we must have the courage to question inappropriate flying actions. In front of students we must always be at our best. We may not always fly the best CAT II approach—it is fairly easy to explain to a student why we were left of centerline due to misanalyzing the winds. But how do you explain to a student why you demonstrated attempting the approach when the weather minimums were not allowable for such an approach?

I was a young, impressionable copilot in Military Airlift Command (MAC), Airlift Mobility Command's (AMC) previous name. I was the third pilot on a static display mission to an airshow. We carried no cargo but we still took three loadmasters and two flight engineers to the event. His was a large crew for an airshow, and all this equaled trouble. The aircraft commander at the preflight briefing said all the right things to ensure the crew maintained their professional demeanor in front of the tax-paying public. For instance, no alcohol near the airplane. One afternoon, I brought some close family friends on board for a tour of the aircraft and much to my surprise there was the entire crew, including the aircraft commander drinking beer on the airplane. (We were not intending to fly that day.) I was shocked. I lost a tremendous amount of respect for that aircraft commander for demonstrating a lack of flight discipline. What is briefed must be practiced, whether it is in the air or on the ground. It all combines to set the tone of professionalism for the aircrew. Since I didn't participate in the crew's drinking activities, I was counseled upon returning from the mission for my lack of MAC-mentality. I also lost much respect for the senior officer in my squadron who advised me to go along with aircrew functions. This is a prime example of senior leadership promoting an unhealthy squadron climate towards flight discipline. This senior leader wasn't concerned with the crew's drinking but about my "fitting in" by lowering my standards. He didn't get it, flight discipline is uncompromising.

Flight discipline is consistency

Flight discipline can come in many forms, but all of them must be practiced with consistency, not just when convenient. A common phrase in flying is "plan the flight, fly the plan." I think this is tremendously simple and valuable advice. Though I was not under the tutelage of an instructor, I was the copilot under the aircraft commander when the following incident occurred. We had coordinated for a fly-over at the Air Force Academy for the noon-meal formation. He had planned and briefed to do one pass and then return to Peterson AFB for landing. The first pass went perfect. On time over target—no problems. Then the controlling agency requested that we do another pass. I had just flown the one pass and my initial reaction was to return to Peterson AFB as planned. We had not planned on doing another pass and therefore should not attempt it due to the many potential traffic conflicts that surround the Air Force Academy. But the aircraft commander was anxious to fly over his alma mater and took control of the airplane. I insisted again that we follow the original plan, but he overruled me. The events that took place ended up being rather comical. The AC had gotten disoriented during the swap of aircraft control and the numerous radio communications. He started to descend and accelerate for the fly-over but unfortunately he was flying over the wrong target. We ended up flying rather low and fast over the base commissary (grocery store). Afterwards I informed him I witnessed Air Force retirees running from the parking lot for fear of an airplane crash. Plan the flight, fly the plan. There is a difference between being flexible while accomplishing the mission and having flight discipline.

A final point for supervisors

The final point I would like to touch on is directed to all supervisors. A common trend is that once a pilot has the minimum hours necessary to upgrade to instructor the upgrade process starts. As this chapter has indicated, being an IP takes more than simply being able to instruct a good instrument approach. Leadership in a flying organization must look at the whole person. What type of role model is this pilot? Character and integrity should be part of the consideration, not just checkride results. When teaching a student to fly, the individual maneuver specifics will get them to pass a checkride in the aircraft that they are training in. After completion of their training, often the student will progress to a different airplane and certainly to a different flying environment. The power settings and entry airspeeds will change. But the role model of flight discipline will apply to every airplane flown. Students never forget.

Chapter review questions

1. What special challenges does flight instruction involve as it relates to flight discipline?

2. Are there any circumstances when flight discipline should take a "back seat" to other training requirements? If so, when and why?

3. What does the author mean when he refers to "human stops?"

4. How can a flight instructor inculcate flight discipline out of the cockpit?

5. How does an instructor become an effective role model for flight discipline?

References

Air Education and Training Command. 1990. *Instructor Development*. Randolph AFB, TX.

Department of Transportation. 1987. *Aeronautical Decision Making for Helicopter Pilots*. DOT/FAA/PM-86/45. Springfield, VA.

10

Communications discipline

No man is an island.
John Dunne

Most experienced pilots are pretty confident communicators. They need to be.

The National Airspace System is an intricate tapestry of routes, radars, regulations, air traffic controllers, aircrew members, inspectors, and equipment. Communication is essential for safe operations. Beyond the communications skills needed to operate outside of the cockpit, there is a special set of crew coordination skills, and written and non-verbal communications that factor into successful flight operations. In fact, communications are a critical part of just about every facet of aviation, and yet we tend to take these skills for granted. We shouldn't. An analysis of over 50,000 self-reports from the ASRS system in 1986 found that over 70 percent of all reports involved some kind of oral communications problem (ASRS Callback No. 86). According to Frank Hawkins, the author of *Human Factors in Flight*

> *"Social, economic, and technological efficiency all depend on effective communication. Its influence is such that without it, loneliness, distress, and death among the aged can result, destructive strike action can occur in industry, and aircraft can crash. It deserves serious attention."* (*Hawkins 152*).

Hawkins points to the universal need for communication, both inside and outside of the cockpit. We will begin our look at communications discipline by considering the significance of even small miscommunications in the cockpit. Consider the following examples.

During an early upgrade training flight in a large commercial transport aircraft, the instructor pilot (IP) decided to "test" the new

201

hire by cutting out the electrical trim system without communicating his intentions to anyone else in the cabin. The new pilot, whose landing technique included trimming off pressure in the landing flare, landed nose-gear first, and the aircraft began to porpoise violently.

During a B-52 training mission in the local traffic pattern, a high-time instructor pilot pulled the nose up 20 degrees, rolled to 60 degrees of bank, and selected full power without warning the crew. The navigators, assuming a catastrophic situation had occurred which required a zoom maneuver for ejection (some B-52 ejection seats eject down), reached for their ejection handles and began to initiate ejection when the IP casually asked how they liked his "Captain Astro" closed pattern.

On another military flight, following weapons release in B-1B bomber aircraft, the offensive system officer (OSO) accidentally "sequenced" the flight management system to the wrong waypoint. The inexperienced copilot, who had flown the same low-level route just two days before, mentioned respectfully to the senior navigator, "The other day when we flew this route we were further north." The obviously irritated OSO replied sarcastically, "The other day you were wrong." A few seconds later, the range controller directed the aircrew to abort the low-level route and go home because they had busted out of the protected airspace.

Communications: Sometimes we don't do enough of it. Sometimes we don't do it correctly, and sometimes we don't do it at all. But aviators generally believe themselves to be excellent communicators—and therein lies our challenge. How can we improve our communications when we ourselves don't think that there is a problem?

This chapter is dedicated to addressing that challenge. By illustrating examples of good and bad communication, outlining research findings, and recommending strategies for improving our communications discipline, I hope to convince you to take a fresh look at your own communication style and habit patterns. Whether you fly a Cessna 172, a Boeing 757, or lead a four-ship of F-16s, your capability and efficiency is directly related to your ability to communicate. Communications discipline is one of the easiest recognized marks of a true professional aviator.

From the aircrew perspective, there are two basic types of communication that take place within the National Airspace System environment. The first is the fundamental communications that take place between pilots and air traffic controllers, and it is here where we will dedicate most of our analysis. The second is communication that takes place within a cockpit or between aircraft in the same formation. Anyone who operates in a team will benefit from the discussion on intracockpit communication. We will discuss each in turn.

Communications discipline also means being able to analyze your own communications patterns and improve them systematically. To this end we will discuss patterns of successful and effective communications which have been identified by exhaustive research, so that you can apply them to your own environment. But let's first take a closer look at the complex environment in which we ply our skills.

The complex airspace environment

The National Airspace System has evolved dramatically over the past fifty years, and this evolution has resulted in some aircrews and areas being served by equipment which is decades old, while others are using state-of-the-art technology. In fact, many air traffic control centers have both types of equipment sitting side by side. What this means is that the system itself can often be difficult to work with and sometimes even unpredictable. Ask any pilot who has flown from the relative isolation of rural Wisconsin into the airspace surrounding Chicago O'Hare how quickly the communication environment can change—from boring to terrifyingly complicated—in just a matter of seconds. Early in my military career, we used to fly a companion trainer aircraft as well as our primary jet to help build flying hours. This "Accelerated Copilot Enrichment"—or ACE—program was a great deal. We got a T-37 twin-engine jet and a credit card for gas. Beyond that, there were few restrictions. Flying out of K.I. Sawyer AFB (a recent victim of military downsizing) in Michigan's Upper Peninsula, we got pretty much whatever we asked for from Minneapolis Center. In fact, the regional controller was usually just glad to talk to someone up there. We would routinely request such things as "a five thousand foot block within a 40-nautical-mile radius of Marquette for airwork"—and get it. But when a friend suggested that we should fly down to catch a Cub's game on the weekend, I got my first taste of real ATC communications. I can still recall the stark terror I felt when I was issued that dreadful clearance as we approached Chicago. It went something like "Superior 30, hold northwest of the...expect the Bellflower Four Arrival at 2345Z...and expect the Localizer Back course to 13." I thought I had just been time-warped into another dimension, one in which they spoke too often, too fast, and in another language. I still do not understand how we ever got into Navy Glenview that night, but I am certain that we botched all three requirements of the initial clearance. The culture shock of going from uncongested—better said unoccupied—airspace into one of the most heavily congested approach-and-departure corridors in the world has never left me, and I have long since gotten serious about ATC communications. I have heard dozens of similar examples from across the United States and around the world.

To stay safe we must know and comply with all of the regulations and procedures, but to maximize the efficiency of this intricate system, we must go beyond merely following the rules and practice sound communications discipline between pilots and air traffic controllers.

The pilot-controller team

It's not a big secret that pilots and air traffic controllers often do not appear to be members of a mutual admiration society. John Stewart, a pilot, flight instructor, professional air traffic controller, and the author of *Avoiding Common Pilot Errors: An Air Traffic Controller's View*, explains.

> *"Pilots and controllers are like two very good teams in the same league. They have a healthy respect for each other, a certain amount of competitiveness, and a truckload of stories about how the other team is a bunch of wimps. Like any other group of individuals who spend a lot of time in the same arena, we try not to allow the other team to take itself too seriously. Spend a little time in either camp and you will hear any group of pilots/controllers make remarks about the other group that, if taken seriously, sound like there is a genuine dislike between the two. But bring these two teams together and you have an all-star gathering that will fight shoulder-to-shoulder against anti-aviation elements who try to cut down either group" (Stewart xii).*

In spite of this view, it is a fact there are some built-in inequalities that have serious implications for communications discipline. The first is the fact that controllers know a great deal more about pilots and the flight environment than pilots tend to know about the air traffic control environment. The rationale for this is fairly simple. A large percentage of air traffic controllers are rated pilots. In fact, many are flight instructors and hold complex aircraft-type ratings. They know both sides of the communication equation. Most pilots, on the other hand, have never visited an air traffic control center or even their local approach control facility, and even if they have, they do not share the enthusiasm for learning the controller's trade that those on the other side of the radar scope do for flight. That is pretty understandable when you think about it for a minute.

My point is simply that we have a lot to learn from each other, and that as pilots we tend to be on the short side of this knowledge equation. To pilots, it all seems pretty simple. We receive and read back clearances, and then go on with our flight. Why do we need to understand the controller's point of view?

Read back, hear back

Part of the answer comes from basic communication theory. If you recall the fundamentals of the "sender-message-receiver-feedback" loop that we learned about in our *Introduction to Communications* class, the problem in aviation communications often comes from a lack of feedback from the controllers to the pilots. Although pilots are required to read back many items to controllers, occasionally the message gets garbled. Bill Monan addressed this challenge in an excellent synopsis of the problem in a 1991 on-line article from the Aircrew Self-Reporting System (ASRS) *Directline*, partially reprinted below with permission. (ASRS *Directline*, March 1991, 1)

All too frequently airmen are reading back wrong numbers, and the ATC controllers are failing to catch the pilot's errors in the readbacks. We call this the hear-back problem. In spite of the fact that the FAA and the industry have actively campaigned for improvement in these areas, the ASRS submissions confirm that hear-back problems in pilot/controller communications continue to be acute.

Causes of communications breakdowns

Why aren't pilots getting it straight? The FARs and ICAO regulations are not that complicated. Pilots are required to read back altitudes, airspeeds, headings, altimeters, transponder codes, runways, and takeoff and landing clearances. An examination of a sample set of

ASRS reports from airmen and controllers identified four major sources for pilot errors in their read-backs.

Read-back problems

1. Similar aircraft call signs. With their hub operations, airlines (as well as military operations) have set a major trap for their airmen. Similar call signs and aircraft all operating in the same airspace, at the same time, and on the same frequency, cause big problems for airmen and controllers alike. "Good for marketing, bad for us" protested one ASRS reporter.

2. Only one pilot listening on ATC frequency. "Picking up the ATIS" and "talking to the company" represented a time-critical gap in backup monitoring during two-pilot operations.

3. Slips of mind and tongue. The typical human errors in this category included being advised of traffic at another flight level and accepting the information as your clearance to that flight level, the classic "one zero" and "one one thousand" mix-up, the left/right confusion in parallel runways, and the interpretation of "maintain two five zero" as an altitude rather than an airspeed limitation.

4. Mindset and expectancy factors. The airmen who request "higher" or "lower" tend to be spring-loaded to "hear what we wanted to hear" upon receipt of a blurred call sign transmission.

The incident set included traffic conflicts, altitude busts, crossing restrictions not made, heading/track deviations, active runway transgressions, and mix-ups of takeoff clearances and parallel runways. Two reports of controlled flight toward terrain were reported.

Hear-back problems

"Why didn't the controller catch the pilot error?" was the questioning theme in the study. While the sources for pilot read-back failures were clearly delineated in the ASRS narratives, hear-back deficiencies diffused into a tangle of erratic, randomly overlapping causal circumstances. But the underlying problem seems to be the sheer volume of traffic: the 9 A.M. to 5 P.M. rush of departures/arrivals; the behind-the-scenes tasks of land-lines, phones, and hand-offs; the congested frequencies with "stepped-on" transmissions; the working of several discrete frequencies; and, at times, the time and attention-consuming repeats of call-ups or clearances to individual aircraft. These activities, together with human fallibilities of inexperience, distractions, and fatigue set the stage for hear-back failures. Indeed, a series of pilot narratives recognized controller overload, working too many aircraft, overwork, and frequency saturation.

These facility conditions provide strong motivations for airmen to drop any "how-the-system-is-supposed-to-work" idealism and adopt a more realistic approach to cockpit communication practices. As a working premise, airmen should assume that during congested traffic conditions, the controller may be unable to hear, or is not listening to their read-backs.

Digging deeper

The report set included a number of aggressively optimistic assumptions on the part of pilots regarding ATC performance. Reluctantly, but more and more frequently, airmen are accepting silence as a confirmation that read-backs are correct. Pilots respond to doubtful or partially heard clearances with perfunctory read-backs, expecting controllers to catch any and all errors.

Airmen seem to hold to the illusion that ATC radar controllers are continuously observing their aircraft as they progress through the airway structure. The reality is that controllers continuously scan the entire scope; they generally do not focus on individual targets. Descent clearances that "seem a little early" or to altitudes that "seem too low" or turns in the wrong direction may well be intended for another aircraft.

Finally, airmen who fail to brief minimum safe altitudes within or near a terminal area or during the approach phase are vulnerable to read-back/hear-back errors leading to "controlled flight toward terrain." Such an event is described in an ASRS report from a shaken pilot who admitted to not checking the charts prior to a night-time descent:

"The dim shape of the mountain came in view...seconds before the 'WHOOP...WHOOP...PULL UP' sounded. We both pulled back abruptly on the controls and climbed."

The ATC controller's report added further details: "The tapes revealed that I had told the pilot to descend to 7,000 feet (6,500 is the MEA) but he had read back 5,000. He got down to 5,700 feet about two miles from a 5,687-foot mountain before I saw him."

Summarized the airmen: "I don't know how much we missed by, but it certainly emphasizes the importance of good communications between controller and the pilots."

The outcome of a similar incident was not as positive. "Reading the tape" was the final administrative step that identified the read-back/hear-back sequence in an NTSB-assisted international accident investigation of a fatal commercial airliner crash.

Time—06:32

Controller: "descend two to four zero zero (Controller meant 2,400 feet). Cleared for the NDB approach..."

Pilot: "Okay, four zero zero."

Tape readout: "WHOOP-WHOOP-PULL-UP. WHOOP-WHOOP-PULL UP."

Time—06:34 SOUND OF IMPACT

Summary and recommendations from the ASRS study

When pilots read back ATC clearances, they are asking a question: "Did we get it right?" Unfortunately, ASRS reports reveal that ATC is not always listening. Contrary to many pilots' assumptions, controller silence is not confirmation of a read-back's correctness, especially during peak traffic periods. Pilots can take several precautions to reduce the likelihood of read-back/hear-back failures:

- Ask for verification of any ATC instruction about which there is doubt. Don't read back a "best guess" at a clearance, expecting ATC to catch any mistakes.
- Be aware that being off ATC frequency while picking up the ATIS or while talking to the company is a potential communications trap for a two-man crew.
- Use standard communications procedures in reading back clearances. "Okay," "Roger," or a couple of microphone clicks are poor substitutes for read-backs.

Controllers can also take steps to safeguard against read-back/hear-back failures.

- Be aware that an altitude mentioned for purposes other than a clearance, such as a traffic point-out, may occasionally be interpreted by pilots as an instruction to go to that altitude.

- Deliver cautionary messages such as "similar call signs on frequency" to help reduce call sign confusion.

The consequences of read-back/hear-back failures vary, but when they occur in the context of high-rate-of-climb/descent operations, ASRS reports frequently conclude: "It was too late to intervene—the aircraft had already passed through an occupied altitude."

Another problem of communication discipline cannot be shared with our ATC brethren, at least not very often or with a straight face. "Lost comm" is a frightening experience, and one that can usually be laid directly at the feet—or the fingers—of the aircrew.

Lost communications

Safe movement of aircraft requires good voice communication between pilot and controller. When this capability is lost, there can be—and often are—very serious consequences. The following event was reported in a 1993 ASRS Directline by an air carrier who lost communications and situational awareness at the same time. (ASRS *Directline*, August 1993)

"...either we missed a frequency change call, or Center failed to pass us to the next sector. Although all three crew members were eating, I can't believe we all missed the repeated calls ATC states they made to us directly and through other aircraft...but through inattention or subconscious reliance on a call from Center to start descent, we continued on at flight level 350. We were nearly at ATL (destination) when we recognized the problem. After a rush to re-establish communication, I made contact with ATL Center and reported overhead ATL at 35,000 feet..."

Although this situation ended in embarrassment, and luckily not disaster, it demonstrates that even three highly experienced commercial airline pilots can fall victim to lost communications through poor crew coordination and cockpit discipline. But lost communications can happen to anyone.

A 1993 study identified some interesting facts about lost communications, including how it happens, how long it usually lasts, and what types of aviation are most prone to have a particular type of lost communication.

How pilots lose communications (ASRS *Directline,* August 1993)

"In addition to simple equipment failures, there are several ways in which pilots can and often do cause self-induced lost communications. Here are just a few.

- Misset the aircraft audio panel.
- Set the radio or headset volume too low.
- Copy and set an incorrect frequency.
- Forget to tune in a new frequency because of distraction.
- "Stuck mike" frequency blocking.
- Trying to communicate on the wrong radio (flip/flop or wafer switch problems)."

Although losing communications is serious enough in its own right, the ASRS study identified an equally serious problem of delayed recognition. An average incident of lost communications lasted 7.6 minutes (ASRS Directline, August 1993, 4), an eternity in densely congested airspace. The shortest incident lasted only 30 seconds, and the longest was one hour in duration. High workload was usually the culprit in prolonged episodes.

Phase of flight

Some interesting differences were found between types of flying operations and when lost communication occurred. Air carriers were most likely to lose communications during cruise flight. This comes as no real surprise. A McDonnell Douglas 1992 Commercial Jet Transport Safety Statistics review showed that air carriers spend approximately 64 percent of their flight time in cruise, and this may provide some clues as to lost communication causes for commercial airliners. Lower attention levels and complacency contribute to many episodes, especially during long legs.

General-aviation aircraft showed a different and far more serious tendency to lose communications during approach and landing phases of flight. Perhaps the reasons for this include lower levels of automation, less experience, and higher workloads in single-pilot aircraft. He or she often is at a far greater risk for task saturation than their airline counterpart. Additionally, many general-aviation pilots find themselves in situations where they may not be aware of the communications environment or requirements, which can lead to situations where pilots enter special-use or restricted airspace not talking to anyone.

Low experience increases risk

Regardless of the type of aircraft flown, there is a significantly increased risk of lost communication when one or more of the flight crewmembers is inexperienced in the aircraft they are flying. For example, the vast number of incidents in general aviation occur with pilots with less than 300 hours of time in type. The trend was also noted in commercial air carriers, and I can tell you from 15 years of personal experience with military training programs, the trend holds true in high-performance military jets as well.

Recommendations

If ever there was a challenge suited for a flight discipline answer, lost communications is it. Attention to detail—the cornerstone of flight discipline, is the remedy for a variety of causes. But let's be a bit more specific. The *Directline* study previously cited made several specific recommendations, which are supplemented and augmented below.

1. Pilots should be aware that there is a significantly increased opportunity for a lost communications event when experience in the aircraft type is low. Continued emphasis on situational awareness will help.

2. Pilots often experience difficulty in returning to an original frequency if there was an error in selection or clearance to a new frequency. A simple and effective aid for pilots is to write down each and every assigned frequency. Should a loss of communication occur at the point of a frequency change, the pilot may easily return to the previous frequency.

3. Pilots should ensure that they are familiar with alternate sources of frequency information, including approach plates, navigational charts, and IFR and VFR Supplements. In addition, many pilots develop their own set of "plastic brains," which often include radio frequencies and changeover points.

4. Thorough preflight planning can help reduce the impact in high-workload environments. Anticipate crunch points and be ready for the increased communications requirements associated with the approach-and-landing phases of flight.

5. Where high workloads contribute to lost communications, such as during approach and landing, adherence to cockpit discipline practices such as the sterile cockpit (no unnecessary communications inside the cockpit), and positional awareness should serve to reduce delays in lost communication recognition.

6. Note facility boundaries on high-altitude charts so you can expect frequency changes instead of missing them.

7. General-aviation types may want to carry hand-held portable transceivers as a backup to aircraft mounted radio equipment. On at least four documented occasions, these radios have been credited with a communications "save."

Some finer points of communications discipline

So far we have spoken about the importance of effective communications and how to stay on frequency. Like in any endeavor, however, there are those who do it better than the rest. To complete the picture of effective communications discipline, we must discuss a few more topics.

Phraseology

Communications between pilots and controllers need not, and in fact should not, be complicated. The pilot/controller glossary in the *Airman's Information Manual* (AIM) spells out the vocabulary of aviation in exacting detail. If communicators on both sides of the radarscope took the time to first learn and then exclusively use the appropriate terminology, there would be no problem with phraseology in aviation. The problem is that we don't. We are undisciplined.

As in any profession, a specialized and unique jargon has grown up over the years of pilot-controller interaction. In the process, unfortunately, much of the standardization we so desperately need has been lost. Take, for example, the simple term "Roger." According to the AIM, it means "I have received and understood your last transmission." Many pilots now use the term inappropriately as a substitute for "yes" or "I will comply." The correct aviation terms for these two desired responses are, of course, "affirmative" and "Wilco," respectively. Many other terms and phrases have suffered a similar fate.

Oddly enough, pilots in the United States, where English is the native language (although my British friends might argue that point), are the worst offenders of poor phraseology. According to Frank Hawkins, "the English language over Frankfurt, in Germany, may be closer to international standards, and so more intelligible, than that over, say Chicago." (157).

The answer to the challenge of poor phraseology among pilots is a relatively simple one. We must all take the time and make the effort to master the terms contained in the AIM glossary and then avoid the temptation to deviate with imprecise aviation jargon or slang.

Written and automated communications

We have spoken very little about written or automated communication in this chapter, but it is a key element of a professional pilot's communications discipline. The pilot who can accurately file a complex flight plan, or who can glean every bit of information from the NOTAM board, is becoming more and more of a rarity. Nothing mucks up the Air Traffic Control system more than pilots who enter a bogus fix or misspell a route identifier or approach fix on a quickly thrown together flight plan. Equally as selfish are the pilots who take off with a short and simple flight plan that they never intend to fly, figuring that they can "air file with Center" once they get to altitude. More and more controllers are refusing to accept such nonsense, and are requiring pilots to contact flight service (FSS) and asking them to hold if necessary until they receive their new clearance.

The point is to take the time required to accurately and thoroughly plan your route of flight and then communicate that information in the standardized flight plan. This is the first and most important step in making the ATC system work for you and at the same time freeing up critical controller time and effort to better serve others.

Other sources of information are also essential to disciplined communications. If all pilots understood how to access all Class L, D, and 2 NOTAMs, and effectively utilized the automated information systems like the automatic terminal information service (ATIS), transcribed weather broadcast (TWEB), and the en route flight advisory service (EFAS), many unnecessary communications could be avoided.

Clearing

Experienced pilots learn to clear using their ears as well as their eyes. This takes a special set of listening skills that tune into the requests and instructions of other pilots as well as you do to your own. A radio call from another aircraft in the visual traffic pattern should automatically direct your eyes to a certain area to locate the aircraft. A call on approach control frequency which cancels an IFR clearance should cue you to expect an incoming VFR aircraft in the near future. This skill can be systematically developed over time by trying to guess the location of aircraft as they make their various radio calls. Obviously, don't let this practice detract from your own situational awareness, but in time you will find radio clearing a valuable tool for your kit bag.

Subtle messages

Pilots and controllers communicate on many levels. Both competence and incompetence communicate, as does a familiarity with local procedures. For example, most serious pilots do not call "level" at an altitude if they are more than a hundred feet off altitude. They take a moment to correct it, and then call in. Likewise, pilots can let controllers (and other pilots in the area) know that they are familiar with local procedures with a simple question or comment such as a short comment on local weather phenomena or other such data that only a local would likely know or comment on. Controllers keep a mental list (and possibly a written one too) on the local aircraft call signs that are used for training or that they have had trouble with in the past. Let them know you are one of the "good guys," and someday when you need it they will be there for you.

Crew leadership and communications discipline

It is true that leadership communications can be a difficult quality to measure, but it would be unfortunate to avoid the subject for the lack of mathematical precision. Indeed, definable actions and attitudes do contribute to the well-managed cockpit, and while it can be uncomfortable to evaluate your personal shortcomings, it can be worse to ignore them. Every good leader is also a good learner, and we can all learn more about effective communication.

An aircrew leader, whether an airline captain or a military aircraft commander, must first learn to avoid the temptation to overcriticize the shortcomings of other crewmembers. While a good debriefing is certainly helpful, excessive criticism can actually lead to reduced effectiveness in communication. Concentrate on self-improvement with your communications—that is where your primary responsibility lies. Help other crewmembers communicate better by setting a good example. Now let's look at the skills and principles of communication that a leader can apply.

Ask the right questions at the right time

Questioning is part and parcel to good CRM. Structured checklists have built-in questions and they should obviously be a minimum starting point. For example, most descent or before-landing check-lists have an "altimeter—checked and set" step. These are simply well-timed questions designed to ensure that you have remembered to ask for and receive an updated altimeter setting before entering the low-altitude environment. Other techniques lend themselves to asking pertinent questions about the mission. Reviewing en route charts is a form of questioning, as is briefing yourself and other crewmembers on approach information. Any abnormal or critical incident should also prompt questions from other members of the crew. Examples of these types of questions might include the following. "What do you think of leaving the gear down?" "Should I shut down number two?" "Is there anything else we need to do?" They are designed to prompt a thoughtful response.

Frankly state opinions

Having strong opinions is typical of aviators, and the old adages of "Don't be so opinionated" and "Keep your opinions to yourself" are often not appropriate for safe and effective aircraft operations. Each crewmember should be encouraged to provide opinions confidently, but in a respectful manner that leaves room for rebuttal and discussion. "I think we have a problem" and "There might be a better way to do this" are two of the most constructive comments a pilot in command can receive.

Honest, considered opinions are the net result of an aviator's career-long training and experience. When those opinions are offered constructively, they can be priceless. If ignored, the results can be disastrous.

Work out differences

One major and pervasive compromise to mission effectiveness and cockpit safety in recent years has come from authoritarian aircraft commanders who cannot or will not consider other opinions. Such a smug attitude has been a measurable component of many aviation tragedies.

Conversely, two potentially deadly civilian crises were managed brilliantly and illustrate how an aircraft commander can seek help in the form of opinions and other assistance and factor them into a successful outcome. United Flight 811 was safely returned to Honolulu after a catastrophic decompression that blew a huge hole in the side of the aircraft. A decisive but assessable captain was the key factor cited in the successful recovery. The aircraft was landed overweight, with only two functioning engines on one side, without flaps and with major structural damage.

The tragic DC-10 crash at Sioux Falls was a triumph in cockpit and crew management by another capable leader. The aircraft was maneuvered without any functioning flight controls, to an airport where emergency support was available to minimize the loss of life.

Neither of these aircraft commanders relinquished their command responsibilities in any way. On the contrary, they enhanced total crew performance by utilizing all of the talents and assets available. Make decisions, but involve others.

As the pilot in command, you are responsible for your mission, crew, and aircraft bottom line! You must make decisions, sometimes without hesitation. When time permits, seek information from your best assets—your experienced and well-trained crew (Kern 1994).

Ground communications

The disastrous collision of two 747s at Tenerife in 1977 highlights the critical importance of communications discipline during ground operations. Frank Hawkins, who cited the Netherlands civil aviation authority comments on the mishap, concluded that:

- *"The crew took off with the absolute conviction that they were clear for takeoff.*
- *The communication procedures and terminology employed were not perfect but were those in normal daily use in civil aviation.*
- *The accident resulted from a breakdown in this normal communication activity and from misinterpretations of verbal messages. Such breakdowns were known to have occurred a number of times on other occasions with resulting incidents, some of which closely resembled Tenerife..."*

Although the majority of pilot communications take place in the air, our communications vigilance and discipline must begin when we first turn on our radio and does not end until we turn it off after the chocks are in—a point this military pilot will not likely forget.

Case study: "I thought you said..."

Our crew was excited about this airlift mission for two reasons: we were going someplace new and doing something different. We had just finished an emergency air evacuation from Howard, where we had possibly saved a baby's life, and we were now taking some of the White House Communications staff to St. Louis. After blocking in our Starlifter, the Secret Service came on board and started asking questions. "How long would we be here?" "Did we have any maintenance problems?" As the aircraft commander, I was trying to be helpful but found my attention being split between my checklist duties for shutdown and answering the agent's questions.

The copilot was reading the checklist over the interphone, and I had one earpiece off my ear so I could hear the Secret Service agent's inquiries. When it came time for the "Scanner's Report," a checklist challenge and reply item with the ground observer, the copilot called for the appropriate response. The Scanner had momentarily disconnected from interphone to install the gear pins, and consequently did not hear the copilot's inquiry. After attempting to prompt the scanner for a response several times, the copilot, now obviously agitated, asked loudly over interphone, "Hey, are the chocks and gear pins installed?" (Can you guess what's coming next?)

The next step on the checklist calls for the pilot to release the parking brakes. Since I had just heard the "Scanner's report" (unfortunately not from the Scanner) and had the next step of the checklist memorized, I released the parking brakes and the aircraft began to roll backwards. I wish I could have seen the Scanner's eyes as the aircraft began rolling downhill towards the truck that he was marshaling towards us for off-loading our cargo! After recovering from his shock, he quickly threw a chock under a main wheel, and inside the aircraft we felt an abrupt stop. (With the engines and APU off, we had no hydraulic pressure available to stop the aircraft with the brakes.) Had a chock not been immediately available, or had the scanner still been directly behind the main landing gear where he had been only a moment before, this series of stupid mistakes could have resulted in a disaster—not to mention the embarrassment of a ground accident with the Secret Service on board! Where had I gone wrong?

My first mistake was trying to do two things at once, diverting my attention from the checklist. The bottom line here is "first things first." By dividing my attention, I had misplaced my priorities. In hindsight,

I'm sure the Secret Service agent would have understood if I had merely said, "Just one minute until we complete this checklist."

The scanner decided to disconnect from interphone because his ground cord was not quite long enough to reach the gear area to install the pins. He should have received clearance but he thought he would only be off interphone for just a few seconds. Murphy's Second Law, ("it will occur at the worst possible moment") warns us about that. The copilot complicated the episode by providing the scanner's response for him in a vain effort to get him to answer. Never say other crewmembers' responses for them, even if it is just to prompt their response. Because I was distracted and the scanner was off interphone, the prompt from the copilot was mistaken for the actual response.

Finally, I completed the error chain by assuming I had heard the scanner's report and automatically proceeded to the next checklist step without the copilot's prompt. By this time I was getting impatient with the situation, the Secret Service agent, and the checklist. I was ready, you might even say anxious, to complete the checklist and deplane. By proceeding with the checklist without the copilot's prompt, I had broken down our last line of defense—checklist discipline. Verbalize all breaks in checklist sequence. In this way, you give other crewmembers an opportunity to preempt a critical mistake. At the debrief, we discussed the sequence of events and how we got into the situation, as well as how we would avoid a similar situation in the future.

A thorough debrief is the most effective prevention tool we have, but an aviator's pride can often get in the way of an open and honest debrief. After we had completed the debrief, the scanner asked, "Do you owe me a case of beer or your pilot wings?" I replied, "Who cleared you off interphone, anyway?" We settled on a six-pack and a valuable lesson learned.

Communications discipline demands precision

Aircrews can be required to communicate mission-critical information dozens of times in a very short mission segment. When it works, the smooth and efficient flow of information rivals the best-choreographed scene Hollywood could produce. When it breaks down, however, the consequences are ugly at best and tragic at worst.

In today's combat or commercial environment, it is not just what you say, but when you say it and to whom. Also important is an awareness of the appropriate "channel" of communication for use. In many cases, the old adage of "keep the aircraft commander informed" has given way to more flexible guidance, "ensure that information is available when needed."

The final 90 seconds of a low-level bomb run can be used to illustrate these concepts. Although few will be as lucky as I was to fly high-performance military aircraft near the speed of sound on the deck, the communications skills and principles used there are as applicable in a commercial or general-aviation emergency situation as they are to military aircrews. As mission priorities shift from, say, avoiding enemy engagements to weapons on target, the source and priority of information must also change accordingly. All crewmembers must realize when this shift takes place and simultaneously feed required information, in amounts that that the intended receiver of the information can handle, to the appropriate crewmember. This may not always be the captain or the aircraft commander, depending on the specialized nature of the situation. All types of aviation have sets of circumstances where the key to success is what we now call precision communications. Precision communications refers to a state of efficiency in the cockpit where the maximum amount of information is transferred in the most efficient manner. It avoids the necessity of repeats, requests for further information, or clarification. At its highest state, even silence communicates.

Take, for example, a crewmember during a night flight who tells the pilot, "We have weather at 12 o'clock." This is an incomplete message that will require at a minimum two more transmissions to complete the message. The missing information on distance and possible options will likely prompt the questions, "How far away is the weather?" and "What do you think we should do about it?" A more precise message might go something like this. "Pilot, we have significant weather at 12 o'clock and 50 miles. Come left 20 degrees for the next five minutes, then proceed direct to the next waypoint. Remember to contact Center for clearance." This type of precise, efficient communications fosters a sense of competence and confidence in the entire aircrew and demonstrates that the individual—and the team—has the situation under control.

The final aspect of precision communications is the selection of the channel for transmission of the message. We are taught from day one of pilot training "not to garbage up the radios." By selecting the appropriate channel of communication, we can avoid this. Most aviators know what is meant by "Navy common" frequency, referring to the Navy pilot's tendency to use 243.0 (UHF Guard) as their backup interplane frequency. But are we aware of our own tendencies in this area? Do we overuse the interphone during critical phases of flight? Are we faithful to the "sterile cockpit" rules we brief?

The goal of precision communications is not to prohibit casual conversation in the cockpit or on the flight deck. We all know that the Super Bowl or the outcome of the NBA playoffs are legitimate topics of discussion for aviators, but let's limit these conversations to appropriate segments of the mission profile. Precise communications are the mark of the professional flyer, and its essence can be captured in a single sentence. Communicate what is needed, when needed, to whom it is needed, in clear and standardized terms through the most effective channel.

A final perspective

What does communications have to do with flight discipline? Simply this: a disciplined mind seeks clarity and precision. Communications discipline extends well beyond following the rules and regulations contained in the *Airman's Information Manual* (AIM) or the FARs. Just like knowing and following the rules of basketball does not qualify you to play in the NBA, if you want to play well in the big leagues of aviation, you must move beyond meeting the minimum standards.

I began this chapter by stating that crisp, disciplined communications was one of the easiest to recognize marks of a truly professional airman. In a very real sense, to others, we are as good as we sound. Many of us can remember our first instructor telling us something like, "No matter how scared you are, don't let them know it by sounding like it on the radio." But communications discipline goes well past presenting an image of professionalism. It makes us more effective—and much safer—aviators. Perhaps controller/flight instructor John Stewart says it best.

> *"This is one of the few professions where people's lives, livelihood, and safety depend on each other's understanding of the spoken word. I consider it ludicrous that some people would climb into an aircraft and launch themselves into the sky without a total understanding of the meaning and limitations of each and every one of those words." (Stewart 67).*

Chapter review questions

1. What communication challenges are inherent in the pilot-controller relationship? Who do you think does a better job?

2. What is a hear-back problem? Have you ever experienced this phenomenon?

3. How does "mind-set" or "expectancy" play in the communications equation? Have you ever heard something incorrectly because it was what you expected to hear?

4. Have you ever experienced a "lost communications" event? List five ways this might happen.

5. Are you guilty of occasional overuse of jargon or slang? Explain the reason phraseology is so important to safe operations.

6. How can a pilot use the radio to aid in clearing? Where are the likely conflicts in your flying environment where this technique might be put to use?

References

Drew, Charles, Andrew Scott, and Robert Matchette. 1993. *ASRS Directline*. Lost Com. Issue 6: August 1993. Internet. http//www-afo.arc.nasa.gov/ASRS/dl6_lost..htm

Foushee, H. C. 1982. "The Role of Communication, Socio-Psychological and Personality Factors in the Maintenance of Crew Coordination." *Aviation, Space and Environmental Medicine*, 53: 1062–1066.

Hawkins, F. H. 1987. *Human Factors in Flight*. 2d ed. Ashgate Publishers: Brookfield, VT.

Helmreich, R. L. 1980. Social Psychology on the Flight Deck. In Proceedings of a NASA/Industry Workshop, Resource Management on the Flight Deck, San Francisco, 26–28 June 1979.

International Civil Aviation Organization. 1989. *Human Factors Digest* No. 1, Circular 216-AN/131, Montreal.

Kern, T. 1994. *The Human Factor Newsletter*. U.S. Air Force Air Education and Training Command, Randolph AFB, TX.

Monan, B. 1991. ASRS *Directline*. Read-back-Hear-back Issue 1, March 1991. Internet. http//www-afo.arc.nasa.gov/ASRS/dl6_read.htm

Stewart, J. 1989. *Avoiding Common Pilot Errors: An Air Traffic Controller's View*. TAB Books: Blue Ridge Summit, PA.

11

Automation discipline

by Christopher Valle and Tony Kern

Tales of two disasters

Automation in the cockpit can take many forms, from the use of a hand-held GPS in general aviation to the glass cockpit of an Airbus 320. Yet despite the huge differences in the level of sophistication, the two central questions for pilots who deal with automated aircraft systems are similar regardless of the aircraft type. "How much control do we relinquish to automated systems?" and "When do we take control back?" In short, it is a question of trust.

This chapter will attempt to help you form your own personal answers to those questions by helping you gain a better understanding of the uses and limitations of automation. Unlike other chapters in this book, the discussion on automation discipline cannot be prescriptive. The human factors associated with highly automated systems are still being worked out by both designers and operators, a point made clear by the differing approaches taken by different segments of the aviation industry. Because firm guidance has not yet been developed relating to automation discipline, this chapter will present an overview of current philosophies, policies, and procedures from industry and the military, along with personal insights from the authors.

A short summary of automation-related incidents brings these questions into sharper focus. A more complete listing of automation-related accidents and incidents compiled in a technical paper by Dr. Charles Billings is included for review at attachment one.

A moonless night in Texas

On November 30th, 1992, *Pyote 14*—a B-1B Bomber from the 337th Bombardment squadron at Dyess AFB in Abilene, Texas—took off on a routine night training mission. The mission profile included a night low-level terrain-following (TF) segment through the rugged mountainous region of southwest Texas. This segment of the mission is flown by coupling the automated flight system to a forward-looking synthetic aperture radar, which displays a painted terrain profile in front of the aircraft. At nearly 600 miles per hour, in the dark, and in rugged mountainous terrain, the pilots are fundamentally reduced to the role of system monitors as the automation flies the supersonic bomber at 400 feet above the terrain.

Approximately six minutes after entering the mountainous low-level route, known as IR-165, the bomber began a slow left turn into rising terrain, stabilizing at nearly 50 degrees of bank. This bank angle exceeded regulated limits for night TF, but the slow rate of turn introduced a high potential for spatial disorientation and may have been responsible for the crew's actions. Neither of these pilots were known as overly aggressive, and in fact, both were models of professionalism and discipline within the squadron.

Although the crew had flown this route many times before, there was no moon this night and in the absolute darkness, the crew may not have been aware of the proximity of nearly 2,000 feet of rising sheer terrain to the north. As the aircraft exceeded the preset radar altimeter bank-angle limits, the terrain following (TF) mode of the autopilot commanded a fail-safe fly-up. This fail-safe, 2.4-G pull-up was designed into the terrain-following system to prevent the autopilot from flying an unsuspecting aircrew into the ground should failure of a component occur, or if a flight limit was exceeded, as it was in this case. The system was functioning exactly as designed this night and would have cleared the terrain if the crew had allowed the fly-up to continue as normal procedure dictated. For reasons unknown, one or both of the pilots disconnected the automated TF system and took action which terminated the automatic fly-up. The pilots rapidly rolled the bomber towards level flight, and began a shallow descent. The aircraft impacted the cliff with a force of greater than 2,000 Gs at nearly 550 miles per hour.

What went on inside the cockpit during the last 30 seconds of this tragic example is anybody's guess. What can be said with considerable certainty is that had the automated system been allowed to function as designed, the aircraft would have cleared the cliff. A second aspect of the tragedy is equally certain, a lack of situational awareness (SA) contributed to the disaster.

A windy day in New York

On February 28, 1984 a Scandinavian Airlines DC-10-30 skidded off the end of an 8,400-foot runway, after crossing the threshold 50 knots above approach speed and touching down over halfway down the runway. The aircraft had a known autothrottle malfunction and on this approach the autothrottle failed to respond to commands. Although only a few minor injuries occurred during the evacuation, it could have been much worse.

The captain of the aircraft was well aware of the malfunctioning autothrottle system. It had been written up for over a month and had even failed to reduce power when commanded on the first leg of the flight from Stockholm to Oslo. Yet he elected to use it to land on a relatively short, wet runway.

As the crew approached their destination in New York, they received a weather forecast warning of a significant windshear. As procedure dictated, they correctly increased their approach speed to compensate for the windshear potential. Although use of the autothrottle was not required, the crew elected to engage it, admitting later that they immediately suspected the system was not functioning as designed. Even with their suspicions, the crew did not disconnect the autothrottle system. In spite of the fact that the aircraft was flying well above the planned reference speed and had an 1,840-ft/min vertical velocity on final approach, the captain elected to continue the approach to a landing on the wet runway using the autothrottle system. As the aircraft came within 50 feet of touchdown, the autothrottle failed to respond to commands to reduce power—just as it had on the previous landing at Oslo. The aircraft finally touched down with only 3,700 feet remaining—skidding off the end of the runway and coming to rest in the water 600 feet past the runway's end.

The NTSB report noted that "performance was either aberrant or represents a tendency for the crew to be complacent and overrely on automated systems." That is an understatement.

So what is the message from these two mishaps? One crew did not trust a system that was functioning perfectly and paid the ultimate price for their mistake. The other crew relied on an automated system that they knew to be faulty and failed to disconnect it even when they had clear indications that it was flying outside of prescribed parameters.

Other case studies add to the confusion of how pilots relate to automation in their aircraft. In 1985 a China Airlines 747 crew watched the autopilot slowly roll the jumbo aircraft inverted at 41,000 feet without taking corrective action. After nearly becoming the first jumbo jet in history to break Mach 1, they eventually recovered at 9,500 feet. Although they left large parts of the aircraft in the Pacific Ocean, the crew diverted to San Francisco and safely recovered the aircraft.

The point of this discussion is that as aviators, we are still working toward effective interaction with these new, automated systems. Many bright folks are thinking long and hard about this challenge, and new research and information on the subject comes out almost monthly. The goal of this chapter is to acquaint you with some of this research, make you a more informed flyer, and assist you in developing your own personal set of criteria for automation discipline based upon your aircraft, your organizational philosophy, and your own set of airmanship skills and knowledge.

Have we overautomated?

In no endeavor has technology been brought to bear more effectively than in the aviation profession, and no profession has more effectively stimulated the advance of technology. In less than one hundred years we have moved from the first flight of the Wright brothers to transporting hundreds of people and tons of cargo in aircraft utilizing the latest advances of computer technology. In the course of this development process, we have learned how to automate nearly the entire process of flying. But have we gone too far? When does the pilot become a redundant component to the automation?

The civilian airline industry led the way with the latest aircraft from Boeing, McDonnell Douglas, and Airbus, which all boast highly automated, two-person cockpits. Many airlines are currently flying these aircraft with great success and the military has just begun operating its first advanced technology heavy-airlift aircraft, the C-17.

Automation strengths and weaknesses

Automation is a technology that works best in predetermined situations which can be planned and programmed for ahead of time. However, technology does not always provide quick and easy flexibility when the external situation changes. Christopher Wickens, a noted industrial psychologist, says automation should be used to perform functions that the human operator either cannot perform, performs poorly, or in which the human operator shows limitations (Wickens 531).

The introduction of automation to the cockpit of current aircraft has greatly benefited aviation. Today's pilot has available at his fingertips more information than could be carried in even the largest map case. However, without proper guidance, this information is of little use, and used inappropriately it can be disastrous. Even when the automation is functioning perfectly, pilots must keep track of what the aircraft is doing or the results can be devastating. Consider the pilots of KAL 007, who apparently allowed their automated navigation system to carry them into Soviet airspace, where they were intercepted and shot down by a Soviet fighter. The results of failing to monitor automation can also be extremely embarrassing, as in the case of a major airline 747 that inadvertently landed in the wrong country (see insert). Degani and Wiener (1994) point out that to operate a complex system successfully, the human-machine system must be supported by an organizational infrastructure of operating concepts, rules, and guidelines.

There cannot be a procedure for everything, and the time will come when the operators of a complex system face a unique situation for which there is no procedure. It is at this point that we recognize the reason for keeping humans in the system, since automation, with all its advantages, remains a coded set of procedures executed by the machine.

Nowhere is there a more dramatic example of human ingenuity than the United Airlines DC-10 that crashed in Sioux City on July 19, 1989. Government investigators determined that the plane crashed because a metal fan disk in the tail engine broke up in flight and metal shards severed all the hydraulic lines that controlled the plane's steering system. When the pilot regained aircraft control he turned to the flight engineer and asked for the procedure in the checklist. The engineer's reply was: "There is none" (Degani 1994). Only human ingenuity, good crew resource management skills, and a little luck saved 184 of the 296 people on board when the plane crashed on a runway in Iowa.

U.S. Jet Bound for Germany Mistakenly Lands in Belgium

The Washington Post, October 1, 1995

by Don Phillips

Section A, Page A01

Reprinted by permission

A (Major US Carrier) Airlines jumbo jet, bound for Germany from Detroit, landed by mistake in Belgium early last month...The investigation into how the Frankfurt-bound plane landed 200 miles away in Brussels so far has traced a trail of missed opportunities to redirect the flight, including the reluctance of the flight attendants to contact the cockpit crew when they and the 241 passengers clearly saw the path the plane was taking on the electronic map displays in the cabin. "The only people on that plane who didn't know where they were were the three guys up front," an aviation industry source close to the investigation said.

Aviation is replete with stories of pilots landing at the wrong airport when two airports are adjacent or at least nearby. But aviation safety officials said they are unaware of a mistaken landing at airports so far apart.

The plane, which was never in any danger, was continually under the direction of controllers who were guiding it to a normal landing at Brussels. Sandra Allen, an FAA spokeswoman, said the FAA's office in Brussels is working with European officials to find out why this happened.

The September 5th incident began when a controller at Shannon, Ireland changed Flight 52's destination in the air traffic control computer. Sources close to the inquiry said the action could not have been done inadvertently because someone would have had to type in the new destination. It is unclear why a controller did so.

Air traffic computers across Europe accepted the new destination, as each country's air traffic computer electronically accepted the assertion that the McDonnell Douglas DC-10 was supposed to go to Brussels. Meanwhile, the cockpit crew failed to follow standard procedure, which calls for crews to routinely crosscheck their location on navigation instruments. The crew had no visual cues to help navigate due to a cloud deck between them and the ground.

Then came another twist. The crew made contact with whom it thought were Frankfurt controllers, and several times addressed controllers as "Frankfurt approach." The Brussels controllers never corrected them.

Passengers and flight attendants, meanwhile, could see their location on an electronic map display in the cabin, installed by (the company) to keep the passengers updated on the flight's progress and to point out features on the ground. Sources said the flight attendants became disturbed by the change of flight plan, sources said, and some of them speculated that the plane had been hijacked. Then, when it became clear that the plane was landing, the flight attendants decided not to contact the crew because of the rule that forbids disturbing the crew on approach to landing except in an emergency. Breaking through the clouds, the cockpit crew saw for the first time the geography of the area and the layout of the airport, and realized it was not Frankfurt. The captain decided to complete the landing rather than go around and head for Frankfurt.

The crew notified (the company's) Frankfurt office and was relieved of duty immediately. Another crew was flown in from Frankfurt, and the passengers arrived in Frankfurt about seven hours late.

The plane's misdirection was reported in the Minneapolis *Star-Tribune* newspaper, but received little notice elsewhere. Most of the aviation community was unaware of it until word of the incident became the talk of a meeting of the International Society of Air Safety Investigators in Seattle.

Automation philosophies, policies, and procedures

Before we can talk about individual automation discipline, we must define what roles some organizations have set for the individual pilot.

This discussion begins with an organizational "automation philosophy" from which is then derived a set of policies and procedures. Automation philosophy can be defined as the overarching view of how automation should be utilized in an aircraft. There is a philosophy toward the development of the system itself as well as a philosophy toward the operation of that system. An airline manager or pilot may not be able to clearly state his airline's philosophy, but a philosophy does exist and can be inferred from procedures, policies, training, and punitive actions (Degani and Wiener 49).

The emergence of flight deck automation as an operational problem has generated interest in the philosophy of operations, partly due to lack of agreement about how and when automatic features are to be used, and who makes the decision, the company or the pilot in command. The nearly unlimited capability of today's Flight Management Systems (FMS) to operate between any range from fully automated to manually controlled has convinced some airlines that it is necessary to formally state an organizational philosophy of operational utilization of automation. However, this philosophy must allow a pilot the freedom to improvise, as no philosophy can address every issue a pilot will encounter. As Earl Wiener made clear in 1987, "automate if you wish, but human operators are here to stay."

Policies are plans, or courses of action, that will dictate what an organization expects on the line. Policies are developed to adhere to the company operating philosophy and provide specifications of the manner in which management expects operations to be performed in training, flying, exercise of authority, personal conduct, etc. (Degani and Wiener, 1994).

Procedures, then, are more exact actions or methods used to carry out the stated philosophy of the organization. Procedures should be designed to be as consistent as possible with the policies (which are consistent with the philosophy). This structure is depicted in Figure 11-1.

Procedures will dictate specific actions to the pilots, but an understanding of the designer's philosophy behind the development of the automation will provide the pilot with the flexibility to confront those situations that are not addressed by procedure. All of this results in improved flight discipline.

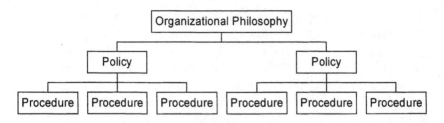

11-1 *The three Ps. Organizational philosophies, policies, and procedures are the universe in which aviators must ply their trade. Failure to comprehend the rules at any level can lead to flight discipline problems for pilots.* (Degani and Wiener, 1994.)

Standard operating procedures (SOPs) are sets of procedures that, apart from being merely specifications of tasks, also serve to provide a common baseline for a flight crew. This starting point is important because the crew may be totally unfamiliar with each other's experience and technical capabilities. The military and airline industry are strong proponents of SOPs and attempts to attain a level of standardization such that if a cockpit crewmember were to be replaced in midflight, the operation would theoretically continue safely and smoothly. While SOPs promote uniformity of operations among large groups, they do so at the risk of reducing the role of the individual human operator to a lower level (Degani and Wiener 1994). When systems operators are reduced to the role of systems *monitors*, complacency is often a byproduct, with potentially disastrous implications. No procedure is a substitute for an intelligent operator.

Human-centered automation

One design philosophy that is gaining momentum in the aviation industry is human-centered automation. This philosophy, developed by Dr. Charles E. Billings of Ohio State University, recognizes that today's aircraft are not always designed to facilitate effective cooperation between the pilot and the machine (Billings 1996). Until now, the philosophy of adding automation to the cockpit seems to be: if it can be automated, do it. According to Dr. Billings, over time this has had the effect of making the flight crew more peripheral to the actual operation of the aircraft, a fact that could have dire consequences if a situation developed which required immediate action by the pilot.

Dr. Billings sums up his discussion about human-centered automation by pointing out that although humans are far from perfect sensors, decision makers, and controllers, they possess three invaluable attributes. They are excellent detectors of signals in the midst of noise, they can reason effectively in the face of uncertainty, and they are capable of abstraction and conceptual organization (Billings 1996, p. 13). Humans therefore provide to the aviation system a degree of flexibility that cannot now, and may never, be attained by computers.

Pilot views on automation

Pilots differ in their views on automation, but one thing they agree on is that too often we are surprised by the actions of automation. Flight crews often are faced with questions such as, "Why did it do that?" "What is it doing to us now?" and "What will it do next?" Some automation surprises reflect an incomplete understanding of either the automation's capabilities and limitations, its intended use, or the aircraft's display (Abbott 1996, p. 33).

A human-factors (HF) study team found differing views regarding training for automation. Some views held that flight crews should be relieved of the burden of fully understanding system operation or the system's underlying design philosophy (Abbott 1996, p. 33). This view would lead a company to a training philosophy in which flight crews are trained to respond primarily in a rote manner (i.e., standard operating procedures). While it is recognized that the use of standard operating procedures (SOPs) is an effective strategy for managing error, the report states:

> *"It is important for flight crews to understand the principles and assumptions embodied in the automation design that affect safe operational use, especially where these principles and assumptions may differ from those of the flight crew. In the absence of this understanding, flight crews are likely to substitute their own model of how the automation works, based on their observations and assumptions of automation behavior." (Abbott 1996, p. 34).*

Pilot decisions

The HF team found that "flight crews differ in use of automation when responding to an abnormal situation, and more importantly, crews may react in ways not foreseen or taken into account during the design, certification, training, and procedure development for highly automated airplanes" (Abbott 1996, p. 34). Observations include situations where crews have either inappropriately continued to use the automation when in an abnormal situation or, if the automation was initially off, turned the automation on to try to accomplish a recovery. The report included specific examples:

- Fixation on following the flight director and ignoring airplane attitude. In one particular case, this resulted in a low speed excursion, after which the flight crew engaged the autopilot to accomplish the recovery.

- Using the autopilot to recover from an overspeed warning rather than resorting to manual control.

- Attempts by the flight crew to engage the autopilot in the moments preceding the March, 1995, crash of an A310 at Bucharest as they attempted to recover from an extreme bank angle resulting from a large thrust asymmetry.

- Engagement of the autopilot by the flight crew of the A300-600 that crashed at Nagoya, Japan, in April, 1994—apparently in response to difficulties in maintaining the glide slope following the inadvertent activation of the takeoff/go-around levers.

The team hypothesized that the action of engaging the autopilot to attempt to recover an unrecognized situation shows that flight crews are becoming less confident in their own airmanship skills relative to the capabilities they perceive to be present in the automation, particularly in a stressful situation. In some cases, where this perception of the automation's capabilities is particularly inaccurate, it can have potentially hazardous consequences.

Operations philosophies

Most pilots will fly a variety of aircraft in their aviation careers. Different design philosophies have resulted in pilots encountering problems transitioning from one aircraft manufacturer to another. This is often referred to as "negative transfer of training" and can have particularly severe repercussions when dealing with highly automated flight management systems. For example, the autopilot disconnect systems in the Airbus A300 and A310 are significantly different than the disconnect systems provided in other large transport-category airplanes (Hughes, et. al, Jan., 1995, p. 60). This difference may have contributed to the loss of an A300 in Nagoya, Japan in 1991. The Flight Data Recorder (FDR) showed the pilot had manually intervened to counter the autopilot. This action in a Boeing or McDonnell Douglas type aircraft would have disconnected the autopilot, but in the A300, the autopilot remains engaged. Further, the autopilot countered the inputs of the pilot, eventually resulting in loss of aircraft control.

One constraint to operating philosophies is senior management or company executives, who spend millions of dollars obtaining these expensive aircraft, and expect to reap a benefit from their purchase. This expectation sometimes trickles down to the line crewmember as a philosophy of "we bought it, you have to use it." Many airlines are avoiding this potential problem by developing formal automation philosophies for pilots. The following companies have published automation philosophies. They represent the best thinking on the subject that industry and the military have to offer. They are reviewed briefly for your comparison.

Delta

Delta Airlines was one of the first major airlines to publish a corporate automation policy. Chapter 4 of the Delta flight operations manual states:

> *"Automation is provided to enhance safety, reduce pilot workload and improve operational capabilities. Automation should be used at the most appropriate level. Pilots will maintain proficiency in the use of all levels of automation and the skill required to shift between levels of automation. The level used should permit both pilots to maintain a comfortable workload distribution and maintain situational awareness." (Bluecoat Newsgroup 1997)*

The following guidelines apply to the use of automation at Delta Airlines:

- If any autoflight system is not operating as expected, disengage it.
- All pilots should be aware of all settings and changes to automation systems.
- Automation tasks should not interfere with outside vigilance.

- Briefings should include special automation duties and responsibilities.
- The pilot flying (PF) must compare the performance of the autoflight systems with the flight path of the aircraft.
- When a pilot conducts a briefing, the level of automation will be addressed. (Bluecoat Newsgroup, 1997)

Delta provides a four-hour course called *Introduction to Automated Aircraft (IA2)* to all pilots that transition to glass cockpit aircraft. The class teaches this automation philosophy and includes accident and incident discussions related to automation.

Cathay Pacific Airways

Cathay Pacific Airways (CPA) has been operating for 51 years as a private airline based in Hong Kong. CPA has over 60 wide-body aircraft, 1,400 cockpit crew, 4,500 cabin crew, and 15,000 employees. CPA insists that pilots not feel pressured to use automation at all times. This airline found it advantageous to provide the following clear policy to its flight crews:

> *"It is the Cathay Pacific Airways (CPA) policy to regard automation as a tool to be used, but not blindly relied upon. At all times, flight crew must be aware of what automation is doing, and if not understood or not requested, reversion to basic modes of operation should be made immediately without analysis or delay. Trainers must ensure that all CPA Flight Crew are taught with emphasis on how to quickly revert to basic modes when necessary. In the man-machine interface, man is still in charge." (Bluecoat Newsgroup 1997).*

> *Cathay Pacific Airways believes if trainees are provided the designer's philosophy, preconceived mindsets can be overcome and knowledge can be acquired faster. Pilots are generally inquisitive and want to know as much about a system as possible. CPA attempts to satisfy this curiosity by answering the question "why?" CPA provides a video to new trainees which features a design test pilot's complete explanation for the automation used (CRM Developers Group 1997).*

United

In 1995, United Airlines revised its pilot training philosophy to a consistent, mission-oriented approach that tends to be more compatible with highly automated aircraft (Hughes et. al, Feb., 95, p. 50). United has taken the approach that pilots don't need to know the systems as well as they did in the past. Since most aircraft today automatically diagnose and correct malfunctions, then inform the pilot of the system's status, United believes that intricate systems knowledge is unnecessary for pilots. United stresses cockpit discipline and following rigid standard operating procedures (SOPs). These SOPs include announcing changes in the status of the autopilot and system modes. United aims at teaching interfacing with the aircraft systems through the flight management system (FMS).

S. William Reichert, United's manager of fleet operations for the A320 and B737-200 through 400 series, says that the key to automation is awareness. He states that operating these aircraft is more a matter of systems management than actually flying the aircraft (Hughes et. al, Feb., 1995, p. 51). United's pilots are trained to follow handbook operations, and would-be systems experts are discouraged from taking nonstandard actions. This philosophy is different from the military philosophy of understanding intricately how the system works, so you are better prepared to react to fix any problems that may occur that might not be in the flight manual.

U.S. Air Force C-17 operating philosophy

The C-17 Operations are guided by Multi-Command Instruction (MCI) 11-217. Section A, General Operating Policies provides the following guidance as the official philosophy of the Air Force toward operating the C-17:

> *"The C-17A is designed to be operated by three crewmembers, two pilots and one loadmaster. In order to perform the same demanding worldwide strategic and theater missions currently flown with larger crews, automation is employed. All technical orders, procedures, checklists, training, and supporting documents are designed to support the human operator. It is the responsibility of the crew to fully understand the operations and limitations of the automation on the aircraft. In flight, the pilot's flying (PF) will determine the most desirable level of automation for a given situation." (MCI 11-217, p. 2).*

MCI 11-217 also provides the following operating guidelines:

- The aircraft commander (AC) has the ultimate responsibility and authority for the safety of the aircraft, passengers, and crew.

- The AC must manage workload, set priorities and employ the available resources, including automation, to maintain overall situational awareness.
- Pilots should use appropriate levels of automation as required by the flight conditions. As the flight situation changes, do not feel locked into a level of automation.

The Instruction also points out the following common pitfalls associated with overreliance, misuse, or misunderstanding of automation:

- Fixating on the automation
- Misprioritizing programming tasks
- Mode awareness
- Assuming automation is programmed correctly
- Overreliance on automation

In addition, the instruction provides a discussion about standard terminology for operating the Automatic Flight Control System (AFCS), including examples of commands and recommended abbreviations. In general, the Air Force expects the C-17 aircrew to let the aircraft perform automatically to the maximum extent possible. The Air Force believes this will decrease the workload of the pilots, allowing them to manage the mission more effectively.

Automation pitfalls

Aircraft designers attempt to anticipate difficulties and dangers associated with the aircraft and incorporate solutions into the design. If they are unsuccessful, the operators of the aircraft will include a warning or caution to avoid the known hazard. However, with all systems there are intrinsic dangers or pitfalls that the operator must remain cognizant of and make every effort to avoid. This section of the chapter identifies pitfalls associated with automated aircraft that all operators should seek to avoid.

In two series of questions conducted four years apart, Captain Harry W. Orlady, a retired United AirLines pilot, sought to identify some of the controversies associated with the "glass cockpit." He identified a number of controversies among pilots:

- The real-life workload that exists under normal, abnormal, and emergency conditions
- The role of the pilot in these new aircraft, including maintenance of the captain's authority

- The existence and, much more important, the operational consequences of fatigue, boredom and complacency that might be caused by these aircraft (Orlady, 1992)

Let's briefly look at each of these.

Workload

One of the stated promises of incorporating automation into the cockpit is that it will decrease pilot workload and allow more time for decision making. After more than a decade of experience, this promise is still largely unfulfilled. Earl L. Wiener, professor of management science at the University of Miami, surveyed 200 line pilots from two airlines flying Boeing 757 aircraft. While the majority of pilots expressed pride in flying the most advanced aircraft in their company's fleet, at least half said they felt automation actually increases workload and, one year later, these pilots showed no shift in their opinion.

While some believe automation may have reduced pilot workload, studies show automation may have merely redistributed workload (Hughes et al, 1992, p. 67). For example, pilots of older aircraft are manually tasked to operate the controls as they fly, while they are cognitively tasked to communicate, process information, and make a decision. The pilots of newer, automated aircraft, performing the same functions, are more cognitively tasked to receive information, identify, and then input that information into the appropriate FMS submenu and engage the automation to perform the desired task. Depending on how workload is measured, either of these situations could be determined to be more or less intensive than the other. The former, manual process clearly is more physically demanding, but the latter example of determining which sub menu is appropriate for the function being accomplished, finding it, and inputting the appropriate information is significantly more cognitively demanding.

Current research has actually identified an increase of pilot workload. Degani and Wiener found that most workload reductions occur when work levels were already low, such as during cruise (Billings 1996, p. 139). The many changes experienced during the historically high workload situations, such as departure and arrival, can actually increase crew activity in highly automated aircraft. Because pilots are forced to focus their attention inside the aircraft to perform these changes, they are not able to look outside for possible conflicts.

Captain's role

The traditional role of the captain or aircraft commander being in command of his aircraft is challenged by automation. As discussed previously the increase in technology over the years has distanced pilots from actual control of their environment. Also, the sometimes-laborious FMS entries require that the pilot not flying (PNF), usually the first officer, make changes to the flight profile, while the pilot flying (captain) monitors the aircraft. By ultimately affecting the course of the aircraft the PNF, in essence, becomes the PF. This shared control of the aircraft weakens and blends established cockpit roles and may lead to confusion about who is actually in control of the aircraft. Confusion is the enemy of discipline.

Pilots can play any of a variety of roles in the control and management of a highly automated airplane. These roles range from direct manual control of flight path and aircraft systems to a largely autonomous operation in which the pilot's active role is minimal. This allocation of functions is represented in the following control-management continuum (Figure 11-2) by Dr. Charles Billings:

The Control/Managemnt Continuum for Pilots (Billings, 1996, p.104)

Management Mode	Automation Functions	Human Functions
Autonomous Operation	Fully autonomous operation; Pilot not usually informed; System may or may not be capable of being disabled	Pilot generally has no role in operating; Monitoring is limited to fault detection; Goals are self-defined; pilot normally has no reason to intervene
Management By Exception	Essentially autonomous operation; Automatic reconfiguration; System informs pilot and monitors responses	Pilot informed of system intent; Must consent to critical decisions; May intervene by reverting to lower level of management
Management By Consent	Full automatic control of aircraft and flight; Intent, diagnostic and prompting functions provided	Pilot must consent to state changes, checklist execution, anomaly resolution; Manual execution of critical actions
Shared Control	Enhanced control and guidance; Smart advisory systems; Potential flight path and other predictor displays	Pilot in control through CWS or envelope-protected system; May utilize advisory systems; System management manual
Assisted Manual Control	Flight director, FMS, nav modules; Data link with manual messages; Monitoring of flight path control and aircraft systems	Direct authority over all systems; Manual control aided by F/D and enhanced navigation displays; FMS is available; trend info on request
Direct Manual Control	Normal warnings and alerts'; Voice communications with ATC; Routine ACARS communications preformed automatically	Direct authority over all systems; Manual control utilizing raw data; Unaided decision-making; Manual communications

(Left margin, top to bottom: LEVEL OF AUTOMATION — VERY HIGH → VERY LOW)
(Right margin, top to bottom: LEVEL OF INVOLVEMENT — VERY HIGH → VERY LOW)

11-2 *The control-management continuum for pilots. Pilots should not only understand their role as it relates to automation, but also understand what to expect from the machine.*

A pilot must be able to operate the aircraft as necessary for safe flight. As previously discussed, Airbus A320 designers have limited the amount of direct control, preventing manual inputs that exceed predetermined parameters. The A320 is only one example where designers have limited the pilot's operating parameters from the design bench. This limited ability to control the aircraft, added to an implied faith in automation, has led to increased levels of complacency in some aviators.

Complacency

Crew complacency is often mentioned as one potential negative effect of automation. The theory is that as pilots perform duties as systems monitors they will be lulled into complacency, lose situational awareness, and not be prepared to react in a timely manner when the system fails. Wiener (1981) defined complacency as "a psychological state characterized by a low index of suspicion." In the NASA Aviation Safety Reporting System (ASRS) coding manual, complacency is defined as "self-satisfaction which may result in nonvigilance based on an unjustified assumption of satisfactory system state" (Parasuraman, et al, 1993).

Other researchers (Parasuraman, Molloy, and Singh 1993) indicated that the major factors contributing to "complacency potential" were a person's trust in, reliance on, and confidence in automation. Crew attitudes such as overconfidence in automation may not be sufficient in themselves to lead to complacency but may set the stage for it. When combined with other factors, perhaps fatigue due to poor sleep or long flights, the combination of the crew's attitude toward automation (e.g., overreliance) and a particular situation (e.g., fatigue) may lead to complacent behavior.

Mode confusion

The flight management system (FMS) on an advanced aircraft is capable of conducting an entire flight without the pilot flying the plane. To do this the flight modes will automatically transition from climb to level off, cruise, descent, and finally approach. In military aircraft this can include low level terrain following and weapons release. Each one of these modes provides different input characteristics to the aircraft, and the pilot must always be aware of the operating mode to remain prepared to return to manual flight. In some aircraft, keeping informed of this mode change requires monitoring a very small display on the glareshield of the aircraft. It is easy to see at night, but can be very difficult to monitor in daylight, especially if the sun is shining on the glareshield. If the pilot takes command of the aircraft manually and assumes the aircraft to be in a particular mode and it is not, the result could be disastrous. The Airbus A300 crash in Nagoya, Japan, is an example of the pilot taking manual control of the aircraft when the aircraft was on an approach. The pilot inadvertently activated the go-around mode, which on the A300 commands full power and a rapid climb away from the ground. The pilot attempted to continue the approach manually and fought against the autopilot, causing the pitch trim to run to the limit. The aircraft became uncontrollable and crashed.

A major area of concern with the rapid development of cockpit systems is that technology has outpaced human ability to fully comprehend automated mode behaviors (Hughes, et al., Jan., 1995, p. 63). Barry Strauch, Chief, Human Factors Division at the National Transportation and Safety Board (NTSB), stated in 1995 that "pilot awareness and understanding of computer modes in modern transport aircraft is a problem" (Hughes, et. al, Jan., 1995, p. 63). A study of mode confusion was conducted by the Massachusetts Institute of Technology (MIT) in 1994. The Aviation Safety Reporting System (ASRS) was searched for incidents related to mode confusion. The MIT team identified 184 incidents of mode awareness problems and broke them down into six categories (Hughes, et. al, Jan., p. 56). The six categories are shown in Figure 11-3 and described in the text that follows.

The MIT study found that 74 percent of the incidents involved confusion or errors in vertical navigation, while 26 percent were problems related to horizontal, or lateral, navigation. R. John Hansman, an MIT professor, noted that newer aircraft provide better feedback on horizontal navigation than vertical. Most advanced aircraft are equipped with a display that overlays an electronic map of the aircraft route, including land-based navigational aids, with the weather. This display provides a clear picture to the pilot of where the aircraft is going and how it will proceed. By contrast, no such picture exists for vertical navigation so the pilot must develop a mental model of how vertical navigation is affecting the flight path.

The following discussion of error categories from the mode awareness study identifies typical pilot errors when operating high-technology, automated aircraft (Hughes et al, Jan., 1995, p. 6).

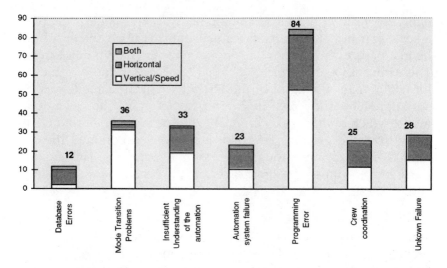

11-3 *Mode awareness incidents. Programming errors far out-distance all other types of automation incidents.* (Hughes et al 1995)

Database error An FMS database error, which causes the crew to either execute an incorrect procedure, or to question the validity of the automatic flight system (AFS) commands.

Mode transition problem Flight crew confusion or aircraft deviation because the AFS executes an unexpected mode transition, or fails to execute an expected mode transition.

Insufficient understanding of the automation Because one or more of the crewmembers do not fully understand the consequences of an action or inaction, confusion or a deviation occurs.

Programming error A partial or full AFS failure that results in crew confusion or an aircraft deviation.

Crew coordination Incorrect data input or mode selector that is not identified immediately and results in an unexpected mode, mode change or deviation.

Unknown failure Any sequence of events initiated by one crewmember that causes one or more of the other crewmembers to incorrectly assume a mode, mode transition, or parameter.

Hansman pointed out that vertical navigation is difficult to comprehend because it involves the use of different combinations of elevator and thrust inputs. Automated systems are capable of maintaining an air speed, a vertical speed, or a flight path angle, in the climb. Each possibility affects the vertical flight path in a different manner.

To preclude mode confusion, Barry A. Strauch of the NTSB advocates a more stringent pilot selection process coupled with advanced training standards that fully expose pilots to the complex capabilities of current and future glass cockpits (Hughes et al, Jan., 1995, p. 63). In summary, Earl Wiener points out that although a key function of cockpit automation has been to help eliminate mistakes, "it is a fallacy to think that automation removes human error when it actually can give the pilot opportunities to make even larger mistakes" (Hughes 1992). It is the goal of designers, operators, and regulating agencies to minimize these opportunities for mistakes.

While at the NASA Ames Research Center, Dr. Charles Billings said, "Automation is there to use, but it must be as simple to manage as the aircraft is to fly" (Orlady 1992). This seems to be the root of current problems in aviation. Unfortunately, early automation was implemented in the cockpit on the implicit assumption that machines could be substituted for humans. The theory of "if it can be automated, do it" pervaded the industry. In the past ten years we have seen the limitations of this theory; the pilot becomes increasingly removed from the operation of the aircraft.

Automation in general aviation: The limits of the GPS

The following information and reports come from NASA's Aircrew Safety Reporting System (Callback 217). It provides light aircraft perspective of the same issues we have now looked at in detail relative to commercial and military high-technology aircraft.

As the cost of Global Positioning System (GPS) units decrease, more pilots are using these devices to supplement their other navigational equipment. However, problems can arise if pilots fail to recognize that GPS is currently designed to be a supplemental— not a primary-navigational aid. A report from a corporate pilot illustrates:

> "I departed on an IFR flight plan with an IFR-approved GPS. I was cleared direct to ABC, at which time I dialed ABC into

*the VOR portion of the GPS and punched 'direct.' The head-
ing was 040 degrees. After a few minutes, Approach in-
quired as to my routing, heading, etc. I stated direct ABC,
040 degrees. They suggested turning to 340 degrees for ABC.
I was dumbfounded. My GPS receiver had locked to ADC,
3,500 miles away [in Norway]! Closer inspection revealed
that my estimated time en route was 21 hours. I did not verify my position with the VOR receiver. I mistakenly, blindly,
trusted a GPS."*

Now that is truly global positioning!

Other reporters have found themselves somewhere other than
where they wanted to be as a result of overreliance on GPS. A general aviation pilot provided this example:

*"I had recently purchased a hand-held GPS and was anx-
ious to use my new acquisition. Without thinking (obviously!), I punched in XYZ VOR and navigated along the
direct route. I did not crosscheck myself and allowed myself to invade restricted airspace. I tuned in 121.5, and received instructions and polite guidance out of my
dilemma. I realize that this is a serious problem and a very
stupid mistake."*

Limits of hand-held GPS

Many hand-held GPS units have an inherent system limitation, as our
next reporter discovered.

*"Flying VFR, using a hand-held GPS for navigational refer-
ence. While en route, position and status seemed fine. According to the GPS position, a "big airport" was getting closer
and closer, but still out of the overlying Class C airspace.
From a visual standpoint, the position was definitely in Class
C airspace. When I landed at ABC, the GPS indicated the location was XYZ [about 40 nm west]. I turned the unit off,
then back on, and the position now indicated ABC.*

*I called the manufacturer, which had received numerous
calls about erroneous positions. A new satellite had been put
in orbit; there were now a total of 26 satellites. My unit only
showed 25. The manufacturer suggested leaving the GPS on
for 45 minutes to acquire the information from the new*

*satellite. I did so, and my unit now shows 26 satellites. The
GPS positions seem correct.*

*Conclusion: Use hand-held GPS as a supplemental reference
only."*

According to the reporter's conversation with the manufacturer,
hand-held GPS units currently in use do not have the RAIM—Receiver Autonomous Integrity Monitor—that is built into installed,
IFR-certified units. The RAIM monitors the actual navaid signal, primarily on SID and STAR routes, to assure that there is adequate signal strength for navigation in the selected mode. If the signal is not
sufficient, an error message will occur. This is analogous to the 'OFF'
flag showing on the VOR receiver when the aircraft is out of range
for adequate signal acquisition. Since the reporter's GPS unit did not
have RAIM capability, there was no way to know that the unit was
providing erroneous information.

Because of the inherent limitations of hand-held units, pilots should
carry and use the appropriate charts as cross-reference material,
rather than relying solely on GPS.

GPS water rescue

When properly programmed and used, GPS has incredibly accurate position reporting capability, which can prove to be a lifesaver—literally.
The next reporter, the pilot of a long-range amphibious airplane on a
ferry flight across the Pacific Ocean, tells a "GPS-to-the-rescue" story.

*"While we were ferrying the aircraft...the left engine began
backfiring severely and would not develop [power]. Engine
#2 was brought to METO [Maximum Except Take-Off] power
attempting to maintain best altitude. An immediate turn
was made for the nearest land, and our 'Pan' emergency
shifted to a 'Mayday' call. After about one hour, descent into
the water was imminent. The night ocean visibility would be
termed zero/zero...and a standard night IFR approach was
set up. After a successful night IFR water landing, we began
taking on water. Seven people escaped without injury into a
lifeboat. Coordination with ATC and very accurate position
reporting with GPS resulted in a very expeditious rescue by
the Coast Guard and a maritime vessel. We were in the
ocean less than one day. Very spectacular efforts by all par-
ties involved."*

It is obvious from these reports that automation discipline is critical to success no matter what your aircraft type, so let's take a closer look at automation discipline and how we can improve it.

Building personal discipline for automation use

Where does all this information leave us with regards to formulating a personal plan for flight discipline and automation? Frankly, it is difficult to make any definitive recommendations that will work in all cases, but there are several areas that emerge from studies that all of us might want to consider.

First, we must recognize the potential for problems. Man and machine are still strange bedfellows. Although Dr. Billings's human-centered automation approach is a huge step in the right direction, it is likely our sons and daughters will be the first generation to realize the benefits of his work. In the meantime, we need to learn to know and respect our automated systems, which leads to my second point.

Whether or not you take the approach advocated by some airlines, that pilots need not understand all the intimate details of the automation on their aircraft, you must understand enough to trust the system while at the same time maintaining a watchful posture. Accident and incident analysis indicate some pilots harbor a secret distrust of automation, but even more frightening is that apparently some pilots no longer trust their own skills. This leads me to my third point.

Until they replace our pilot wings with "automation-system-manager" badges, we must maintain our flying skills and proficiency. The pilots who preferred to trust a malfunctioning autothrottle system over their own skills to land in a windshear are indicative of a cancer of confidence and an unhealthy overreliance on automation.

Finally, until such a time as someone designs a system which will prevent pilot complacency—and that may be a long time in coming—situational awareness is the key to survival. We must find creative methods to keep our heads in the game. Challenge yourself, challenge your crewmates. If the past is any indication of the future, good pilots will die—and likely take others with them—because they lost SA due to overreliance on automation. Don't let it be you.

Chapter review questions

1. What are the two primary questions the author suggests a pilot should ask about automation?

2. What reasons can you think of that would inhibit a pilot from taking manual control of an aircraft that might be suffering from an automation malfunction? Are these justified?

3. What are the strengths and weaknesses of automation?

4. What are the differences between automation philosophies, policies, and procedures?

5. What is mode confusion and how does it illustrate a more general challenge of automation systems knowledge for pilots?

References

Abbott, K., S. Slotte, D. Stimson. 1996. *The Interfaces between Flight crews and Modern Flight Deck Systems.* A Federal Aviation Administration Human Factors Team Report. June 18, 1996.

ASRS Callback 217. 1997. *Going Global with GPS.* NASA Aviation Safety Reporting System, Moffet Field, CA. July.

Billings, C. E. 1996. "Human-Centered Aviation Automation: Principles and Guidelines," NASA Technical Memorandum 110381.

Bluecoat newsgroup. WW Web, http://www.neosoft.com/7Esky/BLUE-COAT/intro.html. The Bluecoat Project was created by Bill Bulfer as a means of opening an ongoing discussion between the engineers who build flight deck automation systems and the pilots who use them.

Degani, A. and E. Wiener. 1994. "Philosophy, Policies, Procedures and Practices: The Four 'P's of Flight Deck Operations," *Aviation Psychology in Practice.* 45–67.

Hughes, D., M. Dornheim, W. Scott, E. Phillips, and B. Henderson. 1992. "Automated Cockpits: Keeping Pilots in the Loop," A special report found in *Aviation Week & Space Technology.* 50–70. March 23, 1992.

Hughes, D., M. Dornheim, W. Scott, E. Phillips, B. Henderson, and P. Sparaco. 1995. "Automated Cockpits Special Report, Part 1," *Aviation Week & Space Technology.* 52–65. January 30, 1995.

Hughes, D., M. Dornheim, D. North, and B. Nordwall, 1995. "Automated Cockpits Special Report, Part 2," *Aviation Week & Space Technology.* 48–55. February 6, 1995.

Murphy, D. 1993. "Simuflite CRM Targets Human Aspects of Flight," *Aviation International News.* 38–39. January 1, 1993.

Orlady, H.W. 1992. "Advanced Cockpit Technology in the Real World," *Journal of ATC.* 47–51. January-March.

Palmer, M.T., W.H. Rogers, K.A. Latorella, and T.S. Abbott. 1995. A Crew-Centered Flight Deck Design Philosophy for High-Speed Civil Transport (HSCT) Aircraft. NASA Technical Memorandum 109171. January.

Parasuraman, R., R. Molly, and I. Singh. 1993. "Performance Consequences of Automation-Induced Complacency," *The International Journal of Aviation Psychology.* 3(1). 1–23.

United States Air Force Regulation 110-14, 1992. Accident Investigation Report, "B-1B Mishap, 30 Nov 92" Dyess AFB, Abilene, TX.

United States Air Force. 1997. Multi-Command Instruction (MCI) 11-217, Volume 5. 15 February.

Wickens, C.D. & J. M. Flach. 1988. "Information Processing," a chapter in E. L. Wiener and D.C. Nagel (Eds.), *Human Factors In Aviation.* 111–155. San Diego, CA: Academic Press.

Wickens, C.D. 1992. *Engineering Psychology and Human Performance* (2nd ed.). New York: HarperCollins.

Weiner, E.L. 1987. Application of Vigilance Research: Rare, Medium, or Well Done? *Human Factors,* 29(6), 725–736.

Wiener, E.L. 1981. "Complacency: Is the Term Useful for Air Safety?," Proceedings of the 26th Corporate Aviation Safety Seminar. 116–125. Denver: Flight Safety Foundation.

12

Disciplined attention

The next best thing to a crystal ball

by Sheryl L. Chappell, Ph. D.

He who...walks in a mist...cannot always discern the right path.
Sir William Napier, History of War in the Peninsula, 1840

(This chapter is edited and modified from an ASRS Directline publication and is used with permission of both ASRS and the author.)

If you had a crystal ball, you would be aware of everything that is happening and is going to happen to your aircraft and the airspace you fly through. Reports to NASA's Aviation Safety Reporting System (ASRS) show many ways that situation awareness can slip away. Some of the lessons learned by others provide a recipe for managing awareness. What we're going to try to do in the next few pages is figure out how to direct our attention in a more discipline fashion so that it's always where it needs to be. Put another way, we'll be learning how to be aware because the consequences of a lapse in awareness can be deadly. The goal is to build skills that result in more awareness; it's the next best thing to having a crystal ball.

From the very first flight lesson, we were taught to "aviate, navigate, communicate," in that order. To aviate, navigate, and communicate, you must be aware of the plane, the path, and the people (crew, passengers, dispatchers, and air traffic controllers). Not only do you need to monitor and evaluate these three things now, but also you need to anticipate what's going to happen in the future and consider contingencies. The current and future state of the plane, the path, and the people are the components of the plan.

The skills for being aware of what's happening now are different than the skills for anticipating what is going to happen later and considering what could happen. These appear in the matrix below and will each be discussed. To many pilots these skills are second nature. They are continuously aware of the plane, the path, and the people and can project into the future and maintain this awareness. However, like all skills, these can be disciplined and refined, and that's our goal here.

Addressing the now

Thomas Huxley said, "The great end in life is not knowledge, but action." His words are particularly well suited to the study of aviation human factors such as attention management. In the end, it is not how much we know, but how well we perform. Attention management is action oriented, and the following areas of concern provide clues to efficient actions in the present.

Monitor

The first skill is monitoring. Unfortunately, we humans have limits to how much we can see and hear at the same time. If we had to put our monitoring goal into one rule, it would be: Be aware of what you need to and ignore everything else. That's real easy to say and probably impossible to do. Let's look at some techniques that can move us toward this goal.

Think of how you focus and direct your attention as you would focus and direct a flashlight. Imagine that everything you are aware of is in the beam of the flashlight. You can hold it steady in one direction and focus the beam (your attention) very narrowly so that you are able to see a small area extremely well. This allows you to ignore all else and concentrate on that small area. Knowing what you can ignore, if even for a moment, allows you to focus on that which you need to be aware of.

Narrowly focused attention can be appropriate when you are solving a difficult problem, as long as someone else is attending to the other plane/path/people issues. The checklist is probably the most common tool for focusing attention. Each crewmember knows what they are to look at, when, and very importantly, they know what the other crewmembers are looking at.

On the other hand, if you broaden the flashlight beam and move it around, you are aware of everything about the plane, the path, and the people. You have the big picture, but less detail in any one area. Well, naturally your job requires you to do both, to focus on a problem and to keep the big picture. This is difficult. There is no way to know when to step back from what you are attending to and move that flashlight around, or move it to a developing problem area for a closer look. Remember, everything has to be covered all the time, so if someone has a flashlight (attention) focused narrowly in one direction, the other crewmember(s) should broaden their beam and keep it moving.

You can point your attentional flashlight in the wrong direction and get sidetracked. There are also things that get in the way of where you're directing the flashlight. Things arise that block your view, both literally and figuratively. These are distractions and they come in many forms. Often in the incident reports, a distraction is the first link in the chain of events that leads to an incident, sometimes embarrassing and sometimes dangerous. More about distractions later.

Monitoring on the three dimensions of the plane, the path, and the people is a balancing act that you, as a pilot, constantly have to perform. The first step towards disciplined attention management is to understand this balancing act. You must:

- Focus on a broad region—keep the big picture.
- Focus on a narrow region—pay attention to detail.
- Focus on the right information—don't get sidetracked or distracted.

Evaluate

It should be pointed out that in addition to monitoring the plane/path/people, you need to evaluate the status of each. The evaluation entails first comprehending what you see and hear. Secondly, you make an assessment of the status of each, the plane/path/people. This leads to an understanding of what the situation is now.

The future

In addition to being people of action, we must plan and anticipate if we are to be successful aviators. The speed at which modern aircraft travel increases the concern we must have for both the near-term and long-term future of our flight. The techniques which follow are designed to help you in this critical arena.

Anticipate

A key to maintaining situation awareness is anticipation; stay ahead of the airplane. If you project what is going to happen later, you will go a long way toward that crystal-ball view. It's like having the answers before the test. Anticipating simply involves projecting the current situation into the future. Most of the time everything follows the laws of physics and the prediction is very accurate. Standard procedures allow you to anticipate what other crewmembers will do in a given situation. A crew that is skilled in managing their situation awareness has a shared vision of what's going to happen in the next few minutes and on into the future. Anticipation is particularly important for high-workload situations. If you as a crew know what each person's responsibility is ahead of time, the awareness level can remain high on each person's part, even when a lot is going on.

Consider contingencies

Sometimes things happen that cannot be anticipated. These can be aircraft malfunctions, ATC clearances, or simply a normal event at an unexpected time. These are the things that simulator checks are made of. Playing the "what if" game has a tremendous advantage in the management of situation awareness. As a common example, when properly briefing an approach, the entire crew is made aware of the required flight path for the missed approach. Should a missed approach be executed, each crewmember has a shared awareness of the sequence of actions, the airport environment, and the navigational information. They collectively know the "what, where, when, and who" of the missed approach procedure.

There is no substitute for thinking things through ahead of time and dividing the tasks and the information so that both become manageable. Using the flashlight analogy again, this allows focus on the big picture and the necessary details, no matter what situation you are presented with.

Plan

The plan is comprised of the current and future states of the plane, the path, and the people. This plan is the foundation all crewmembers are building their situation awareness upon. The plan is constantly being updated based on the awareness activities. As a crew monitors, evaluates, anticipates, and considers contingencies, they continuously modify the plan. Ensuring the entire crew has the same shared plan will ensure that they have a shared situation awareness.

Traps

Below are some traps that can take away situational awareness, if you allow yourself to fall into them. Knowing they are there will hopefully make them easy to avoid.

Focus on the right information at the right time

Keeping the priorities straight is a constant challenge, as this NASA safety report describes.

> *"After we exited the runway, the first officer asked me a question about the ground control frequency, and I looked down at the airport diagram which was on the yoke in front of me. When I looked up, I saw the runway markings for the approach end of runway 7L in front of me. I then looked right and observed a wide-body aircraft approaching our intersection on his takeoff run on runway 25R. I slammed on the brakes and we came to a stop about 20 feet short of the runway. Two flight attendants were out of their seats, but fortunately no one was injured, although I did have a planeload of concerned passengers. My airline has been emphasizing 'situational awareness' lately, but although I was familiar with LAX and well aware of runway 24L, I momentarily lost track of where I was while I dealt with the question about ground control frequency. This brief lapse could have been fatal and it underlined the importance of knowing where you are at all times, and above all, control the aircraft first and worry about the incidentals once that is accomplished."*
> *(ASRS Report 135526)*

Be sure to take your awareness vitamin before every flight. Even those who have had the situation awareness vaccination can have lapses.

If something doesn't look or feel right, it probably isn't

As humans we are aware of many cues from our surroundings for which we cannot always identify the origin. These cues are very real. Don't ignore them, even when they only manifest themselves in a feeling of uneasiness. Excerpts from the cockpit voice recorder prior to the tragic accident in Cali, Colombia, emphasize this point. The flight crew turned their aircraft into a mountain.

> *First Officer: "Uh, where are we...goin' out to*
>
> *Captain: "Let's go right to, uh, Tulua first of all. OK?"*

First Officer: "Yeah, where we headed?"

A few seconds later, Captain identifies Tulua.

Captain: "Just doesn't look right on mine. I don't know why."

Two minutes later they impacted a mountain.

Watch out when you're busy or bored

Studies of humans performing many different tasks show us that we will be less likely to detect something when we're busy attending to something else. We will also be less likely to detect something when we're not attending to much of anything. During times of low and high workload, try to compensate for this human characteristic and be more vigilant. Work out crew procedures to keep each other in the loop during these times. Predetermine roles for high-workload times, especially abnormal situations.

Habits are hard to break

As highly trained aviators, you likely have developed very complex habit structures. These enable you to perform all the tasks required to skillfully fly your aircraft. There are times when these habits can get in the way of safety. If you are required to perform a task differently than you normally would, watch out because the habit pattern may take over without your even realizing it. The best way to combat this natural tendency is to create a barrier so that you prevent or at least are aware of what you are doing. For example, when receiving an aircraft with a failed generator, one airline directs its crews to put a coffee cup over the flap handle so that later, during the approach, the flap handle will look and feel different and alert them not to lower the flaps according to the normal landing checklist. This procedure was adapted after many instances of pilots falling into their normal habit patterns during the high workload approach phase and failing to use the nonnormal checklist.

Expectation can reduce awareness

Above we discussed the importance of anticipation. The downside of anticipation is that it can bias your hearing or seeing what is really there. It is very common that the reports to the ASRS contain the phrase "we heard what we expected to hear." This trap often comes in the form of published "expect" altitudes on arrival charts or familiarity with an airport, resulting in an altitude deviation. The following report is a typical example.

"I anticipated the next crossing restriction to be FL220 at MAYOS and programmed it into the legs page of the FMC As we descended through 25,800 feet, Washington Center issued the following: 'Your discretion to FL240, expect to cross MAYOS at FL220.' However, I anticipated and heard the following: 'Your discretion to 240, cross MAYOS at 220.' Very routine, however incorrect. Fatigue and anticipation had led me to hear what I wanted to hear. The captain working the radio read back the clearance as he had heard it, correctly. I once again heard what I wanted to hear...Moral of the story: same old thing!! Stay in the loop!! and keep the communication flowing!!" (ASRS Report 141158).

When you are expecting something, double-check to make sure that it really was the way you expected it to be.

Things that take longer are less likely to get done right

If you're doing something over a period of time, it is less likely to get done correctly, and this is especially true in the cockpit with all the things going on. Fuel cross-feeding is an example that most of us are familiar with. The problem is that you get interrupted with other tasks during the time that you're cross-feeding so that the time seems shorter, or you might even forget that you have the cross-feeds on. Take special precaution when a task takes a long time, is subject to interruption, or is something that you can't do right away and have to remember to do later. Fuel cross-feeding, checklists (especially before-start) and contacting the tower at the outer marker are examples of things that have shown up in ASRS reports as not getting done right or not at all. We'll talk later about creating reminders that will help alleviate a lack of awareness for these tasks.

Reliable systems aren't always reliable

We know logically that all the systems we rely on to get an airplane from point A to B can fail. We practice this stuff in the simulator. However, research has shown that people actually stop crosschecking reliable systems (Chappell 1997). When the system fails, it can go undetected. This is especially true in glass cockpits where systems are very reliable and failures are difficult to detect. The only cure for this is to force yourself to double-check information against other sources. The following report illustrates this point.

"What I failed to notice was that by inserting the arrival in the FMS, the computer dumped the crossing restriction I had

*inserted just a few moments earlier...Through about 17,500
feet, Approach Control asked if we would make the BUMBY
restriction (10,000 feet) and it was immediately obvious that
we would not. The cause, I believe, was a combination of
cockpit management workload during the approach phase
coupled with an overconfidence in the FMS to present valid
descent profile information. I allowed myself to get too busy
during the descent to make essential crosschecks to confirm
the FMS was working as advertised. The correction: always
double-check the FMS data against other available navigational data to ensure that your programming is correct and
that the aircraft is following accurate FMS guidance. Overconfidence in the FMS and increased workload in the cockpit during bad weather and approach preparation are no
excuse for sound pilotage and the maintenance of situational awareness." (ASRS Report 272508.)*

It's hard to detect something that isn't there

Probably the hardest task we have, being pilots and being human, is
to detect something that isn't there. Much of what we need to be
aware of is the absence of something. You'll probably notice that the
engine fire bell/light isn't on, but harder to detect is that the other
crewmember didn't say, "after-takeoff checklist complete" or that the
green arc didn't move to reflect the new crossing restriction you
thought you entered correctly. Sometimes even serious aircraft malfunctions can be manifested in the absence of a subtle cue, at least at
first. The only way to detect something that isn't there is to specifically look for its absence. These checks have to be built into your flying techniques, your personal checklists. It takes attention discipline.

Automation keeps secrets

Although an entire chapter is dedicated to automation discipline
later in the book, it is worth mentioning how it relates to attention
discipline. The information in the glass cockpit is sometimes less obvious than in the traditional cockpit. A simple error in a numerical
entry, if not caught at the time of input, can be nearly impossible to
detect and correct. The following report of an altitude deviation illustrates this vulnerability.

*"This situation resulted from three crew errors: 1) My first
priority was data entry rather than situational awareness. 2)
I entered the crossing fix incorrectly in the legs page,*

and 3) my first officer did not detect my data entry error be-
fore I executed the command. Cross-checking data entry be-
fore execution is company policy and part of my cockpit
briefing." (ASRS Report 254092).

Distractions come in many forms

Crew distractions are a serious impediment to safety. Probably the
most documented case was the Eastern AirLines L-1011 crash in the
Florida Everglades that was the result of the crew's preoccupation
with a landing gear problem. This brings us back to the juggling act
between focusing on a specific area and keeping the big picture.
Distractions result when the attentional flashlight beam is too narrowly focused and not moving. Many things pop up in that beam of
light that get in the way of seeing everything that is going on. A list
of distractions compiled by Capt. Monan (1978) from 169 reports to
the ASRS show some examples of distractions that led to a variety of
safety incidents. (See Fig. 12-1.)

The ASRS incident reports describe many cases of pilots being faced
with aircraft malfunctions that distract the situation awareness of the

Type of Distraction	Number of Reports
Nonoperational activities	
Paperwork	7
Public address	12
Conversation	9
Flight attendant	11
Company radio	16
Flight Tasks	
Checklist	22
Malfunctions	19
Traffic watch	16
ATC communications	6
Radar monitoring	12
Studying approach chart	14
Looking for airport	3
New first officer	10
Fatigue	10
Miscellaneous	2
Total	169

12-1 *Distractions by type of contributing factor. There
are many causes of distraction in flight operations. Notice
that many of the "high-risk" activities are normal flight
tasks.*

crew. Sometimes the malfunctions are big and obvious; sometimes they're small and illusive. Someone has to attend to the malfunction and figure out the appropriate course of action. What often happens is that everyone is engaged in solving the problem and no one is flying the airplane. No one has the big picture, the wide, sweeping beam of attention. Capts. Sumwalt and Watson (1995) took a further look at ASRS reports, examining 230 reports of in-flight aircraft malfunctions. The attentional demands on the flight crew during the resolution of the aircraft malfunction caused adverse safety consequences in 38 reports. The safety consequences included altitude deviations, course/track/heading deviations, nonadherence to other ATC clearances, and noncompliance with Federal regulations or company procedures.

In short, attentional discipline has a direct impact on multiple other areas of flight discipline. As such, it may be an excellent starting point for overall flight discipline improvement.

When a malfunction occurs, it should trigger a "red flag" for a heightened sensitivity to a potential loss of situation awareness. Sometimes distractions come from something that has already happened and is over. Many incidents have been the result of a crew dwelling on something that has previously happened and neglecting the current situation. You have to recognize that you're doing this and shake it off. Think and talk about it later, on the ground.

How to build a crystal ball

There are a few tricks that you can use to get and keep the awareness you need to fly safely. You can use "plane, path, people" as a checklist. Take a moment to assess the current state of each. What are the plane, path, and people doing now? What is likely to be the state of each later? Finally, consider all the "what if" possibilities for each. If you periodically run this checklist, you'll find that your awareness has increased. Most of the surprises will go away and the ones that pop up will be more manageable.

Manage crew awareness

Crew procedures are designed to focus attention while keeping the big picture by dividing the awareness responsibilities. When functioning as a crew, you not only have to concern yourself with what you're doing, but also with what other people are doing.

You need to check that the other crewmembers do certain things that fall into their area of responsibility. You also need to check that they do not do certain things that are inappropriate or unsafe. Crew-shared awareness is high when doing a checklist. Attention is focused on each item as one crewmember reads and another checks. It's obvious what's being looked at and by whom. We've developed other techniques, e.g., for handing off the job of listening to ATC, saying, "You've got the radio." Many of the other things we do in an airplane are less structured. It's these other situations that cause a crew to misunderstand who's aware (or not aware) of what.

A study by the National Transportation Safety Board (1994) showed that monitoring/challenging failures were identified in 31 of the 37 accidents reviewed. Crewmembers failed to monitor and challenge the errors or the lack of awareness of the other crewmembers. As a crewmember, you have to watch each other for what actions are taken and what actions aren't taken.

What do they know that I need to know?

As team members we need to utilize all our sources of information to be aware of everything we need to be. Many of those sources are other people's eyes and ears. The following ASRS report shows how information from the tower, the flight attendants and the relief pilot were critical in detecting a tail strike and making a prudent decision to discontinue the flight.

"Shortly after takeoff tower reported they had seen sparks from the rear of the aircraft...From the cockpit the rotation had seemed normal to all three crewmembers and nothing abnormal was felt by the pilot flying...Aft flight attendants...said they had heard scraping on takeoff...With all systems normal and no other adverse information, it was decided by the captain and company maintenance it would be OK to continue to our destination. After leveling off at 25,000 feet, the captain asked the relief pilot...to inform the flight attendants what the noise had been on takeoff. On return to the cockpit, he informed the captain that the flight attendants in the aft had also heard a loud metallic bang or crashing sound on rotation. The captain then called this flight attendant to the cockpit for more information. After receiving this new information, the captain felt it would not be prudent to continue over water not knowing if there was damage to the aircraft

fuselage. A request for return to JFK and fuel dump was received... Upon inspection, paint was missing from the tailskid...It was found that the cargo bins had come loose... I believe that on takeoff the cargo bins shifted aft, causing a slight movement of the center of gravity to the rear, causing tailskid contact." (ASRS Report 243137).

Use all the sources of information you have available to maximize your situation awareness.

What do I know that they need to know?

Periodically ask yourself, "Do I know something my other crewmembers don't that they should know?" If the answer is yes, then tell them. If the answer is that they don't need to know, but they should know that you're keeping an eye on it, then tell them that you are. When something takes your attention away from what the other crewmembers are expecting you to keep an eye on, tell them that too. There will be times when, despite your crystal ball, you will have a reduced level of awareness due to fatigue, distraction, or some other factor. Let the other crewmembers know when this is the case so that they can back you up more carefully.

What do none of us know that we need to know?

The other question to ask yourself is, "What are we as a crew not paying attention to?" If everyone is looking at the same thing, then something's getting missed. If you are unsure whether another crewmember is maintaining awareness of something, be sure to clarify. The request "keep an eye on that for me" comes in handy.

Create reminders

A powerful way to ensure disciplined awareness is to create reminders. There are many that are employed by flight crews, both formally and informally. Checklists are formal reminders. Some people have developed informal reminders, such as turning the checklist upside down on the yoke clip when it is interrupted, as a physical reminder that it has not been completed. Other reminders include selecting the radar test pattern when cross-feeding fuel and putting the nose landing light upon being cleared to land. These are obvious visual reminders that are in the scan of normal flying activities. Reminders can be aural as well. Some pilots have the technique of selecting the audio for the outer marker when they have been instructed way out on the approach to contact the tower at the outer marker. This gives them a reminder that they don't have to look at during a busy time in the flight.

As we discussed above, things that take longer, are subject to interruptions, or can't be done until later, are less likely to get done right. Creating reminders for these things is probably the best, if not the only, defense against forgetting them. Reminders work for other things as well. Reminders should be unique and consistently used for the same thing. That's why the string around the finger never worked.

Summary

So, in summary there are a few key things to do to manage your situation awareness:

- Focus attention on details while keeping the big picture.
- Anticipate; stay ahead of the airplane.
- Consider contingencies; have a plan for the "what if" situations.
- Predetermine who will watch what in busy times.
- Have a plan for handling distractions, especially malfunctions.
- Use all your team members for awareness.
- Create reminders.

OK, so the bottom line is: Be aware of where your attention is and is not. Don't fall into the awareness traps. Just like money, situation awareness is very hard to get and very easy to let slip away. Periodically stop and ask yourself, "Is there something that we're not aware of that can bite us?" If you do take on the challenge of disciplined attention and learn to manage your situation awareness, you'll have the next best thing to a crystal ball.

Chapter review questions

1. What three areas of focus does the author suggest for developing better monitoring skills?
2. What are the "traps" that can lead to poor attention discipline and a loss of situation awareness? Can you recall instances when you have fallen into any of these?
3. According to the study of distraction causes done by Captain Monan, in what circumstances is a pilot most likely to

become distracted? Can you think of others from your personal experience?

4. What three questions can help you manage crew awareness?

5. What reminders do you use to help you maintain awareness? How many of the techniques suggested by the author might you use? Can you think of any other reminders that would be helpful for your type flying?

References

Chappel, S. L. (1997). The effects of experience and automation on failure detection. Proceeding of the Ninth International Symposium on Aviation Psychology. Columbus, OH: Ohio State University.

Monan, W. P. 1978. Distraction—A Human Factor in Air Carrier Hazard Events. NASA Aviation Safety Reporting System: Ninth Quarterly Report, 2–23. Moffett Field, CA: National Aeronautics and Space Administration.

National Transportation Safety Board. 1994. A Review of Flight Crew-Involved, Major Accidents of U.S. Air Carriers, 1978 through 1990. Safety Study. (Report No. PB94-917001 NTSB/SS-94-01). Washington, DC: National Transportation Safety Board.

Sumwalt, R. L., and A.W. Watson. 1995. What ASRS Incident Data Tell about Flight Crew Performance during Aircraft Malfunctions. The Ohio State University Eighth International Symposium on Aviation Psychology, 758–764. Columbus: Ohio State University.

13

Killing conditions

Common scenarios for breakdowns of flight discipline

Death be not proud.
John Gunther

The purpose of this chapter is to drive home the survival lessons inherent in the term "flight discipline." It is designed specifically to make the theoretical positions offered by this book real, because they are. They can be quantified in gravestones. If the publisher would allow it, I would write this chapter in red. It is hoped that through vicariously experiencing the fates of failed aviators, you will be able to recognize warning signs on the road to failed discipline. I am limiting this discussion to four areas: the unexpected weather encounter, spatial disorientation (specifically as it applies to transitioning between visual to instrument conditions) low-altitude flight operations, and midmission changes. There are certainly more areas worthy of discussion as "killing conditions," but I would prefer to highlight these four in hopes that the reader will recognize them as truly lethal, and not lose them in a lengthier discussion.

Analysis of accident databases points to specific conditions—or scenarios—which are particularly lethal to aviators. I have selected the scenarios which have distinct flight discipline implications and present them, not as an all-inclusive list, but rather as a set of memory joggers should you find yourself in a similar condition.

Lethal scenario 1: The unexpected weather encounter

Pilots must plan for the unknown. The very term "unexpected" should become a misnomer in flight operations, especially as it relates to weather, one of the most destructive and fickle of aviation hazards. Yet time and time again, we read the mishap report of a noninstrument-rated general-aviation pilot entering conditions he was unqualified to be in, a commercial pilot who gets caught in a microburst or windshear on short final, or as in the following case, a military flyer who sticks his nose somewhere he wished he hadn't.

Case study: My "hail-lacious" low level

by Lt. Norman Metzer, USN (*Approach* 1996)

"It was my second hop in a fleet A-6 squadron operating out of NAS Oceana. I was lead with the safety officer as my BN (bombadier-navigator) and the operations officer was on my wing. We took off for a low-level through the West Virginia mountains. Although the Dash-1 (weather sheet) read clear and a million, with no forecast thunderstorms or precipitation, the weather deteriorated about halfway through the route, and my wingman lost sight of me.

We climbed through the cloud deck and rendezvoused at the (low-level route) exit point. The worsening conditions should have been our first indication that the weather sheet was not worth the paper it was printed on. I should have knocked it off before the weather got bad.

We weaved in and out of developing thunderstorms, staying VFR en route to the target area. Our target time was uneventful (broke even on bets).

The off-target rendezvous flexed to a lower altitude because of lower ceilings (than expected or forecasted). The join-up went well, and we departed the restricted area for home. We could see that there was no way around the line of clouds between us and Oceana. My BN (bombadier-navigator) reported that on radar the weather looked like a squall line of rain showers, so we headed straight home to save gas. We found out later that the (wingman's) radar didn't work, so they were operating under the 'join-up-and-shut-up rules.'

As we entered the clouds, my BN said we should be through in about three miles. As soon as he said that, the visibility went to zero in thick clouds and heavy precipitation. My wingman lost sight again and took a cut away.

A split-second later, all hell broke loose. It's hard to say how large the hail was since it flattened on the windscreen. As usual, the A-6's characteristic radio squeal in the weather, combined with the extremely loud noise of hail banging against the windscreen and the moderate to severe turbulence added to our problems. As I checked our airspeed, I saw it was close to 300 knots. (An A-6 thunderstorm penetration airspeed is 230-280 IAS.) I pulled the power.

The nightmare was over as quickly as it started, and we were in VFR again. In 20 seconds, we went from good formation flight to lost sight and heavy hail. We could see some damage to our aircraft, but it flew fine. The flight assumed individual call signs and returned for straight ins. Our wingman sustained only slight damage, most likely from his cut away at lost sight and immediate reduction in airspeed.

All of this happened within 35 miles of Oceana. When we climbed out of the jet, we could see all of the damage that we couldn't spot from the cockpit. There were no anticollision lights left. All of the fiberglass antenna covers were badly damaged. The drop tanks looked like someone had taken a ball-peen hammer to them. The biggest concern was the engines. The leading edge of the intakes were destroyed. We felt lucky that both of the engines kept running.

I learned that day not to push my comfort level trying to hack bad weather. I am thankful that our A-6 Intruder proved to be truly 'all weather' and stood up to what we put it through."

There are several lessons in this scary story. First, if the weather begins to look different than forecast, assume the worst. Weather prediction is still very difficult, especially during the summer months with all of the moisture, heat, and convection.

Secondly, realize that radars don't see ice unless it is coated with water. A hailstorm like this one is likely to produce no more—and perhaps less—significant returns than a rain shower. This holds true for both airborne radar and those used by the air traffic controllers, so they aren't the magic solution either. If in doubt, and you even suspect that there may be hail or heavy turbulence in a cloud bank, slow to the appropriate airspeed prior to entry. As you saw in this example, these encounters are often quick and violent.

Finally, realize that most of us don't fly aircraft as sturdy as the A-6 Intruder. An encounter like this one could completely destroy a lighter aircraft. Treat all weather with respect, and be particularly careful when you see unforecasted phenomena. Sometimes, the result is not as benign as merely a damaged aircraft and ego.

Case study: What you see isn't always what you get (NTSB 1995)

On 12 January 1995, a Cessna 208B crashed into a ridgeline about 14 miles short of its destination airport in Oakland, California. The pilot was operating an on-demand air taxi service on a routine flight from Visalia to Oakland. Although there were a few clouds and some light precipitation reported along the route, the weather was of little concern to the experienced pilot. The 63-year-old had been flying for years and had collected over 25,000 hours. He held all of the ratings required and had filed an IFR flight plan, which was required because of the forecast clouds and the fact that it would be night when he arrived in Oakland.

The flight progressed normally, perhaps too normally, until the pilot prepared to make his descent. Cruising at 7,000 feet MSL the pilot asked for a descent and reported the airport in sight from 30 miles away. Although a ground station was reporting two cloud layers (1,500 scattered, 5,000 broken) between him and the airport, the pilot opted for a visual approach. I think you can see what's coming.

The aircraft first hit a tree and then impacted a 1,500-foot ridge in a wings-level attitude. The accident investigation board found no evidence of mechanical failure, and there were no emergency radio calls to indicate anything other than controlled flight into terrain. The pilot was not available for comment.

The weather in this example was just as much a factor as in the previous case study, but in a different fashion. The pilot here assumed that since his slant-range visibility was good 30 miles from the airport, it would remain good for the rest of his descent. A pilot with this much experience should not have made such an elementary error. One has to wonder where the origin for this undisciplined action may have been. The pilot had probably gotten away with similar decisions in the past. His experience level might have convinced him that he could not make such an elementary mistake. The combination of these factors most likely turned a highly experienced pilot into a poor decision maker.

One could say that there was at least one common factor between the two mishaps—a lack of respect for the weather.

Lethal scenario 2: Spatial disorientation—transitioning between instrument and visual conditions

One of the most difficult tasks in aviation is transition from visual to instrument conditions, or vice versa. It requires a pilot to smoothly switch control and performance references while maintaining a stable aircraft. The stability aspect is particularly critical when going from visual to instrument conditions, as this can quickly lead to spatial disorientation.

Moving in the other direction, from instrument to visual conditions, can be just as deadly, as many pilots abandon their instrument cross-check prematurely at the first sight of the ground. This mistake has killed many a pilot and passenger in the final stages of an instrument approach. If you add acceleration or deceleration to the equation, which is often the case on takeoff and approach to landing, you are literally flirting with psychological disaster.

Our first study looks at the problem of spatial disorientation, which haunts the first few seconds of transition from visual to instrument conditions. It is written by a Naval aviator who's "been there" and "done that."

Some "autogravic" illusion thing

by Lt. Commander Tom Ganse, USN (*Approach* 1996)

"Now launch the Alert fighter!" The not-so-alert pilot shook himself awake and instinctively scrambled to launch at an unlit remote site.

Full afterburner flung him down the dark runway until he aggressively pulled the nose up into the black moonless night. The aircraft arced through the darkness until, 30 seconds later, it crashed 2.5 miles from the departure end.

Investigators used recorded flight data and ground scars to determine that the pilot achieved a maximum climb angle of 35 degrees before he began a 0 to 0.5-G push over through the featureless sky. The aircraft hit the ground 30 degrees nose low, wings level.

In another incident, a two-seat fighter leapt forward accelerating down the runway at full power. Once airborne, the pilot hugged the deck, gunning for maximum airspeed. He hauled the back on the stick at the runway's end, pitching the jet up into a steep climb at more than 300 knots. The aircraft quickly penetrated the winter overcast. A few seconds later, the pilot began pushing forward on the stick. The jet scribed a near-perfect parabolic arc through the thick, gray clouds, ending up in a smoking hole two miles from the takeoff point.

In a third mishap, weather was reported as 300 overcast, tops at 4,500 feet, with visibility at six miles when a demonstration pilot manned his showbird. His maximum performance takeoff catapulted him to 300 knots under the dense cloud layer before he pulled the stick into his lap. The fighter quickly pitched 45 degrees nose up and rocketed into the overcast. Once in the clouds, the pilot pushed the nose over, maintaining a near zero-G trajectory until his jet crashed 30 degrees nose low and wings level at 540 knots on the extended centerline just past the departure end of the runway.

The pilots in each of these mishaps experienced a sustained acceleration along the longitudinal axis while using outside visual references, followed by a transition to instruments during an acceleration along the vertical axis. They each responded by pushing over with no regard for the actual horizon reference—the predictable response for a pilot experiencing somatogravic illusion.

That's a big word for us tactical types. In fact, the first time many of us heard about it, we described it as "some autogravic illusion thing." Several learned intellects have written about the topic. I'll try to make their research more digestible to the average aviator.

Somatogravic comes from somato meaning "body" or "of the body" and gravic, which means "pertaining to the gravitational force." Somatogravic illusion occurs pilots respond to inner-ear sensations, that is, fly by the seat of their pants. The situation is more likely to occur during a transition from visual references to instrument flight and is aggravated by high-G maneuvers...Even though it feels like our bodies are being forced down by gravity, we can't distinguish this force vector from a net force vector experienced during accelerations. Pilots can be trained to overcome any disparity between the two by concentrating on, and believing in, flight instruments. (This takes considerable discipline.)

When you can't suppress the conflict between what your orientation feels like and what your intellect tells you it is, you have vertigo. We are most sensitive to this hazard when transitioning to instruments after losing visual reference. It is much harder to correct perceived orientation by using flight instruments if you are already disoriented.

Somatogravic illusion, then, is a fancy word that means your seat-of-the-pants orientation is wrong...this perception can be overwhelming when you are deprived of outside references, especially if you have not previously established your instrument scan. What makes it so lethal...is that any pushover results in further acceleration and an increased sensation of pitch-up. The pilot then feels he needs more forward stick, which further aggravates the situation....Simply stated, you are at risk when you accelerate and aviate using visual references, then lose those references before establishing an instrument scan exclusive of them.

Keep in mind, unforecast or unexpected IMC is bad news, and even if it doesn't kill you, it can shake you up pretty good. We conclude this scenario with an ASRS reporter who, though qualified for instrument flight, knows the difference between "legal" and "proficient." His decision embodies how flight discipline works in this deadly scenario.

"The flight was conducted in VMC. Turn to final approach course was a sharp descending turn from VMC into IMC. I immediately got disoriented and started hyperventilating. After a short period of time that felt like forever, I decided to abandon the approach and advised the tower controller. I calmed down (subsequently)...and successfully completed an ILS approach and landing. Though I am legally current and have a significant amount of "real" instrument time given my level of experience, I plan to grab an instructor and

go get some more, particularly with the VMC to IMC transi-
tion. (ASRS Callback 1995)

Lethal scenario 3: Flying low

I guess it stands to reason that you would run into things more often flying close to the ground, but there is something more to the
low level environment that seems to entice aviators into poor flight
discipline. The mishap files are filled with reports of pilots who
know the regulations which prohibit flight below certain altitudes,
realize that they have not been trained to operate in the low-level
environment, and yet still insist on scud running, buzzing, and even
attempting low level aerobatics. In other cases, the pilots have been
trained and are operating in an approved area for aerobatics, but still
push themselves and/or their aircraft past its limits.

In the United Kingdom, where they track mishaps slightly differently
than in the U.S., unauthorized low-level flight is the third leading cause
of mishaps and fatalities (UK Air Statistics 1996). No matter where these
incidents occur, the results are typically tragic, as in the following cases.

Case study: Low-altitude antics lead to tragedy

On June 14, 1995, the pilot took off into the clear summer from an El
Reno, Oklahoma, airstrip in his Walter W. Bell Lazer Edge for a bit of
sport. The pilot was going to put the aircraft through its paces on this
day, planning some low-altitude aerobatics. Witness testimony to the
mishap investigation team describe the unintended consequences.

The airplane executed a "snap roll" to the right, recovering in a 90-
degree right bank and "nose-low" attitude. The airplane then "snapped
to the left," with a wings-level recovery, "but the nose [was] extremely
low, in a 30–40-degree nose-low attitude." During these maneuvers,
the airplane had been "rapidly losing altitude and was about 400 feet
above the ground." Witnesses then observed the airplane's nose
"pitchup" and heard an application of engine power. The airplane then
"broke to the left, doing a 3/4 or more spin, before impacting the
ground and exploding into a fireball." Examination of the flight control
system did not reveal evidence of a flight control failure prior to impact.

The NTSB listed the probable cause of the accident as "the pilot's failure to maintain control resulting in an inadvertent stall/spin. Factors
were the pilot's performance of aerobatic maneuvers at the low altitude (NTSB 1995b).

A second example occurred three days earlier in Bath, New York. The pilot, flying a Berg Steen Skybolt, was practicing aerobatic maneuvers when he too lost control. Witnesses reported that shortly after takeoff, at an undetermined altitude, the pilot attempted an acrobatic maneuver. Review of a videotape taken of the accident sequence revealed the airplane rolled left and entered into a spin. The airplane completed four revolutions before recovery was initiated. The airplane stopped rotating just prior to impacting trees. Post-accident examination of the airframe and engine did not reveal any anomalies.

The NTSB determined that the probable cause of the mishap and fatality was the pilot's failure to maintain sufficient altitude/clearance to avoid trees while performing aerobatic maneuvers. Related factors are the low altitude at which the maneuver was initiated and the delayed recovery/pull-up.

Special skills are required

The low altitude environment is no place to practice maneuvers until you have perfected—and I mean perfected—them at altitude. And this is only after you have received professional flight training and are operating in an authorized area and at authorized altitudes. As a pilot trained in both high-speed low-level operations, which include aggressive maneuvering for combat operations, and in aerobatics, I must tell you that even after thousands of hours in these environments, I retain a healthy degree of respect for the ground. Learning when to start climbs, initiate descents, where to look before and during a turning maneuver, are all topics of considerable importance. Unless you have received formal training, stay out of this environment. Special skills are required for safe operations down there.

The low-altitude environment has a built-in capability to lead a pilot into trouble. The longer you fly at low altitudes, the more comfortable you become, and the more likely you are to try flying even lower. Additionally, if you are using wide scan visual references to maintain your altitude, (and you better be!) you will likely find it is more difficult to fly at 500 feet than at 200 feet. That is because the visual cues are more pronounced the lower you get. When you combine these two factors, you end up with a deadly combination, a more comfortable—and perhaps complacent—pilot, edging ever closer to the ground.

If all of these issues aren't enough to convince you it's dangerous down low, there are even more factors to consider. First, wind and thermals off the ground are significant players below 500 feet, and large raptors and scavengers such as hawks, eagles, and buzzards love to swoop around the tops of ridge lines. An encounter with one of these big birds can ruin your day. In addition, man-made structures such as power lines or cables crossing valleys and canyons are nearly invisible to the pilot. Even towers can appear out of nowhere if the pilot is not actively and continuously scanning in front of the aircraft.

In summary, when pilots decide to operate in the low-level environment, they are increasing their risk factors dramatically in terms of the aircraft, the environment, and themselves. Plan and fly accordingly.

Lethal scenario 4: Midmission changes

Replanning in flight is demanding, and when it is coupled with an event that requires it in the first place, such as bad weather, low fuel, etc., it can flat out overwhelm the best of pilots if not handled correctly. Although this subject is thoroughly discussed in our chapter on planning, I want to review the dangers inherent in midmission changes to re-emphasize the hazards. In one of the most classic examples of how a midmission change can effect flight discipline, we begin with a case study of a military crew with a few minutes of extra time on their hands.

Case study: A few minutes to play

On October 7, 1992, a USAF Air National Guard cargo aircraft took off from its home base in West Virginia on a local training mission. Normally, the aircraft would depart on an IFR clearance and practice instrument approaches, patterns, and landings. This particular mission was scheduled as an "Engine Running Crew Change" and did not include a low-level training portion of the flight. As such, the crew would be expected to fly minimum IFR altitudes in the local-instrument traffic pattern. In the case of this airport, that would be 3,000 feet MSL unless on a published portion of the approach.

The changes started early for this crew. When they requested their IFR clearance for their traffic pattern activity, they were told to expect a delay and made the determination to depart VFR and request their IFR clearance after becoming airborne. This was change number one, and it would set the stage for a domino effect that would lead to disaster.

The crew departed to the west and climbed out to 3,000 MSL. Upon leveling, they contacted Approach Control and requested an IFR clearance. Once again they were denied, and told to expect a ten-minute delay due to inbound IFR traffic, which had priority since they were already on an IFR flight plan. Change number 2.

The crew—commanded by a Lieutenant Colonel Instructor Pilot—had already deviated from the initial plan, so why not a little more, eh? Flexibility is the key to airpower after all. The crew decided to stay VFR and "proceed west." They told Approach Control that they would be back in contact shortly.

Approximately ten minutes later, the aircraft impacted a set of high-tension power lines at 381 feet above the ground. Although they remained airborne for another minute and a half, the final result was a crash and explosion, killing all six crewmembers and destroying a civilian house, trailer, and cinder-block shed.

This incident resulted from a change in the mission almost before it began. The following case study shows that it can also occur at the end of a flight profile.

Case study: Descent into disaster (Hughes 1995)

The mishap flight was the first leg of a three-sortie, routine resupply mission carrying equipment, rations, mail, and passengers. The crew departed on schedule and proceeded along the flight-planned route. Once airborne the aircrew requested a phone patch with a command post for a routine operations report. This report was relayed by an airliner because the mishap aircraft was unable to establish direct radio contact.

The mishap crew requested and was given clearance to descend from flight level 220 to 200 with further clearance to descend in visual meteorological conditions VMC for landing at their first stop. The distance from the last reported intersection to the mishap site was approximately 185 miles and the elapsed time to fly this distance was estimated to be 61 minutes. During this leg the flight path, type of descent, airspeed used, and weather encountered could not actually be reconstructed by the accident investigation board. However, the mishap crew should have penetrated an overcast cloud deck at 8,000–12,000 feet and another broken deck at 1,500–4,500 feet. The visibility could have been as low as 1/2 mile—not exactly VMC conditions.

The mishap aircraft was last seen visually passing over a peninsula at 300–500 feet, then it initiated a right turn towards the open sea. Shortly thereafter, observers and local residents heard a loud crash from the direction the aircraft had traveled. Two individuals rushed to the shore and observed what they assumed was the tail section floating off the coast. The wreckage floated for approximately 15 minutes before sinking out of sight. The aircraft was destroyed and the crew and passengers, a total of 21 personnel, were fatally injured.

In attempting to reconstruct the accident, several pieces of information become relevant. The mishap crew's last contact with a ground agency was through an Eastern Airlines aircraft, when they departed flight level 220 and were cleared to descend under VMC to landing. The crew had to penetrate several layers and scattered clouds could be found as low as 300–500 feet. The distance from the last reported intersection to the crash site was 185 nautical miles and was covered in 61 minutes. This suggests that the aircrew maneuvered significantly to descend to a point where they could maintain visual contact with the surface.

The aircraft was last seen over a peninsula near the field of intended landing. Weather for that area was changing constantly with rain, fog, and low clouds obscuring visibility to as low as 1/2 mile at times. The aircraft flew along the peninsula and then made a right turn out to sea. The turn could have been made for a number of reasons. As the aircraft proceeded along the peninsula, the aircrew should have been able to observe the area of the terrain on the left. In all likelihood weather at the intended point of landing was worse than over the water. The right turn could have also been made because of a decision to terminate efforts to reach the airport and turn away from the mountainous terrain of the mainland. The turn also could have been made to place the aircraft in an area of perceived better weather. The crew could have been maneuvering for another attempt to approach the airport or to remain in visual conditions to obtain an IFR climb clearance from a controlling agency. Either of the latter three would also explain a decision to remain at low altitude.

When the aircraft turned to the north, the aircrew probably did not have a definite horizon for visual reference because of poor weather conditions. The aircraft could have been flown into the water because of visual illusion such as a false horizon or spatial disorientation. Aircrew attention could have been channeled outside the aircraft in an attempt to maintain visual contact with the surface. Conversely, attention could have been focused inside the aircraft because of some aircraft system malfunction. Any of these factors could have resulted in an undetected descent and bank. Although unlikely, the crew could have been faced with a mechanical problem that could not be resolved in the existing weather conditions and available altitude. The aircraft's initial impact with the water was in a left-wing low attitude.

The above factors could have resulted in this attitude. Initial impact caused the aircraft to violently pitch down and rotate to the left.

The crew was very experienced except for the copilot who had only about 182 hours in the C-130. He did have, however, an aviation degree and a civilian pilot's license. The copilot may have been hesitant to take action and express his opinions based on his relative inexperience. The pilot was a highly motivated, mission-oriented flyer who may not have recognized his own limitations in this situation. The weather conditions were probably a factor in obscuring the horizon and causing a loss of perspective cues. There are three human factors that may have contributed to this mishap.

1. The decision to operate the aircraft at low altitude in marginal weather conditions.

2. The loss of situational awareness which can be reconstructed by the use of two scenarios: The loss of a true horizon by cloud decks creating visual illusions or the loss of outside visual cues resulting in spatial disorientation while maneuvering.

3. The distraction of the crew through channeled attention on inside or outside factors.

The point to be made here is that all of these factors merged because of a relatively small change in the plan. The crew did not expect to encounter problems with a VMC descent. When they did, their planning and risk management fell apart. There is no evidence of problems within the aircraft that would have caused channeled attention, therefore, it is highly likely that poor weather conditions caused an obvious focus of attention leading to a neglect of flight instruments. There were other indicators of a crew predisposed to poor flight discipline.

The pilot probably violated AFR 60-16 visual flight rules in the descent through the overcast cloud layer. The pilot operated the aicraft at low altitude in marginal weather conditions for an undetermined reason. Other violations include an overdue aircraft inspection which the pilot should have caught during an inspection of the aircraft maintenance forms during preflight, an overdue flight physical on the copilot, and the fact that the aircraft commander was carrying a concealed weapon loaded with a full clip. In addition, the alternate destination was invalid due to insufficient reserve fuel to reach the designated field. Quite a combination.

How can a sequence of small events have such tragic implications? The answer may be found in many pilots need to assert control over a flight mission. As controllers, pilot's function very well. But when the situation begins to become slightly more chaotic—when things change—two phenomena occur. First, the pilot and crew are thrown out of their comfort zone, and the stress levels rise. Secondly, because pilots are not typically accustomed to admitting they are having a problem, they look internally for a solution. Instead of realizing that a change in circumstances represents a significant change in risk, and requires what amounts to a new mission planning session in flight, they choose to remain in dangerous waters while they seek to work out solutions by themselves.

The "4C" approach to changing situations

My father used to tell me, "Son, you gotta have a plan." I recommend the "4C" approach to address the challenges of midmission change. I admit that it is a modified version of the "I'm lost" checklist that we are taught in the early stages of pilot training (Climb-Conserve-Confess), but I believe it works. Simply put, the "4C" checklist for midmission change says to Confess-Climb-Conserve-Consult. Let's look at each step in a bit more depth. A complete analysis addressing the multiple challenges of midmission change is presented in Chapter 15.

Confess. Admit to yourself and the rest of the crew if applicable, that the situation has changed, and that you must make a new plan. This establishes the increased risk level in your mind and sets up the next steps as appropriate first actions.

Climb. If you are close to the ground, or if a higher altitude is advantageous in terms of either fuel or weather, climb. This gives you a greater margin of safety if you should somehow become too involved in the mission change and get distracted or lose situational awareness.

Conserve. Conserve fuel, as the recent change may require diversion to an alternate or more time to work the problem.

Consult. Get the help you need. Look at all available resources to professionally replan the situation. Talk to ATC, flight service, or other resources to let them help you. Pull out the charts; do a thorough review.

A final perspective on lethal setups

Safety experts are fond of saying that each accident is the result of a "chain of events." This is referred to in different ways, but the essence of the argument is that if you can break the "error chain," you avoid the mishap. While I fully subscribe to this theory, I believe it leaves out an essential prerequisite for mainstream pilots. In order to break an error chain, we must be able to recognize it as it develops, which is far more difficult for us when we have a handful of airplane and do not know what the future holds than it is for post hoc accident analysis.

Certain conditions set pilots up for the lethal mistake, and we must learn to recognize, respect, and if possible, avoid those conditions. In turn, these conditions lead to the big buzzwords in aviation safety: lost situation awareness, distraction, and poor judgment. The purpose of this chapter was to isolate four of the most deadly conditions an aviator can face, and present them in such a way as to strike respect—if not fear—in the hearts and minds of the reader. It's a funny thing about flyers; they tend to get comfortable with hazardous conditions and like to see what is around the next corner. A short quote from the US Army Flightfax addresses this tendency.

> *"Each new plateau of risk, when first attained, seems to be the last; but, as we grow accustomed to it, a new horizon beckons.*
>
> *What insulates us from fear as we approach the danger is simply habit, the familiarity of a point we have reached and all the points we have left behind.*
>
> *Until one steps too far, it's often hard to tell the difference between recklessness and skill." (US Army 1997 p. 6.)*

And then of course it is too late.

Any time you have an unexpected weather encounter, are transitioning between instrument to visual conditions, are flying in the low-altitude environment, or experience a significant midmission change of circumstances, you had better be on top of your game and concentrate on the business at hand.

Chapter review questions

1. What are the four lethal scenarios identified in this chapter? Do you agree that these are the top four? What other candidates for this list can you think of?

2. What is the somatogravic illusion and how can we use flight discipline to avoid it and to get out of it?

3. Which is trickier, going from visual to instrument conditions, or vice-versa?

4. Why is the low-level environment conducive to poor flight discipline? How does it lull a pilot into ever-decreasing altitudes? What special skills and training are required to effectively master it?

5. What is the "4C" method for addressing the challenges associated with in flight re-planning? Can you think of times when all of the steps might not be appropriate?

6. What is the difference between recklessness and skill? Why do we have a hard time identifying this point?

References

Ganse, T. 1996. *Approach.* Vol 41, No. 5. September–October p. 24–25. Naval Safety Center, Norfolk, VA.

Hughes Training. 1995. *Aircrew Coordination Training Workbook.* Hughes Training Division, Dyess Air Force Base, Texas. pp. 15–17.

Metzger, N. 1996. "My Hail-Lacious Low-Level." *Approach.* vol 41, No. 5. September–October pp. 12–13. Naval Safety Center, Norfolk, VA.

NASA Aircrew Safety Reporting System (ASRS) 1995. *Callback #192: It's Almost Summertime.* NASA, Moffet Field, CA. p. 1.

National Transportation Safety Board. 1995. *NTSB Report Brief #46413.* Internet. http://www.ntsb.gov/Aviation/FTW/95A244.htm

National Transportation Safety Board. 1995b. *NTSB Report Brief #FTW95LA244.* Internet.http://www.ntsb.gov/Aviation/FTW/95A244.htm

National Transportation Safety Board. 1995c. *NTSB Report Brief #BFO95LA062.* Internet.http://www.ntsb.gov/Aviation/FTW/95A244.htm

U.S. Army. 1997. Flightfax. vol 25. Fort Rucker, AL.

14

Flight planning

Discipline at ground speed zero

by Pat Barker
and Tony Kern

Plan the flight and fly the plan.
Old aviator's saying

No plan survives initial contact.
**Helmuth von Moltke, Chief of the Prussian General
Staff**

Flight planning is certainly one of the most underrated components of flying. Perhaps it is because it is not as glamorous as other parts of the flying experience. Perhaps it is because most of us are not adequately trained in the finer aspects of flight planning. Or perhaps it is because it is hard work to flight plan thoroughly and accurately. Whatever the reason, one thing is clear. Underestimating its importance is dangerous.

Knowns and unknowns

Disciplined flight planning does more than provide a greater margin of safety. It should also improve the efficiency and enjoyment of your flight operations, by allowing you to maximize the resources at your disposal, and by eliminating nagging fears that something has been left undone. Simply put, thorough planning answers the known elements of a flight, identifies the known unknowns, and it frees up your mind to handle the unknown unknowns. Did I say, "Simply put?" Let me explain.

279

The knowns of a flight are concrete and unchangeable aspects of the flight, such as times, distances, fuels, terrain, runway lengths, approach procedures, etc. The known unknowns are less predictable phenomena such as winds, weather, air traffic control congestion, and the like. The unknown unknowns are the tricks that Murphy has up his sleeve for your flight. For example, the engine bearing that is about to go, the closing of the airfield you intend to land at, the food poisoning from your breakfast at the *Hangar Queen Cafe*. Think of flight planning as the task of taking as much guesswork out of the cockpit as possible. It's not complicated, but it is work.

How often have we stepped into the cockpit thinking, "I've done this dozens of times; if something happens I'll figure it out," or "It's clear and a million; let's just get out to the airplane and do it." Recently I was participating in an Internet discussion on IFR flight planning. I was dumbstruck by an aviator's comment in response to a query about how much time should be spent in IFR flight planning. He wrote that his plans typically took about 10 minutes' time "unless the weather was lousy," in which case a few extra minutes would be prudent. While we hope this attitude is only found in a minority of aviators, it surfaces often enough to be of concern. Planning your approach procedures into, say, Atlanta while talking to the approach controller who is vectoring you about in deteriorating weather conditions is hardly solid flight discipline.

Haven't we all had "one of those days" where we just couldn't get it together—when suddenly you have less fuel than you should have because somewhere in your planning 2+2=8? Or perhaps you are conscientiously performing your pattern work at 1,000 feet while everyone else seems to be flying a few hundred feet above you and you begin to wish you had reviewed the VFR Supplement a little better? It happens more often than you think. The experiences of this B-52 crew show—at the expense of pride, perhaps—just such a day.

Case study: A tanker with a view

On the air-refueling portion of a standard B-52 training mission, a B-52 pilot team was convinced that they were about to link up with "their" tanker situated 5 miles off the aircraft's nose. On the lower deck of the bomber, working without the benefit of cockpit windows, the two navigators repeatedly announced "their" tanker was, in fact, ten miles away to the east of their 5 o'clock. The closer the

B-52 neared the pilot's tanker the louder the interphone conversation became. Ignoring the navigators' incessant pleas (to be specific: "shut up, I have the tanker in sight"), the aircraft commander closed to precontact position, approximately 50 feet below and behind the "tanker." Downstairs, the navigators sat back in their seats and folded their arms in silence.

At this time the co-pilot remarked over interphone, "Hey, look at all those windows." They had, in fact, come up behind a Delta airliner. At about the same time, Fort Worth ARTCC noticed the problem, too. In response to the controller's question as to why the B-52 was 20 miles away from the tanker, the pilot sheepishly responded, "Roger, Luger 16 turning right at this time."

What can we do to prevent occurrences like this one? Can we anticipate these days? In the above case probably not. What started out as an honest mistake turned into a fierce matter of pride for the aircraft commander. The airplane off his nose was going to be his tanker no matter what anyone said. Perhaps, though, we can make ourselves more prepared through a more disciplined approach to flight planning. We all make mistakes, but why set ourselves up to fail or set the stage for a minor error to become a major mishap?

This chapter offers insight into the planning process based primarily on the five pillars of airmanship (see Chapter 1): know yourself, know your aircraft, know your team, know your environment, and know your risk (Kern 1997). While "checklists" are offered for what one might do before and during a flight, they are in no way the final word on flight planning or anticipating change. They are merely guides for the flight-planning virtuoso.

Flight planning as an art

Flight planning is much, much more than "plugging and chugging" numbers and filling out FAA or DOD paperwork. It's a physical and mental manifestation of flight discipline. Planning, then, evolves from every aspect of flight discipline. In the case study that follows, the aircrew missed two opportunities to adequately plan the approach to a strange field. The first was during normal mission planning, the second was in the traffic pattern after they realized that this was not a routine approach and had executed a go-around at the prompting of the navigator. See how many failures of flight discipline you can identify in the next case study.

Case study: Short planning—long landing (Hughes 1997)

The C-130 was on a regularly scheduled "channel mission," a normal rotation for resupply. The weather at the destination airport was VFR with a temperature of 75 degrees. The airfield is in a nonradar environment, and the crew was told to expect the VOR/DME approach. The crew descended late and arrived overhead the Initial Approach Fix (IAF) at 9,000 feet. With the runway in sight they requested a left 360-degree turn to lose altitude. The copilot executed the left turn and began a shallow descent, rolling out on a three-to-four-mile final, but they were too high for a safe approach. The crew entered a left VFR pattern for another attempt. The pilot chose gear up and flaps 20 percent on the downwind leg. When rolling out on final, he directed 50% flaps and gear down. He touched down at approximately midfield and 26 KIAS fast on the 5,437-foot runway. The aircraft departed the runway and continued through a culvert and perimeter fence before impacting the center of a highway interchange. Seven of the ten persons aboard were able to egress with various injuries, three were fatalities, and the aircraft was totally destroyed by postcrash fire.

The crew consisted of two relatively highly experienced pilots, both with over 2,000 hours, and more than 1,000 hours time in type. The navigator, flight engineer, and loadmasters were also highly experienced.

The airport is located in high terrain that rises sharply to about 7,500 ft. MSL in two quadrants. The FAA has included this airport in an FAA advisory circular that requires special qualifications for the pilot in command (PIC). The single runway is 6,073 feet long with a displaced threshold that leaves only 5,437 feet available for landing with a 225-foot overrun. On the approach end, the terrain rises to 100 feet above field elevation within one-half mile and to 700 ft. above the field within 1-1/2 miles. The "day-only" runway is marked as a nonprecision instrument runway, has no approach lighting, no distance remaining markers, and no visual approach slope indicator (VASI) system. The Airfield Suitability and Restrictions Report (ASRR) addresses the visual illusion created by the terrain, and the resulting tendency to fly a high, steep approach with a high sink-rate at touchdown or a long landing. The plan view of the DOD approach plate contains cautions and a warning about terrain and required climb and descent rates.

The pilot had not been to this airport before, and the crew did its premission planning on the day prior to the flight. They did not review the required airport qualification video located in the planning room. Before arriving at the airfield, the copilot asked the pilot for the flap setting for landing. Theoretically, this should have been discussed during the normal mission planning day activities, but was not. The pilot said, "We'll probably fly a 50% flap approach and 100% flap landing." He checked the approach plate and said, "Six thousand feet, yeah, we'll do that at six thousand feet." This comment was not challenged by anyone on the crew. The computed the landing distance for a 100% flap landing was 4,200 ft, and 5,500 feet for a 50% flap landing. After adding the 500 feet required by Multiple Command Regulation (MCR) 55-130 for RVR adjustment, the minimum runway requirements were 4,700 feet for 100% flaps, and 6000 feet for 50% flaps. There was no discussion about the RVR adjustments to landing data. In short, the pilot's plan would force him to land in the first 737 feet or he would run off the runway. No one bothered to mention that to him. Let's take a closer look at the sequence of events to see where the failures of discipline occurred.

When the crew contacted the approach control, they were instructed to descend to 12,000. They set up their navigational aids for the approach and agreed that they may need a left 360-degree turn to lose altitude for the approach. Soon after they were cleared to 9000 feet, the copilot said that he had the runway in sight and that they needed to "drop very quickly." The pilot requested flaps 50% "on speed." The copilot did not hear or acknowledge this call due to radio traffic. When the runway was in sight, the copilot, with concurrence of the pilot, asked for a left 360-degree turn.

During the turn, the pilot mentioned that he would not descend too low because of terrain. The navigator informed the crew of the highest terrain in the quadrant. The pilot continued the turn and reported that he was intercepting the inbound course. As he rolled out on final, the pilot again asked for 50% flaps and called the runway in sight. The copilot confirmed the runway, set the flaps to 50%, and informed tower that they were on a 3.5-mile final.

The crew continued with the before landing checklist, set flaps 100%, and was cleared to land by the tower. The copilot and navigator never called their items complete on the before landing checklist. The pilot expressed some frustration about being high on final, the navigator suggested they make a left-hand VFR pattern, and the pilot initiated the go-around. So far, so good. They were struggling, but still making good decisions.

As the crew entered the visual pattern, they did not run any checklists. The gear was retracted and the flaps set at 20% while rolling out on downwind. The crew discussed the terrain, and the copilot attempted to put the pilot more at ease by coaching him around the visual pattern with comments such as, "there you go," and "looking good, looking good." This may have distracted the copilot from running the standard checklists and/or checking critical information such as landing data. The unplanned VFR pattern had broken their habit patterns and the crew was breaking down procedurally.

The pilot started the base turn early, maintained a continuous turn to final with 45 degrees of bank at 160–165 KIAS and the flaps at 20%. He descended only about 100 feet in the final turn. He overshot, and was once again high and hot on final. The pilot called for gear down and the copilot confirmed flaps 50. The pilot then stammered, "We'll take a fast landing is what we'll do if anything...flaps to 100,...ahh...Just leave the flaps at 50, so I can get the thing down." The lethal decision had been made, and no one on the crew challenged the pilot's action.

As they continued the approach, the copilot, stated, "OK, gear's down and flaps are at 50. Everything is set. Keep going. Looking good...148...138." Although the copilot did not run the checklist, he did point out that the gear and flaps were taken care of and was monitoring the airspeed and approach. The flight engineer was silent. Flight data recorder analysis indicated that the descent rate on final exceeded 2,000 fpm and the airspeed reached 170 knots. The pilot's next words were, ". . . can't slow down any." The aircraft crossed the threshold at approximately 164 KIAS and 150 feet above the field. It was time for another decision—force it down or take it around.

Touchdown occurred 2,025 feet down the runway at 151 KIAS. The nose gear touched down and the throttles were moved into the ground range within two seconds of touchdown. Six seconds later, the throttles were moved into reverse and maximum brake pressure was applied. In spite of the attempt at maximum braking, physics would determine when the aircraft would come to a stop; it wasn't where the crew intended and three died as a result.

Although all of the information on the airport was available to this crew to preplan and carefully analyze the approach and landing at this field, they did not access all available resources, not the least of which was a video that would have given them exacting details of the runway and terrain environment. Once they realized the difficulty factor involved with this approach, they had the opportunity to discuss it at some length in the traffic pattern, or request a holding pattern to work through the challenge. Instead, the crew chose to put themselves in a time-constrained box and then failed to recognize the danger of the second botched approach. When the pilot made the decision to land hot and long, there was no more time to plan for anything other than an emergency egress route from the wreckage.

Disciplined planning means knowing yourself

Understanding yourself is a lifelong endeavor, and probably the most difficult thing anybody can be asked to do. It is however, essential for disciplined flight planning. "The human physiological and psychological systems are far more sophisticated than any machine ever invented. Beyond this is the problem of self-image. We all like to think of ourselves as great aviators" (Kern 1997). In short, aviators prefer to analyze the mechanical things that will take them into the air rather than the aviator(s) within. After all, everything we need to know about the airplane we can glean from the manual and from talking to other pilots. We can even call up the manufacturer. Learning about ourselves is a far more daunting proposition.

Medical airworthiness

Are you feeling okay today? Do you feel something "coming on?" Tired? Hungry? Can you anticipate how any of this is going to affect your performance? How much have you had to eat lately? Proper nutrition and exercise is an important part of flight discipline, and taking them into account during mission planning accomplishes two tasks. First, it reminds you to take better care of yourself. Secondly, it identifies increased risk if your self-analysis comes up a bit short of ideal. A quick look at your physical state should be the first action taken in the self-analysis of your medical airworthiness. Equally important is your psychological state of mind.

Psychological airworthiness

How much concentration and attention do you have today? Are you distracted by things beyond your ability to compartmentalize and fly with the appropriate focus? Before you answer this question during planning, think of all the things that could go wrong on your flight, not the milk run you have planned. Judging your psychological state is a tough one, because many of us fly to get away from earthbound stress. This is healthy, but only as long as you retain the ability to focus and manage your attention. One area of particular concern in your psychological airworthiness is any hazardous attitudes you might be harboring or are at risk of falling victim to.

Dangerous minds

Make a special search for any lingering hazardous attitudes (see Chapter 7). Are you likely to fall victim to "get-home-itis?" Are you pushing yourself, your aircraft, or the mission too hard? Are you flying for the right reasons? Are you taking unnecessary risks? Are you primed for the antiauthority, machismo, and invulnerability hazardous attitudes? Are you set up for resignation, complacency, or excessive professional deference?

Good stress can be dangerous, too. Have you just fallen in love, bought a house, or gotten recognized for some great achievement? Any deviation from your psychological equilibrium can be a potential player in upsetting your ability to perform in the airplane. Be aware and be wary. The best time to think these items through is during flight planning. It won't happen automatically. You need to make a conscious decision to ask yourself the questions.

In summary, maturity is one of the hallmarks of good airmanship. Can you admit to yourself—and to others—that you are not fit to fly today? Do you have the willpower to sit yourself down if you find any of these issues to be significant?

Subtle implications and personal minimums

What if you are struggling with the go-no go decision and you decide that things aren't perfect (they seldom are if you are honest with yourself), but you've made the call to give it a shot and take off. There are subtle implications of discipline here that you must take time to think about before you get into the air. Does your honest self-assessment of physical and psychological airworthiness warrant setting some slightly more forgiving personal minimums for this flight? I have taken many an aircraft with a slight headache or runny nose. Other times I have been distracted by personal deadlines, commitments, or relationships. But I make allowances for these conditions, and I recommend that you consider doing the same. For example, if I am flying with someone else I will admit up front that I am not 100% today and by doing so put them in a more watchful posture. Other concessions can be made as well, concerning G tolerances, weather flying, or the extent of complexity you are prepared to accept in a nonemergency situation. By thinking these items through before strapping on the aircraft you have mastered some of the finer points of disciplined flight planning.

What scares you?

Fear is a funny thing, but I tend to trust it. If you harbor secret apprehensions about specific flight maneuvers or conditions, train them out with the aid of an instructor—or avoid those conditions like the plague. Although rocker David Bowie advised us to "turn and face the strain" (from *Changes*—I still own it on vinyl LP), it is best to steer well clear of areas your intuition tells you might be unprepared for.

Disciplined planning means knowing your aircraft

Would you willfully put your life in the hands of someone you barely knew? Most likely not, yet it happens all the time in aviation when we take to the sky in an aircraft we don't thoroughly understand, a common denominator in many aircraft mishap reports.

General aircraft knowledge

Aircraft knowledge should be one of the more straightforward aspects of flight planning. We have defined the study of aircraft rather thoroughly, and most flyers know their aircraft systems and procedures well. Flight planning is the application of this knowledge and equipment to the planned flight objectives, with a careful eye on what might happen as well as what you have planned. In addition to the generic knowledge of aircraft type and equipment, flight planning is an excellent time to consult the maintenance history of the specific tail number you are going to fly.

Specifics on the individual aircraft

Every aircraft is different, and most pilots will admit to having personal favorites for one reason or another. When flight planning, it is worth the time and effort to get as much information as possible on the specific aircraft you are going to fly to avoid any nasty surprises. Questions that you might ask maintenance personnel or other pilots who have flown the aircraft recently are:

- Has there been any major maintenance done on this aircraft lately?

- Has anyone refused to take her up? If so, why?
- Is there anything special about this aircraft that I should know before flying it? For example, odd noises, vibrations, etc.
- Are there any new parts on the aircraft that weren't there when it flew last? Of course, a thorough review of the aircraft forms should tell you if there are any required inspections due or overdue, but it is always a good idea to ask when the aircraft had its last major overhaul or "phase inspection."

Fundamental planning criteria

Some fundamental questions should be answered on all mission-planning efforts. Here are just a few of the basics.

- How can I communicate (radios, transponders, etc.)?
- How can I navigate (compass, TACAN, VOR/DME, GPS, etc.)? Do I have appropriate publications and charts to get me to where I am going, and are they current for the duration of my trip? If not, where can I get updated ones along the way?
- How high can I fly (pressurization, oxygen requirements, service ceilings, etc.)?
- How long can I fly (fuel, oil, speed, daylight, etc.)?
- How equipped is my aircraft to handle weather (antiicing, backup attitude indicators, weather radar, etc.)? Do I need to take extra equipment or supplies along, such as gust locks, tie downs, wheel chocks, oil, etc.
- Most importantly, do these items meet the demands of the flight I am planning or might have to perform in the event of an emergency?

In a nutshell, disciplined planning entails knowing what your airplane can and cannot do, and juxtaposing these facts against the projected demands of your flight.

Disciplined planning means knowing your team

Everybody benefits from good teamwork but few fully understand its nature and intricacies as it applies to a complex, and often hostile, environment of flight. Books have been written on cockpit/crew resource management (CRM) and I highly recommend you read them. But for our purposes we will limit our discussion to three primary areas: team formation and participation in the planning process, briefing, and interpersonal relationships. The flight mission actually begins the first time a group who intends to fly together meets.

Team formation and participation

In a landmark study of airline captains and crews, Robert Ginnett discovered that what a captain did in the first two or three minutes of getting his or her crew together was an accurate predictor of overall crew performance in flight. Effective captains made formal and clear introductions, set up expectations for the planning and briefing process, and generally established norms for the crew from the outset. In short, effective captains made it clear that they were in charge but would expect participation from everyone on the crew. Less effective captains were typically more *laissez faire* and did not set up specific duties, and in some cases did not even formally introduce themselves to crewmembers they did not know.

What this means for us is that flight planning should involve everyone, and the pilot in command should establish expectations early in a friendly—but formal—assignment of duties. For a couple of buddies who are planning a short hop for lunch in nearby town, this may be something as simple as "Bob, I'll fill out the paperwork if you can get us some takeoff data and weather. Let's meet back here when you are finished and brief the flight." Or, in commercial or military operations, it will involve the use of a formalized planning and briefing guide or set of procedures. In all cases, getting off on the right foot is essential to disciplined flight planning and effective flying.

Briefing

All flights, even if you are flying solo (perhaps especially solo) should be proceeded by a formal review and briefing. This accomplishes several purposes. First, it insures that all minimum safety considerations have been performed. Second, it gives you a final opportunity to seek additional information or make last minute adaptations to the plan. Finally, it puts the pilot and crew in the correct frame of mind before stepping to the aircraft. It says, "Everything's done; let's go fly."

The briefing, at a minimum, should consist of an overview of the flight to include:

- NOTAMs, weather, and special instructions for the departure and arrival airfields

- Ground operations: interior and exterior inspection checklists and preflight responsibilities, taxi routes and communication procedures, airfield obstructions and lighting

- Takeoff procedures: to include how you would accomplish an emergency return to the field, abort items should be discussed

- Routings, altitudes, airspeeds, and fuel reserve, high terrain for the route of flight

- Specific practice maneuvers: such as stalls, slow flight, or other special flight objectives

- Arrival to include planned runway, approach, and specific pattern work

- Crew coordination: including transfer of aircraft control, clearing, radio procedures, and in-flight checklist responsibilities. This should be accomplished even if you are just flying with a passenger.

- Emergency procedures: be sure to cover all emergency procedures which have **time-critical actions** involved as well as how you plan to handle egress from the aircraft on the ground and any physiological incidents.

- Questions: Be certain that you leave enough time to answer any questions that might have been prompted by the briefing or the planning session.

Depending on the complexity of the flight, a good initial briefing to someone you haven't flown with before can take anywhere from fifteen minutes to one hour, depending on the thoroughness and the questions brought out. Remember that this is not just for show; the preflight briefing is your single best opportunity to break any error chains that may have started before you even get to the aircraft. But even the best of briefings cannot overcome poor teamwork.

Interpersonal relationships

Poor interpersonal relationships can kill a flight team, often literally. In some instances it can be caused by poor communications skills; at other times it can be more systemic.

According to one pilot from a major carrier, it can be common practice in some airlines for captains to crosscheck the names of his crew with a list of strikebreakers from previous airline strikes who now fly with the airline. If "scabs" are found on the crew, they will be ignored by the other crewmembers. This is an example of a systemic problem which can be quickly isolated and corrected. Other teamwork problem areas are harder to identify, such as crew competency, strengths and weaknesses, assertiveness levels, fears, outside distractions, and medical airworthiness of your flight team members, who may have become adept at hiding these very items from view. In short, you should attempt to know everything about your flight team that you do about yourself, but it is unlikely that you can build that kind of openness and trust immediately.

Perhaps the best we can do is to open the lines of communications, ask the right questions, and admit some of our own frailties. I have had aircrews admit more personal weaknesses to me after I mentioned that I only got six hours of sleep last night or that I just got over a cold, than at any other point in the planning process. Even if you must search deeply to find something, admitting that you are somehow less than perfect somewhere during the planning session will often bring out other crew limitations.

The positive side of conflict

Don't confuse conflict and negative feedback with poor teamwork. The last thing a good pilot in command wants is a bunch of "yes men" telling him how well he is doing. We can pat ourselves on the back, we don't need it from others. There must be an environment for open and honest communications, which often means debate and well-intentioned conflict. The pilot in command is still in command, but he or she is a fool to worry about it. Mission planning and briefing is the perfect place to establish this environment. Tell the crew you plan to take all inputs pro and con to any decision, but you reserve the right to make the final call. Worry about conflict resolution at the debrief on the other end of the flight.

Disciplined planning means knowing your environment

Thorough planning means understanding the conditions and environment in which you plan to fly. Your environment includes three areas: the physical environment, the regulatory environment, and your organizational environment. Look carefully at each of these to catalog all factors that might become players on your mission. Physical environmental concerns should include weather, terrain, and airfield diagrams. Regulatory environment issues could include special-use airspace, terminal control areas, required equipment and pilot ratings and qualifications. Organizations also provide an environment in which we operate. Are you likely to encounter any unusual pressures to perform this mission? Are there any other organizational issues, such as on-time takeoffs or maintenance personnel problems, that could affect your mission? Disciplined planning means addressing, or at a minimum, being aware of such issues.

Workload management

Thorough planning should include identifying crunch points, discussing and planning workload management strategies, and identifying any potential pitfalls that you might be likely to encounter.

Flight planning is the ideal time to identify crunch points in the mission, and find methods for balancing the workload to cope and maximize efficiency. This applies whether you are flying alone or with a crew. Tale advantage of planned slack times in the flight. Get the weather early, figure out your landing data, review the approach and landing procedures. Play "What if . . ." a little bit to sharpen your situational awareness and emergency-procedure confidence for the environment you plan to fly through. If the trip you are making involves more than one leg, take time to analyze the implications of the "turn" at the intermediate point, something the crew in the next study could have accomplished somewhat better.

Case study: A failure to prepare (USAF 1996)

The following example shows a tragic breakdown in flight discipline that cost the lives of eight military aviators and one passenger on a military airlift mission at Jackson Hole, Wyoming. See where you can identify the mistakes that could have been corrected with disciplined planning.

Friday afternoon had been a hectic one for mission planning, but all too typical for an airlift wing with a high operational tempo. Two days earlier the crew learned that their joint (i.e., Army-Air Force) High Altitude, Low Opening (HALO) airdrop mission to Pope Air Force Base had been modified. Now a new leg had been added—a presidential support mission. They were to leave their base in Texas, pick up cargo and one passenger at Jackson Hole, proceed to John F. Kennedy International airport in New York, and after crew rest fly to Pope Air Force Base, North Carolina, for the joint forces airlift mission. At 10:30 A.M., came an additional change. The mission had been pushed back 14 hours. The copilot spent most of the day planning out the modified mission and left at 2:30 P.M. The pilot, who held important additional staff duties such as section commander and executive officer, spent most of the day at those duties rather than flight planning. He only managed to spend fifteen minutes with the navigator and copilot. The navigator was still hard at mission planning at 5:00 P.M., though some noticed that by this time he was visibly frustrated while working with his flight computer.

His reactions hardly raised an eyebrow in all the wing's activity. Eleven crews were already deployed to Southwest Asia and were getting ready to swap out with eleven fresh crews from the home base. Other commitments to Germany and special alert missions, combined with a shortage of navigators and flight engineers, stretched the wing's available resources. There was little possibility, or thought, given to finding another crew for this mission, even though this crew had some very junior crewmembers at key positions.

The aircraft commander was an experienced—but otherwise busy—pilot who had become an instructor three months before. Joining him was a relatively experienced flight engineer with 500 hours' time, and a new copilot with just over 60 hours in type who was not yet qualified to fly at night in mountainous terrain.

Rounding out the crew was a young navigator with about the same total experience as the copilot but less than 20 hours in type since his qualification. It was also the navigator's first flight in the Air Force as a mission-ready, qualified crewmember without an instructor. Despite the obvious shortcomings in experience, on paper this crew was qualified—if only barely—to fly the mission. And in an airlift squadron like this one, stretched to limit by high operations tempo, "qualified is qualified."

In all the rush to put together the mission, a few items were overlooked. First of all, the 14-hour delay meant that the landing and takeoff into Jackson Hole would be at night rather than the daylight mission originally planned. A month prior to this mission, Headquarters, Air Mobility Command, had forbidden its crews to fly into or out of this field at night. The crew was not aware of this higher headquarters restriction, nor were they aware that Jackson Hole was included in a special FAA pilot advisor circular that required special pilot certification. Though FAA advisories aren't binding upon the USAF, it would have been nice to have this information available to the crew. It might have given them pause to think and plan in a bit more disciplined manner.

Unfortunately, the crew was likely built with the joint HALO mission in mind, which, though complicated, was nowhere near the complexity that resulted when the presidential-support mission was added at the last minute.

While these organizational omissions were not necessarily the crew's concern, it was certainly within their responsibility to look at the approach plate for Jackson Hole (in *FLIP Terminal Low Altitude Procedures,* Volume III) and discover that this airfield had special departure instructions, including a specified climb gradient. Although one would have had to look at the index of the publication to see that Jackson Hole had a specific departure routing, the ramifications of high terrain in the vicinity were glaring. The approach plates for Jackson Hole warn aircrews not to circle east of the field, which, coincidentally, was where the crew planned to fly their departure. It is believed that the aircrew was unaware of any of these restrictions during the planning or briefing. Additionally, climb performance of the aircraft itself was an issue. If the crew had looked closely at the terrain and departure out of Jackson Hole, they would have discovered that the aircraft was not assured terrain clearance had they lost an engine after takeoff. This at least should have raised a warning flag to the crew to take a closer look at the surrounding terrain.

To a great degree the crew—particularly the pilot in command—set themselves up for a fall. The pilot, freshly qualified as an instructor, had a solid flying reputation. Perhaps that would have been enough to compensate for the inexperience of his copilot and navigator had he spent time thinking about the upcoming flight. Better yet, squeezing an extra hour out of his schedule to sit down with his young cockpit team members and go over their planning procedures or crosscheck their work might have been time well spent.

Whether the pilot's excessive duties—or his misprioritzation of them—took him away from the flight planning process is not important. What matters is that he left the entire planning of a very demanding mission alone to two barely qualified individuals.

The flight into Jackson Hole was relatively uneventful—though the pilot took special note of the mountains on the way in that would soon claim his life. However, the lack of adequate planning became apparent prior to departure from that remote field. While waiting for the cargo to be unloaded and additional fuel to be added, the flight planning room found the pilot, copilot, and navigator updating their flight plan and getting the most current weather.

Back in the airplane for the flight to JFK, it became obvious that neither the pilot nor the copilot had looked closely at the departure during flight planning. When the navigator asked the pilot what navaids to set for the departure, the pilot replied, "You tell me what makes sense because I am not really sure of what the course is and all that stuff out here." (All direct quotations are taken from Tab N of the AFI 51-503 report.)

To the extent that the departure was addressed in the cockpit that evening, the pilot indicated that he wanted to keep a healthy climb rate "just to get climbing away from all those mountains." What that "healthy climb rate" might be was anyone's guess at that moment, since no one had taken the time to calculate it.

Both the navigator and pilot discussed the importance of a left turn after takeoff most likely based on noise abatement procedures rather than terrain avoidance. It must have been assumed that the climb would take care of clearing the mountains. In retrospect, the copilot's silence on the issue during the departure discussion is almost deafening.

Between engine start and takeoff, the crew's earlier lack of planning manifested itself in a further loss of situation awareness. The "formal" cockpit pretakeoff briefing was rushed—only the takeoff was addressed in the cockpit briefing, and not the departure. It appeared from the voice transcripts, that the pilot uncharacteristically neglected to refer to the standard wing briefing format, which, had it been used, would have given the copilot an opportunity to say something that might have broken the deadly chain of events. As it was, it appears that the copilot may have exhibited excessive professional deference (see Chapter 7), not feeling inclined to challenge or question the incomplete briefing. Apparently nobody thought to crosscheck what the inexperienced navigator was saying, either.

During the taxi out to the runway, the crew almost experienced a ground mishap. Twice the pilot complained of bright runway lights and once of bright cockpit lights. The combination of the rushed takeoff and decreased visual acuity nearly resulted in a ground collision with another aircraft. The crew was clearly annoyed that they were being taxied so close to other aircraft. In retrospect, it might have been a blessing if they had hit it.

Following this harrowing event, the pilot finally asked for a departure heading, to which the nav responded, "left turn zero-eight-zero." "Left to zero-eight-zero," remarked the pilot. "That's going to be my terrain clearance heading, too?" "Yeah, that's affirmative," replied the navigator. "The sooner we can turn, the better." Neither response was sound, as there is towering terrain to the east. In fact, many local aircraft who fly in and out of the airfield regularly prefer to stay in "the hole" to perform a 360° climbing turn in order to get high enough to clear the mountains into which the aircrew had chosen for their "terrain clearance heading."

At 10:46 P.M., the aircraft took off. At less than 300 ft AGL and one mile south of the field, the heavy, lumbering C-130 began a climbing left turn towards the east. The pilot handed control of the aircraft over to the copilot. Shortly thereafter the pilot remarked quizzically, "My radar altimeter just died." What he took to be an instrument malfunction was in reality the last chance for the crew to avoid disaster. At 10:50 P.M. the aircraft slammed into the side of a mountain at 10,392 ft MSL. There were no survivors.

Disciplined planning means knowing your risk

Obviously, the crew in the previous case study did not have a good grasp on the relative risk of flying a heavily loaded aircraft out of Jackson Hole at night. Although there are several examples of aircraft and crews who perish in controlled flight into terrain (CFIT) coming out of the weather on an approach to landing, this is one of the few I know of where a professional aircrew lost situation awareness so quickly after takeoff. This unfortunate chain of events can be clearly tracked back to inadequate flight planning and briefing, but there are many other factors which can increase risk.

Risk management

Risk management, much like CRM, is the subject of volumes of research and findings from high-powered statistical analyses. The goal of all risk management is the same, however, to reduce risk to the lowest possible level while accomplishing your task with maximum efficiency and results. That doesn't sound so hard. Thankfully, there are a few experts we can look to to help us sort this out.

The insurance industry would go broke if they were not accurate risk assessors. They use a very simple equation to calculate risk and it goes like this:

Risk = probability of loss × cost of loss × * length of exposure

This is helpful. For example, we know that flying approaches and departures in mountainous terrain and bad weather is risky, so we try not to do it unless we have to, and then we minimize our exposure and maximize our training and concentration. This is risk management as its most basic form, and you will probably not be surprised to hear me tell you it should be done extensively during flight planning.

Identifying risk

Risk is found everywhere an aviator looks, but the magnitude of the problem should not cause us to shield our eyes. On the contrary, there are several approaches we can take to identify serious risks before and during planning. There are many sources from which a hazard can be identified. One of the best is simply talking to your buddies whom you fly with. The safety types call this "a group process with functional experts directly from a common environment." In our case, we are most likely talking about other aviators with common experience. If this process is used in a positive way, these discussions can be productively funneled into hazard identification. "What if...?" every situation as far as you can go. Ask questions until there are no more questions to ask. What should come out of this process is a picture which not only identifies common risk factors—but also possible methods for addressing these challenges. "Hangar flying" is uniquely and ideally suited for hazard identification.

A second source for hazard identification are the local safety representatives. The following is a partial list of assets that they have at their disposal, along with some suggestions for possible use, taken from *Redefining Airmanship* (Kern 1997).

Mishap Reports come from a variety of sources. They can be found within the organization, from the NTSB, and from various military sources. Obviously, a "missionized" or specific hazard identification is the best, and often your safety representative would be happy to help you construct a complete list based on your individual aircraft and mission type.

Military Inspector General (IG) reports: The IG, a military inspection team, claims that "we're only here to help you," and they do, by providing important feedback and written documentation on military hazards and local procedures to combat them. Everybody in the military gets a visit, and all reports are official records and are archived, so they are easily accessible.

Mishap and Incident Databases: The Department of Transportation, and various military safety agencies retain extensive databases. All information is not releasable to the general public, but many reports and statistics are. One of the most useful and informative databases is the NASA Aviation Safety Reporting System (ASRS), which has literally thousands of incident reports available to anyone who asks. They can even be purchased on CD-ROM with an easy to use search engine to identify common areas of hazards for your particular type of aircraft or location.

Surveys: Safety experts often use surveys to identify hazards, and you can, too. Design your own questionnaires. Target an audience and ask some very simple questions related to such topics as "What will our next accident be? Who will have it? What will cause it? When will it happen?" Don't ask questions like, "Does the boss support safety?" The answer is always the same, and does nothing for the improvement process.

Controlling risk

Before you control risk, you have to understand that there are some alternatives to risk. It can be accepted, reduced, avoided, spread, or transferred. The trick to managing risk is in knowing how much we are accepting. Until you exercise a process that identifies and assesses risk, you don't know what you have. Consequently, you can't plan for or control what you don't know.

Summary of risk management

There are three rules that must be understood in order to make sound decisions related to risk during flight planning.

1. Do not accept unnecessary risk. We accept risk all the time without knowing it. The trick is identifying and exposing risk, then breaking it down into component parts, and managing the parts. Risk properly managed is acceptable.

2. Make risk decisions at the appropriate level. Who in your organization should accept risk? During flight planning, the buck stops with you, the pilot in command.

3. Accept risk only when benefits outweigh costs. Obviously, this is related to the previous two rules. We accept a lot of risk,

but we must know what the benefits are in order to balance the risk. If we do it wrong, the mission becomes either too risk aversive which jeopardizes mission accomplishment, or too risky, which jeopardizes safety. Neither is desirable, and both can seriously degrade your performance.

Formal risk-management tools

As the science of risk management becomes more exact, several off-the-shelf products are becoming available to assist the pilot and crew with decision making and risk management. One of the most useful products on the market is the Flight Safety Foundation's *CFIT Risk Assessment* checklist. It is a very precise and systematic tool for measuring the relative risk factor of a CFIT incident. It is reprinted by permission and included at Appendix B for your use.

A final perspective on flight planning

Discipline is doing the hard stuff, and flight planning is the price of the ticket to slip the surly bonds of earth. Although preflight activities are often looked at as a necessary evil, real pros take a more positive approach. If I may, allow me to use an analogy from military history, where it is said that amateurs talk about tactics while professionals talk about logistics. In our case, I believe that amateurs tend to focus exclusively on flying, while true professionals improve their abilities by planning more effectively.

Chapter review questions

1. Explain what the author means by "known unknowns" and "unknown unknowns." Have you ever experienced an unpleasant surprise as a result of an unknown?

2. If flight planning is so important to safety and effective flying, why do so many aviators short-change it? Are these legitimate reasons?

3. How does the author use the Airmanship Model from Chapter 1 as a guide for flight planning?

4. What is a personal minimum? Are these fixed or changeable?

5. How can you use fear as an ally in flight planning?

6. What sources of hazard identification do you use in your type of flying? Are these sufficient?

References

Hughes Training Incorporated. 1995. *Aircrew Coordination Workbook.* Hughes Training Inc., Dyess Air Force Base, Texas.

Jensen, R. S. 1995. *Pilot Judgment and Crew Resource Management.* Avebury Aviation Publishers: Aldershot, UK.

Kern, T. 1997. *Redefining Airmanship.* McGraw-Hill: New York.

NASA ASRS. 1995. The AeroKnowledge CD-ROM. Trenton, New Jersey.

Skolarsky, Col. 1996. United States Air Force 51-503 Accident Investigation Report on C-130H Mishap at Jackson Hole Wyoming. Barksdale AFB, LA.

15

Chaos theory?

Structuring change in the cockpit

by Pat Barker and Tony Kern

Man has a limited biological capacity for change.
Alvin Toffler, *Future Shock*

Things change. Carl von Clausewitz, the oft-quoted but little understood Prussian general, called it fog and friction in 1832 (Clausewitz
1976) . Most aviators today call it Murphy's Law. Whatever the term,
rarely does an aviator sit back during a flight and observe that
things have taken place—to paraphrase the evil emperor from *Star
Wars*—"exactly as I have foreseen." Although there are many ways
to deal with uncertainty in aviation, one comes eventually to two
major considerations. We can either attempt to reduce uncertainty,
or accept uncertainty as a fact of life and function within it.

Reducing uncertainty was the goal of the last chapter, which was
dedicated to disciplined flight planning on the ground. This chapter
takes on the next challenge, functioning well amidst change in flight.

When things change

Because we are human and base actions on an enormous amount of
sensed and stored input, we act—or perhaps better said react—to an
intense series of feedback loops which can create a decidedly unpredictable, even chaotic, situation. In many circles, the inability to
structure this phenomena has given way to a relatively new term—
chaos theory. I believe it applies well in aviation.

Chaos has come to refer to a system of beliefs, or frameworks, describing the behavior and structure of many natural systems (Schmitt 1995). We need not go into a extensive discussion of chaos and complexity here—authors like James Gleick and Uri Merry have already done an admirable job—but we might consider just how our intuitions are predisposed to think of what is "okay" and what isn't (Gleick 1987, Merry 1995). Our perceptions of normalcy have a big impact on our in-flight decision making, and nowhere is this skill more critical to an aviator than when things change midstream.

Two reasons why we react poorly to change

Have you ever wondered why birds that are suddenly caught in wind gusts never seem to crash? I'm serious. Take, for example, barn swallows, who swoop and bank at incredible rates reacting to changes in the flight paths of bugs, wind conditions, and other barn swallows chasing the same insects. How do they do that? Darwin tells us that somewhere in the past several hundred millennia, the barn swallowlike bird-things that did crash never got a chance to reproduce, and those that didn't crash passed along their genes to create what is now a marvelously adaptive flying machine.

Since humans did not start flying until, say, a century or so ago, we still have several million years of catching up to do. In short, humans have not evolved to handle fast-paced change, hence Toffler's belief (see Chapter heading quotation) which is valid and safe, at least for the next few million years.

Add to this unfortunate state of affairs the culture that humans developed which decidedly favors stability over change. Just look at our language; the default condition is orderly and neat. We have words like synchronous, stable, periodic, and regular. Abnormal conditions are described by simply negating the basic word. Something messy is now irregular—something erratic is unstable. In short, our intuitions, thanks to Newton and his cohorts, are fundamentally linear.

The real world, however, is nonlinear. A prominent mathematician recently wrote, "Calling natural phenomena 'nonlinear' is like calling the bulk of the animal kingdom nonelephants." (Beyerchen 1992). If this makes your head spin a bit, it is just further validation of our point—we don't really understand the nature of random change very well, and in aviation, we pay for this lack of understanding with blood.

In the math textbooks we learned from as kids, small inputs into systems yield correspondingly small outputs, and gradual inputs could be expected to produce gradual change. We were taught that the world was based on mathematical principles, and it probably is, but not the kind of traditional Newtonian math I grew up with. For all of Newton's steadfast reliance on laws and principles, he was also secretly a closet alchemist, trying to turn lead into gold. Perhaps he too knew that everything was not as simple as it seemed.

We have come to expect—albeit subconsciously—the same cause-and-effect relationships in real life that we learned in school. What *really happens,* though, throws us for a loop. Small inputs into our lives, bodies, and the in-flight environment might indeed produce small outputs, but they can just as easily have huge, seemingly out-of-proportion results. Examples from history include the assassination of a relatively unimportant archduke in a small country in Europe, which led directly to World War I—the bloodiest war humans had ever known. Or when a young black woman, Rosa Parks, decided she would no longer sit in the back of a bus, she spawned a huge social movement that is still in action today. Small events—big consequences, and in aviation these can, quite literally, explode in our faces.

Small inputs, big outputs

Aviation is no stranger to such phenomena. The case of the B-52 crew in the case study that follows began with what seemed random minor events but quickly escalated into a life-threatening situation. It describes an aircrew whose situation descended from a calm and certain combat delivery over hostile territory to a chaotic and lethal chain of events that was at least partially self-induced. The explanation for the apparent contradictions in capabilities speaks to the heart of what this chapter is about. Simply stated, the crew could not handle either the rate or the magnitude of the change. The apparently random events were nonlinear, and the result was not the sum of its parts.

Case study: Coincidence, chaos—or both?

(Rouse 1991; Kern 1997)

During Desert Storm, the crew of a B-52 took off from a forward-operating location on their fourth combat mission. Tragically, this would be the last mission for this aircraft and crew. As the case study progresses, we will see a chain of apparently unrelated and random events that culminate in the fiery crash of the aircraft and crew just 15 miles—three minutes—short of their final destination.

Although there were many factors involved with the mishap, in the final analysis the accident revolves around the B-52's electrical system and a series of seemingly unrelated maintenance factors and aircrew actions. It is a classic illustration of the need for a structured approach to change in the cockpit.

During preflight inspections, the aircrew was not informed of two recurring maintenance problems—a critical lack of information that would come back to haunt them later in the mission. The first was a number-one fuel quantity gauge failure, which had been written up five times previously. The second bit of unknown was the recent troubles with the number five engine, which had failed to "subidle" conditions at least twice on recent flights. Neither of the malfunctions was documented in the aircraft forms.

These two existing—but unknown (to the crew)—aircraft malfunctions would prove instrumental in the gradual unwinding of the situation. They are both significant in that they affect engines with generators. The first, a simple fuel gauge malfunction, would be misdiagnosed by the crew. The second malfunction would cause an underexcitation relay to cut the number five generator off line at the worst possible moment, ample proof that chaos has some pattern, perhaps designed by a leprechaun named Murphy. As the B-52 took to the sky, the sequence of events resulting in its demise had already been set into motion.

It didn't take long for the first malfunction to manifest itself. Between the first and second air refuelings, the pilot noticed a fuel gauge fail indication and that the number one main tank fuel quantity indicator was reading zero. After taking a quick look at the checklist, the copilot reset the fuel quantity circuit breaker in violation of flight manual guidance, and the gauge seemed to work properly—for the moment.

Matters got worse when the number three generator dropped off line, meaning that it was no longer supplying AC power to the aircraft systems. The copilot attempted to reset the generator twice, but it failed to come back on line. The aircraft commander did not consider this malfunction to be especially worrisome but was aware that it could lead to further problems.

The crew completed their combat delivery and turned back towards home, the day now wearing into evening. During a second air refueling, the number one fuel gauge began to act up again. The crew pulled and reset the circuit breaker once more. When the number one main fuel tank quantity dropped 7,000 pounds below its number four counterpart, the pilots began to discuss the possibility of a fuel leak but took no checklist action. Although the flight manual points out that "the most positive indication of a fuel leak may be an abnormal change in the pitch or lateral trim requirements," the pilots did not take advantage of this cue because they failed to disconnect the autopilot to check for a difference in feel. Following this discussion, the copilot checked off for his turn to rest.

Shortly thereafter, the aircraft commander noticed that the number five engine was displaying "abnormal indications." The engine readings were somewhat lower than the other seven, and this caused the aircraft commander enough concern that he had the copilot awakened to come back into position so that they could work the problem as a team. The crew discussed the indications: Low readings on the number five engine instruments across the board; the engine would only respond to decreasing throttle movements and not throttle advances. The copilot testified that at this time the number five generator was still on line. They determined that they would not elect to shut the engine down, and with that, the copilot went back into the bunk to rest again.

As the crew approached their destination, and the aircraft entered a holding pattern—the number one and two engines suddenly flamed out. The copilot returned to his position and the aircraft commander informed him that engines numbers one and two had flamed out because they "had run out of fuel." After a short disagreement about the situation with the fuel system, the copilot immediately began to work the problem without referencing the flight manual. This was a critical error because the amplified checklist contained in the flight manual directs a step specifically designed for the case of flameout due to fuel starvation—which was highly likely given the fact that the two engines flamed out simultaneously. This checklist step directs some specific throttle actions designed to "purge the fuel control unit of air and the engines of fuel," which would significantly increase the chances of a successful airstart. As luck would have it on this day, the number two engine restarted, but number one—the engine with a generator on it—would not.

The situation now was growing rapidly worse. The crew was preparing for a six-engine approach, considering engine number five, which had failed to subidle conditions, as unreliable. Ironically, the most crucial product being produced by the number five engine was not thrust, but the electrical power flowing from its generator. Although the number five engine hovered precariously close to the generator cutout speed, the crew seemed not to appreciate the possibility of its loss. A good deal of time was spent preparing to land with the loss of two engines, but none on the possibility of having to rely on a single generator. The copilot did advise the crew to reduce the electrical loads after seeing high amperage readings on the two remaining generators, which were running off of the number five and number seven engines, but apparently the copilot did not recognize the impending loss of the number five generator due to low rpm.

Unknown to the crew, the aircraft began to operate on a single generator shortly after they began their descent out of the holding pattern. The number five engine fell below generator operation speed and tripped the underexcitation relay. Interestingly for our discussion on complexity, the master caution light did not illuminate to indicate the new malfunction, as it should have—a new failure, and another example of chaos in action. Had the crew been aware of the single generator operation, they may not have elected to lower the flaps, which greatly exceeded the capacity of the lone remaining generator, causing complete AC power failure.

The pilot immediately called to advisory control, "Sir, (we are) declaring an absolute emergency at this time; we have no power to the aircraft and I need minimum vectors to get me home." After moving a preceding tanker out of the way, the controller gave the stricken B-52 an initial heading for the approach. "Hulk 46 heavy, fly heading 170." When Hulk 46 responded, they made the final mistake in the long sequence of events. The pilot of Hulk 46 replied "Sir, I'm gonna need no-gyro vectors."

A "no-gyro" approach is required when the aircraft has lost the capability to determine its heading due to a failure of the aircraft heading system(s). In this case, the pilot still had some heading information available, including a heading indicator and a magnetic standby compass. The no-gyro approach is completed by beginning and stopping turns based upon a radar controller's instructions and is frequently a very abrupt approach, as the pilot attempts to precisely "turn left/right" and "stop turn" in accordance with directions.

When the B-52 lost all AC power, it also lost the capability of the electrical fuel boost pumps to provide positive fuel pressure to the engines. Although the engine-driven pumps should have continued to provide pressure, they are susceptible to cavitation, which can result in flameout. To help prevent this eventuality, a critical, three-step "Fuel Management without Boost Pumps" checklist is included in the emergency procedures section of the B-52 tech order. It was neither referenced nor remembered. The third step in the procedure reads as follows: "Maintain aircraft in as nearly level attitude as possible and avoid abrupt changes in speed or direction." To emphasize the point, a WARNING was included immediately below the checklist step, which read:

"Changes in flight attitude or acceleration forces may cause main tank boost pumps to become uncovered allowing air to be drawn into the system, thus causing engine flameout."

Over the next several minutes, the radar controller responded to Hulk 46's request for a no-gyro approach by providing two turns and stop turns to align the aircraft with the runway, and the crew experienced total engine flameout at low altitude over the Indian Ocean, less than three minutes from a would-be landing.

The case study above serves notice to us all that there are, in fact, no unrelated events. In many cases we may not see the relationships, but given the right chain of events, as the chaos theory pundits put it, the disturbance caused by a butterfly's wings can cause a hurricane on the other side of the world. The lesson in all of this is that small events can have big consequences, with lethal implications for aviators.

If one example is not enough to convince you of the "unfairness" of the cause-and-effect equation in aviation, consider these. Recently, a young Civil Air Patrol Cadet on a T-43A (B 737) aircraft ignored the steadfast rule for all guests on military aircraft—"Don't touch nothin!" The inquisitive would-be fighter jock reached up to the navigator's panel and turned off the Inertial Navigation System. Since at that moment the aircraft was using the INS for its flight control reference, the 737 pitched over fifteen degrees and began a rapid descent which, luckily, was quickly caught by the pilots. A similar situation occurred a few years ago in the same type of aircraft. An accidental press of a button by an inexperienced navigator student sent a T-43 into an uncommanded split-S. Little buttons, big results.

While pressing little buttons can have surprising consequences, more common are the "little changes" in aviation that become major considerations, and often in a big hurry.

Reacting to change

An analysis of incidents and mishaps resulting from in-flight changes indicate that the concept of time plays the pivotal role in the success or failure of an action taken as a result of a mission change. In a multitude of cases, the inability to structure a changing situation revolves around the pilot's or aircrew's use of available time.

An aviator's performance is often degraded by a perceived or actual need to hurry or change the tasks at hand. Time-related pressures, for example, might arise from any number of factors, including pressure from ATC to expedite taxi, takeoff, climb, descent, etc.; maintenance or weather delays; or quite simply the desire to get home and get to sleep or watch your daughter's soccer game. In addition, the phenomenon of "temporal distortion" comes into play, where our perception of time is destroyed by a significant emotional event.

In both the case of the CT-43 at Dubrovnik (See Chapter 1) and the C-130 at Jackson Hole (Chapter 14), the urge to "hurry" can be traced well back into the planning phase, where major or multiple mission changes occurred. At times it seems as if change is the norm in aviation. Major in-flight mission changes—to include replanning an entire mission in the air—are not uncommon. Aviators, as humans, are usually predisposed to bristle when changes occur too rapidly to handle smoothly, and we even have physiological manifestations of this stress. Perhaps the hair on the back of the neck stands up, muscles tighten, or sweat breaks out. Things are moving too fast, we can't stop the aircraft, we are rushed, and we are angry.

Doug Edwards, an Australian safety expert calls these moments "stress steps" and postulates that we should expect them, realize their debilitating effects, and even use them from time to time. Certainly recognition is a key aspect of monitoring ourselves in the moment of change. The NASA ASRS office also recognized the problem with time-constrained change, and further points out the human tendency to exacerbate the problem by rushing ourselves for no apparent reason.

The hurry-up syndrome

In 1993, the NASA ASRS center conducted an extensive study of errors caused in part by a perceived need to "hurry." The following paragraphs summarize and expand the report to include military examples of the same type.

Once the pilot or crewmember feels some need to expedite a situation or quickly change plans, a phenomenon which has become known as the "hurry-up syndrome" occurs. Aviation's worst disaster, the terrible KLM/Pan-Am accident at Tenerife, was due in great part to schedule pressure problems experienced by both flight crews. Two Boeing 747s, one operated by KLM and the other by Pan Am, collided when the KLM flight was taking off and the Pan Am flight was taxiing on the runway. Both aircraft caught fire and were destroyed—there were 583 fatalities and only 61 survivors. The Air Line Pilots Association (ALPA) conducted an 18-month, three country investigation of this accident, with an emphasis on the human factors of flight crew performance.

ALPA found that the KLM crew had strong concerns related to duty time, specifically that they would be able to return to Amsterdam that evening and remain within their duty time regulations. They also expressed concern about the weather and its potential to delay the impending takeoff—the cockpit voice recorder indicates that the KLM captain said, "Hurry, or else it [the weather] will close again completely."

Pan Am's crew was equally concerned with potential weather delays. They were detained for more than an hour due to the KLM flight crew's decision to refuel—the KLM aircraft and fuel trucks blocked the taxiway, thus preventing Pan Am's departure. These schedule-related problems set the stage for the catastrophe that followed (ASRS 1993).

In a recent ASRS study of 125 time-pressure aviation incidents, 63 percent had their origins in the preflight phase of operations, where the conditions for a mistake later on downstream were established. For example:

> "...inbound flight was late and we were rushed because of the scheduled-out, time-report-card mentality...It turns out that the clearance I got on ACARS was for the inbound flight. The squawk was incorrect, the altitude was wrong, and so was the departure frequency." (ASRS 1993).

Target seen but not hit

In the late 1980s, a Strategic Air Command B-52 bomber about to take the active runway was stopped short by the tower when it was discovered that the pilot forgot to file a flight plan. He had become sidetracked in base operations by squadron matters and rushed out at the last moment to catch the crew bus. The resulting delays brought on a hurried eight-hour mission with two refuelings and a low-level bomb run, which was basically replanned in flight by the navigators. Preoccupied most of the flight, the pilot was not quite "with" the airplane the whole day. The copilot was added on the crew late and was not much more "into" the flight than his boss, the aircraft commander. He had not been present during the previous days' flight planning to study the low-level bombing route, and had received a cursory explanation of what the target area looked like just prior to takeoff.

Since the aircrew was dropping live weapons—Mk-82, 500-pound bombs—extra care had to be taken to positively identify the target before dropping high explosives upon it. This is something we take pretty seriously. A minute prior to bomb release, the navigators thought they had the target on radar, so they asked the pilots for a visual identification of the target. This safety step is required for the navigators to complete their bomb release checklist. The pilots could not see the target area.

Thirty seconds prior—again, no visual tally...fifteen seconds...ten. At the last moment—five seconds prior to the release point—the copilot called out that he had identified the target. A flurry of activity ensued, knobs turned, switches flipped madly, and 2,500 pounds of ordnance dropped from the B-52. Indeed, the copilot had identified the target, but he neglected to say at the time that it was almost a mile to the right at the aircraft's two o'clock position when the bombs were released. Hurried mission changes, omissions, and rushed briefings manifested themselves in this glaring error in crew coordination. Luckily—except for some injured egos—no one was hurt by the error.

Changes can occur even prior to takeoff, which can catch a crew off guard. The captain on this commercial airliner was saved in this case by an alert tower controller.

"...we were busy with checklists and passenger announcements while changing to tower frequency. Tower cleared us

for immediate takeoff, and even though we had not finished our checklists, I taxied our aircraft into position and started to advance the power for takeoff... After about 1,000 feet of takeoff roll, tower canceled our takeoff clearance...(we) asked the tower why we had our takeoff clearance canceled. The F/O said (that) we're not on the runway. At that point I realized we had started our takeoff roll on an active taxiway." (ASRS 1993)

Yikes! If these types of errors can occur to highly trained professional crews employed by the military or major carriers, maybe we should look into this change-related error phenomenon a bit deeper.

Errors facilitated by changing conditions can be made by one individual, or they can be made by the flight crew as a collective unit. The majority (68 percent) of errors appeared to be collective, according to the ASRS study. Collective error on the part of a three-person flight crew is well illustrated by the following report:

"I am relatively new at this position as Second Officer...We had a tailwind which precludes reduced power [for takeoff] in this aircraft, but they [Captain and F/O] didn't notice and I was so rushed that I didn't back them up and notice. So we took off with reduced power...We were just in too big of a hurry..." (ASRS 1993).

Doing something wrong—or maybe not at all

Human errors may be categorized as errors of commission or omission. Errors of commission are those in which crewmembers carried out some element of their required tasks incorrectly or executed a task that was not required and which produced an unexpected and undesirable result. Errors of omission are those in which the crew member neglected to carry out some element of a required task.

Errors of commission

Sixty percent of human hurry-up errors in the ASRS study were errors of commission. In the following example, the flight crew erred in not adequately examining the airport surface chart.

"Takeoff was made from displaced threshold instead of the beginning of the paved runway. I feel some contributing factors were: Not studying airport runway chart closely enough to realize. We had an ATC delay and were at the end of our takeoff release time..." (ASRS 1993).

In another example of a commission error a young student pilot in South Carolina landed after a bumpy ride. In his eagerness to get the airplane parked before the deteriorating weather got any worse, he completely forgot to contact the ground controller. Thinking he had clearance to taxi back to parking, he rolled across the active runway just as a twin-engine aircraft was taking off. No mishap occurred, but the pilot learned a good lesson about feeling rushed.

Errors of omission

In 38 percent of instances in the ASRS study, pilots made errors of omission. In the following report, the flight crew neglected an important element of preflight preparation—with annoying and unnecessary results.

> *"Got a pod smoke warning from central annunciation in cruise en route between Fresno and Ontario...Diverted to BFL...no evidence of fire...we found a placard, which showed the pod smoke detection system as deferred and inoperative...We were pressured to hurry, and in the process, failed to check the aircraft log prior to departure." (ASRS 1993.)*

What led to the error?

In each incident report, one or more contributory or causative events promoted a "hurry-up" error on the part of one or more of the flight crew. Based on 309 citations from 125 reports, high workload was cited in 80 percent of all incidents, while problems involving physical or motivational states were next with 74 percent of incidents.

ATC may contribute to the "hurry-up" mindset by requesting an expedited taxi or an intersection departure by issuing a "clearance invalid if not off by..." or other time-sensitive requirements. Of course, ATC personnel are similarly under pressure to maximize traffic flow. In this example, the flight crew clearly felt pressured by ATC:

> *"Our inbound aircraft was late arriving, and upon receipt of our ATC clearance for our outbound leg, we were informed we had a 'hard' wheels-up time. Needless to say, we were rushed...about 100 yards before reaching the end of the runway we were cleared for takeoff on Runway 12...I taxied onto what I thought was 12R, but what was actually Runway 17." (ASRS 1993.)*

Time compression leads to poor discipline

What types of incidents result from change and hurry-up errors? Deviations from Federal Aviation Regulations and/or ATC clearances are most common, while deviation from company policy or procedure was next. Fully 48 percent of hurry-up errors resulted in a deviation from ATC clearance or FARs. Twenty-one percent of reports resulted in deviations from company policy or procedure. There were also a number of runway transgressions, like the one that caused the mishap at Tenerife. Runway transgressions were mentioned in 17 percent of all reports. Other incidents mentioned included landing below minimums and altitude deviations. Hurry-up errors result in failures of flight discipline. So what do we do?

Managing change: Operating in the fourth dimension

Napoleon said that the only resource the logistical wizards in his *Grande Armee* could not replenish was time, for "once lost, it is lost forever." Certainly, decisions made in the dynamic environment of modern flight must be timely, but what does that really mean? The following questions outline a method for utilizing time to tip the scale in your favor. If you begin to feel the "stress steps" mentioned earlier in this chapter as a result of some impending flight change, ask yourself these questions.

The critical importance of time and tempo: Four critical questions

There are four questions that can help structure in-flight change by taking better advantage of time, our most valuable resource in decision making. If you take a moment to ask yourself these questions, you are well on your way to efficiently managing in-flight change.

Do I need to act immediately?

If the answer is "yes" you are literally betting your life, and possibly the lives of others, on your training and preparation to handle the situation. Realize this before you act. Most immediate decisions are made with precious bits of fragmented information, and your individual abilities at that moment in time will have to carry the day. An immediate decision to act isolates the actor from outside assistance.

Will it help—or hurt—to delay this decision?

If you decided that immediate action was not necessary, ask yourself and the crew if any decision is necessary in the near term. Many scenarios are actually improved by delaying a decision, while under different conditions waiting merely narrows the options available. For example, options available on a transoceanic flight usually increase as one approaches land. Conversely, some conditions—such as approaching nightfall—can force decisions to be made. Several years ago, a B-1B landed with the nose gear up on the lake bed at Edwards Air Force Base after many hours of attempting every potential fix known to man. They might still be up there trying to connect circuits with paper clips if the tech reps and command post had a way to keep the sun from going down. But the aircraft commander said "enough is enough" as daylight waned and made a safe landing on the Edwards lake bed. There are many ways to "stretch time," including the 4C method mentioned in an earlier chapter ("climb-conserve-confess-consult"). Know your best endurance altitudes and speeds. If you are lucky enough to have an air refuelable aircraft, sharpen your A/R skills. On the flip side of the coin, don't allow your options to be narrowed by bad weather, low fuel, nightfall, etc.

What help do I need?

Whether or not you decide to act fast, it is imperative that you make every effort to access all available resources. If you have forgotten which electrical bus a critical piece of equipment is on, an emergency situation is not the time to hide your memory lapse. Many an aviator has died following that old saying, "I'd rather die than look bad." Swallow your pride, get help—and get it early. The "experts" may have forgotten too, and they may need time to find the information you need.

What is the worst-case scenario?

This final question will help project you into the future and help you guard against the lost situation awareness that often accompanies the distraction of an in-flight change. Additionally, by considering the worst case, your course of action may be more conservative and give you a greater margin of safety. If the B-52 crew in the earlier case study would have asked themselves this single question, they would all likely be alive today. Bottom line—leave yourself an out.

Tempo: The final element

One final word on time—the fourth dimension in aviation. Once you have structured your actions and made that quality decision, don't let events accelerate beyond your control. Manage the pace—or tempo—of the actions you take. Even the right thing can be done too fast or too slowly. Mishaps occur when humans lose control. Remember Toffler's admonition from the front of this chapter. We aren't very good at this change thing. I don't have the statistics to prove it, but I would bet my next month's flight pay that impulsiveness has killed far more aviators than procrastination. Rushed checklists often get started but not finished; good intentions and plans of action are not followed up. Pace yourself and keep this in mind—change begets change. As soon as you get one situation under control, get ready for more.

More hints for managing change

Beyond better time management, there are some other fundamental techniques that we can use for a more structured and disciplined approach to in-flight change. Most of these are part and parcel to our everyday flying activities but tend to get lost in the distraction of in-flight changes.

Don't forget the basic four

We are taught from our earliest training days the four-step process for handling an emergency.

1. Maintain aircraft control.
2. Analyze the situation.
3. Take appropriate actions.
4. Land as soon as conditions permit.

These steps, at least the first three, are sound guidance for any in-flight change. It should not be assumed that flying the aircraft will automatically jump to the forefront of your mind when a midmission change occurs. On the contrary, many mishap reports can be traced to the moments of distraction immediately following a curve ball thrown to the pilot or crew on an otherwise normal mission. A senior training officer for a major air carrier told me recently of an observation he made while flying in the jumpseat of his company's aircraft. He related that after the two-person crew was given a single point change to their flight plan, both pilots "played with the flight management system for 14 minutes without ever lifting their heads to look outside the cockpit." That is putting a lot of trust in ATC to clear your flight path. Only after a conscious step has been taken to keep a firm hand on the tiller should you move on to the next step of active analysis.

Analysis of an in-flight change should be thorough enough to identify its ramifications on your overall flight objectives but still not detract unnecessarily from flying or monitoring duties. To relate back to the issue of time, don't let a change push you into something you are not prepared for. Slow down and request a holding pattern if necessary. Remember, it's not the change that will kill you; it's the ground or another aircraft.

When you determine what actions to take relative to the change, do a quick risk analysis. Does this modification to the original plan add significantly to the risk level of the aircraft and crew? If the answer is yes, either reconsider your plan or admit the increased risk to put everyone on a more watchful posture.

Use the checklist

Just because we pick up some other planning materials to assist in our decision making, this does not mean we should throw down the checklist. More times than not in mishap analysis of change-related accidents, if an aircrew had taken the time (there that word is again) to run their normal procedures checklist, they would have identified, and hopefully prevented, the cause of the mishap. No one intentionally lands gear-up unless there is a serious aircraft malfunction; they just forget. Checklists prevent forgetting; it's as simple as that.

Keep an eye on capabilities

When things change in flight, many options may present themselves. Some of these options will exceed the capabilities of either you, your crew, or your aircraft. Stay within your comfort zone. In all likelihood, the change itself has decreased your performance somewhat by negating some of your disciplined flight planning and possibly increasing your stress level or that of your crew. Don't make the situation more hazardous by accepting—or worse yet, proposing—a course of action that will require you or your aircraft to perform perfectly to pull it off. Don't paint yourself into a corner.

A final perspective on in-flight change

A former commandant of the Marine Corps, General Alfred Gray, understood change better than most of us. In fact, he understood it so well that he advanced the idea of creating chaos so his troops could exploit it. Don't get me wrong here folks. I am not advocating that we adopt the Marine's philosophy of intentionally creating confusion. I am, however, strongly advocating that we take their approach to understanding it.

General Gray said, "Chaos—learn to love it. Create it, revel in it, swim like a fish." Even for the best aviators, that may be going a bit too far, but the idea of being comfortable, even confident with the rapid onset of change is definitely a goal for all flyers to aspire to. Discipline comes from confidence, and confidence comes from preparation. We may not all be Marines, but we may want to adopt their legendary approach to discipline.

Pat Barker, captain in the U.S. Air Force and course director of a new course titled Foundations of Aviation Excellence, collaborated on the writing and development of this chapter.

Chapter review questions

1. Is chaos a fair word to use when describing the rapid onset of change in the complex aviation environment? Why or why not?

2. Why do people resist change? How does this play out in the cockpit?

3. What four questions should you ask when operating in the "fourth dimension?"

4. In addition to managing time, what other techniques are effective for structuring in-flight change?

5. How comfortable are you with change overall? Do you feel you are an effective manager of in-flight change? Why or why not?

References

Beyerchen, A. 1992. "Clausewitz, Nonlinearity, and the Unpredictability of War" *International Security* 17:3.

Clausewitz, C. 1976. *On War* : Princeton, New Jersey: Princeton University Press.

Gleick, B. 1987. *Chaos: Making a New Science.* New York: Penguin Books.

Hughes Training. 1997. *Aircrew Coordination Workbook.* Dyess Air Force Base, Texas: Hughes Training, Inc.

Kern, T. 1997. *Redefining Airmanship.* New York: McGraw-Hill.

Merry, U. 1995. Coping with Uncertainty: Insights from the New Sciences of Chaos, Self-Organization and Complexity3. Westport, CT: Praeger.

Rouse, D. M. 1991. Report of Aircraft Accident Investigation B-52G #59-2593. United States Air Force. Feb 3.

Schmitt, M. 1995. "Chaos, Complexity and War: What the New Nonlinear Dynamical Sciences May Tell Us about Armed Conflict." Paper written for Marine Corps Combat Development Command. Quantico, VA.

16

Flight discipline in action

The credit belongs to the man who is actually in the arena.
Teddy Roosevelt

Flight discipline is at once attitude and action. It is difficult to describe but easy to recognize. It permeates everything we do as aviators, and yet is seldom discussed unless it is recognized as lacking. This chapter is designed to help us recognize flight discipline in a positive environment, for examples of excellence abound if we care to look and know what to look for. The following case studies are designed to show you what to look for. The first case study illustrates events of solid flight discipline in the face of a serious emergency. The final two case studies juxtapose two similar military missions from our recent past, one a huge success, the other a humiliating failure. The difference was discipline in planning and action.

As you read these case studies, reflect upon what we have learned in previous chapters, and see flight discipline played out in actual events.

Case study: Prepared to use all available resources

Sometimes it pays to have thought of things in advance, like when you lose your only engine ten minutes into your flight over open ocean. The pilot and his wife had just taken off from a coastal California airport and were flying on an IFR flight plan in VMC conditions. His wife was the only passenger on board and she occupied the right front seat. Takeoff and departure went like clockwork, and the aircraft was in a cruise climb configuration approaching 6,000 feet on a clearance to 10,000 feet. The ocean below and blue skies above gave every indication of peaceful serenity, it was one of those days that makes a pilot know why he took up flying in the first place. But all that was about to change.

Passing through 6,000 feet the pilot noticed the vacuum gauge needle was flickering just above the green arc. Abnormal certainly, but not an abort item just yet. The vacuum system might just have a small leak. Besides, it was such a beautiful day, but the pilot must have been running through his time and distance equations in his head, and moments later he needed them. The pilot relates what happened next.

> *"The engine then made a grinding noise for about three seconds and the propeller came to an abrupt stop. I immediately decreased airspeed 35 knots to 85 knots IAS, turned toward our departure airport, closed the cowl flaps, and alerted Departure (control) of the engine failure and my course deviation." (ASRS 1993).*

This pilot was on top of the situation. He not only ran his "engine-failure-in-flight checklist" flawlessly, but had the presence of mind to alert the controlling agencies early enough so that they could clear outbound traffic away from his recovery flight path. More importantly, he did not forget rule one—maintain aircraft control. His initial actions were precise, and it would pay dividends for the rest of the recovery. But there is more to a forced landing than one checklist and a turn towards the field. Let's see how they handled the remainder of the life-threatening emergency.

The pilot's next actions exemplify the discipline required to take the next step in any emergency procedure—analyze the situation and take appropriate actions. The situation was obvious, he had lost his one and only engine, but what actions should he take next? This pilot chose to get more help, and he knew right where to find it.

> *"I then passed to my wife the emergency procedures laminated cards to read aloud to me while I turned the fuel valve and mixture control to off, maintained airspeed, and alerted ATC that I was shutting down the transponder."*

By utilizing his passenger as a checklist reader, the pilot accomplished two critical aspects of flight discipline. First, he has managed his workload by creating a team. He could now expend more of his time and concentration on flying the aircraft. A near perfect long distance glide was essential in this situation, as ditching was the only other option if the aircraft could not get back to the coastline. Secondly, the pilot eliminated the chance of forgetting a step in the checklist by trying to do too much in too short of a time. He could now divide his time between keeping the airport in sight and monitoring the airspeed, and he had time to think about the finer points of an emergency recovery. "I instructed my wife to insure her seat belt and shoulder harness were tight."

As the aircraft glided in towards the coastline, the air traffic controller handed them off to the airport tower. The pilot informed the tower he was an emergency, a fact the tower already knew and was prepared for, as evidenced by an immediate clearance to land on "any runway." The onshore breeze they had taken off into just moments before was still a factor. The active was runway 24, which meant a longer gliding distance to set up for a landing into the wind, although this was certainly not required in an emergency like this one. But the pilot's disciplined early actions allowed that luxury. In fact, he even had to increase his drag a bit to get it down to a normal glidepath.

> *"My altitude, airspeed, and progress to the airport were good and I elected to fly an abbreviated downwind and base to 24. I lowered the gear on final to help lose altitude. I maintained 85 knots until short final, when I lowered the flaps to enable me to slow the airplane and land in a normal landing attitude."*

The remainder of the episode was uneventful, and the aircraft rolled to a stop on runway 24. The "crew" had functioned perfectly, and good preparation had led to good fortune. It is clear—exceedingly clear—that his pilot had thought through his response to such a situation before it actually occurred and was confident and certain in his actions. He didn't miss a beat. To what does he attribute his success?

> *"Twelve minutes had passed from the time the engine shuddered to a grinding halt and we rolled to a stop on Runway 24. I believe the successful outcome of this flight can be attributed to the fact that I remained focused on the task at hand, flying the airplane, and used all of the resources available to me, including my wife, who read the emergency procedures. I also believe that I was very lucky to have been presented a situation which was equal to my level of skill."*

Twelve minutes of disciplined action, but likely a lifetime of preparation. This example illustrates that discipline means being able to admit your limitations and actively seek additional resources, in this case the passenger and a very helpful air traffic control system. Preparation, sound fundamental flying skills, workload management, and attention to detail made the difference between a walk back to operations and treading water in the Pacific Ocean waiting on the Coast Guard. A tip of the hat to a disciplined pilot and wife team.

Our next two case studies look at a different type of life-and-death situation, where the U.S. military puts it all on the line. The first case is an attempt to rescue American hostages; the second is the beginning of the operations to evict Iraqi forces from Kuwait.

Perhaps nowhere in aviation is the concept of flight discipline as important as it is in the military. Thankfully, most flyers do not have to strike deep into enemy territory to rescue hostages or take out early warning radar sites on their flight missions, but there is much to be learned from those who do. The following case studies come from two examples which are flown by similar type aircraft with similarly trained crews. One results in failure and national embarrassment, the other is the first bold stroke of a hugely successful military operation.

The first account is taken from multiple sources who have analyzed the disaster that was Operation Eagle Claw, the United States's attempt to rescue hostages held by Iran in 1980. The second example comes from a pilot who flew one of the very first missions of Desert Storm.

As the tales unfold, look at the critical aspects of planning, communication, and personal flight discipline. You will see that the difference between victory and defeat is often the narrowest of margins, where a single oversight or undisciplined act can turn the tide. Although these are military examples, the same holds true across the spectrum of aviation operations.

Operation Eagle Claw

On November 4, 1979, a mob of Iranian students overran the Marine guards at the United States embassy in Tehran and seized the compound and over 100 hostages. They demanded that the United States send back the Shah of Iran, who was in New York to receive medical treatment, to stand trial (and presumably execution) for "crimes against the state." Because the United States's position did not allow bargaining with the terrorists, President Carter needed options. Less than a week later, the military was beginning to put a daring rescue plan together. The plan that was conceived was called "Operation Eagle Claw" to symbolize the strength and power of the United States swooping in to pluck the hostages from danger. Less than six months later, the plan would fail horribly, leaving twisted and burning aircraft, highly sensitive classified materials, eight American corpses, and a good bit of American pride in its wake.

What went wrong at "Desert One," the code name for the first staging point located several hundred miles into Iran? How could America's most elite military forces, backed by the full weight of the presidency, fail before ever reaching the primary destination? Before we go into a human-factors analysis of what went wrong, let's take a quick overview of the plan.

The plan

Operation Eagle Claw was to be a joint operation, that is to say involving more than one service of the military. The tip of the spear was the newly-formed and ultra-secret Delta Force, under the command of Colonel Charlie Beckwith, a decorated, tough-as-nails, Vietnam veteran. In military parlance, these guys were real "snake-eaters." The Delta Force was planning on nothing less than invading a foreign country, storming the Tehran embassy, defeating the terrorists, freeing the hostages, and escaping with them back to friendly control. All in a day's work—or maybe two, with a little help from a few AC-130 gunships and other airborne assets. While Beckwith's boys were razor sharp and ready for action, the transportation in and out of Iran would have to come first, and it would be supplied by the Air Force, Navy, and Marines.

The first stage of the operation called for a desert rendezvous at Dasht-e-Kavir, Iran, a rural desert location with a straight stretch of highway that would be used for an initial C-130 landing sight. It was prepared in advance by a special team from the CIA, who had covertly flown into Iran on a Twin Otter and surveyed the sight and even imbedded landing lights into the sand, which would be activated by the 130s on their initial approach. Eight Air Force C-130s, some full of equipment and commandos, others would haul fuel to resupply the Navy helicopters were being flown in from the aircraft carrier Nimitz by Marine pilots. The plan called for a rendezvous and refueling at Desert One, and then a final insertion into Tehran by helicopter, all under the cover of darkness. After the refueling and transfer of the Delta Force to the helicopters, the 130s would fly out of Iran, and be refueled by Air Force KC-135 tankers staging out of Diego Garcia, an island in the Indian Ocean. Navy fighter and attack aircraft were standing by just in case things got out of hand with the Iranian Air Force.

It was a complicated plan, something that military commanders have tried to avoid since the days when Alexander the Great fought on these same sands some twenty-three centuries earlier. Due to the nature of the operation, some complexity was required. However, when the flames erupted at Desert One, there were no fewer than four on-scene commanders trying to direct various parts of the operation. This resulted in unnecessary confusion and chaos, and is indicative of an organizational failure of discipline—a lack of unity of command.

Execution

On April 24th, the 173rd day of "the hostage crisis" as it had come to be known in the United States, the plan was put into action with a short message from the commander of the joint task force: "EXECUTE MISSION AS PLANNED. GOD SPEED" (Kyle, 235). With those words, C-130s launched from Misirah, Oman, and eight RH-53 helicopters launched from the USS *Nimitz*, just southeast of the Strait of Hormuz in the Persian Gulf. Flying low to avoid Iranian radar, and "blacked out" to minimize visual detection, the aircraft began their 500 mile-trip to Desert One. The 130s were to arrive first, followed closely by the helos.

The first chink in the plan occurred when the lead MC-130 encountered a reduced visibility condition caused by suspended sand particles in the air—a condition the Iranians call a "haboob." While the MC- and EC-130s were capable of navigating through the dust, the Air Force component commander had some concerns about the helicopters. He directed that a SATCOM message be sent immediately, warning the Marine helo drivers of the condition so they could better prepare their formation for the dust storm penetration. But the Air Force personnel could not send the message on the new SATCOM equipment, which was located in the back of the plane. The person assigned to send the message lamented "I wasn't able to send the message about the dust. It was too dark back there, and I couldn't

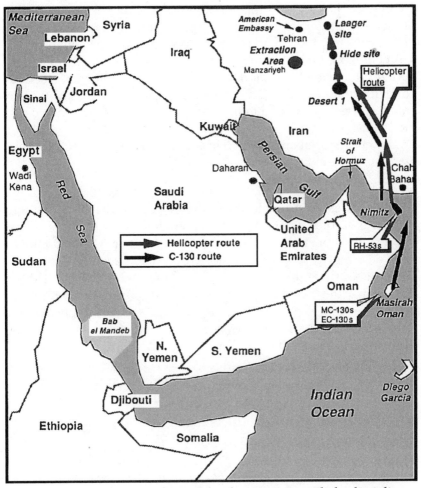

16-1 *Plan of operations for "Eagle Claw" mission. Flight discipline began to fail in the planning stages of Operation Eagle Claw. An overly complex plan and insufficient preparation led to disaster at Desert One.* (Kyle 1990. Reprinted by permission.)

make out the code-word matrix" (Kyle 251). Ironically, the Air Force personnel had been given different instructions than other members of the task force on how to use the SATCOM system. A "code-word-matrix" was designed to encode critical pieces of information, but it was unnecessary because the SATCOM system itself was secure. All that was necessary was to key the microphone and pass the message in clear text English. A key piece—perhaps the key piece—of information had failed to be passed due to poor planning and discipline.

Back in the helicopter formation, they were experiencing problems of their own. After six weeks of continuous testing and maintenance, the eight RH-53 Sea Stallions were supposedly in excellent mechanical condition. Only 140 miles into Iran, helicopter number six experienced a blade inspection method (BIM) warning light, a serious condition that could indicate impending failure of one of the aircraft's nitrogen-filled aluminum rotor blades. It was determined that the aircraft was no longer airworthy, so helo number eight picked up the crew and abandoned #6 in the Iranian desert. The problems were just beginning.

As the helicopters entered the first of two haboobs, formation flying became extremely difficult due to the decreased visibility. The situation was further complicated by communications restrictions that did not allow the helo crews to talk to each other. The lead pilot finally determined that he could not proceed, and made a U-turn and landed about 300 miles short of Desert One. Although he had expected that the remaining helicopters in the formation would see his maneuver and follow his lead, only one other chopper actually did. The result was a leaderless group of five continuing on in the dust towards Desert One. Although the two aborting helos eventually re-launched and tried to join the formation, the integrity of the formation, which was designed to provide mutual support, had been destroyed. According to Colonel James Kyle, the Air Force Component Commander

> *"...the lead choppers were now 175 miles from the desert rendezvous and in a world of hurt. This was nothing like the mission for which they had planned and trained. They were battling a long unrehearsed instrument flight and had no secure radios to hold it all together—it had turned into a blood-curdling experience." (266).*

Things went from bad to worse, when one of the RH-53s lost one of its two hydraulic systems. Although this was a grounding item in peacetime training operations, the crew would continue to Desert One if possible to assess the damage and evaluate the airworthiness of the craft. In addition, the helicopter carrying the senior Marine lost sight of the formation and experienced loss of a primary flight instrument. Not wanting to risk collision, and still having to navigate over a 9,600-foot mountain range, the crew elected to abort the mission and return to the *Nimitz*. One source reported that the copilot on the aborting aircraft vehemently opposed turning back. He reportedly stated "We must go on . . . this is the Super Bowl!" He was apparently told to "sit on his hands" as he was not the aircraft commander (Kyle 270). Once again, a lack of communication within the formation meant that this decision was made without knowledge that there had already been one abort and a second chopper was questionable for the rest of the mission because of the hydraulic problem. In a nut-shell, the minimum requirements for continuing the mission past Desert One were now seriously in question.

The end of the plan

Back at Desert One, all of the special operations 130s had arrived but were running low on fuel because the helicopters had been delayed by the weather. The 130s kept their engines running during ground operations, as they did not want to risk shutting them down and not getting a restart. Therefore, the longer the delay, the more critical became the fuel state. Once the six remaining Sea Stallions did arrive, they were quickly marshaled into refueling position—which was directly behind the C-130s—and the fuel was rapidly transferred. But even as the preparations moved forward for the second phase of the operation, it quickly became apparent that one of the RH-53s would not be able to continue. After analyzing the single hydraulic failure that had occurred en route, it was determined that one of the remaining six helicopters was not airworthy. Without six helicopters, the mission had to be aborted. But the worst was yet to come.

One of the C-130s was now getting critically low on fuel, which might require a KC-135 to come into Iranian airspace for an emergency refueling. If this occurred, there would be no hiding the fact from hostile radars, and the situation would likely escalate quickly from there. Before the C-130 Hercules could move however, the helicopters directly behind it had to taxi clear, as the run-up of the big C-130 props would likely bury the helos in sand. As the first helicopter lifted off to move out of the way, it became lost in the blowing sand and rammed the stationary Hercules, immediately engulfing both aircraft in a ball of flame. Just prior to the impact, the copilot reportedly told the pilot "you're not going to clear it." In the immediate aftermath of the disaster, eight Americans were killed.

There have been several detailed and controversial analyses of this operation, but our focus here is on a few key human factors around which all debates swirl. The first fatal flaw came in the planning process and resulted in extremely poor communications discipline throughout the mission. A new satellite communications capability—called SATCOM—was to be utilized on the mission, yet the capability and equipment was so new that it was not installed on the helicopters or C-130s which would be used for the mission themselves until just before the mission. This did not allow the crews or commanders to practice with the technology. As a work-around, ground communications personnel were added to some of the aircraft to run the SATCOM equipment, but they did not fully understand aviation jargon and on occasion were unable to effectively rely on critical pieces of information. The lesson here is that new technology is not always better. This holds true whether we are analyzing the use of Satellite Communications on operation Eagle Claw—or deciding whether to use a new Global Positioning System for navigating our next Cessna 152 cross-country. The bottom line is that technology is only as useful as the human is capable of making it. A failure of discipline here can set the stage for future problems, as seen in the case study above. Simply stated, the aircrews and command and control elements of the mission did not practice the way they would play.

A second human factor was the failure of anyone associated with the mission to realize that things were getting out of hand to manage the change. If anyone did notice, the communications security should have been broken to attempt to get the mission back on track. There is an analogy for us here as well. How many times have we seen situations where aviators are afraid to speak up when they first perceive a problem because they feel that it is not their place to do so. This not only jeopardizes mission accomplishment, but often safety as well. When crewmembers did speak up, as with the reported comments by the copilots on two of the problem helos, they were not listened to. The comment from one aircraft commander to the copilot to "sit on his hands" is clearly contrary to good cockpit/crew-resource-management (CRM) practice.

To summarize the discipline issues involved with the failure that was Operation Eagle Claw, it can be said that poor communications training and discipline were present from the start, which led to questionable teamwork, decision making, and judgment by individual flyers. When combined with incredible bad luck, the stage was set for disaster.

In the next case study, the planners and aircrews had the benefit of hindsight, and somewhat better technology to guide them. It is written by a crewmember who actually flew the mission.

Case study: Black Knight: A mission that couldn't afford to fail

by Corby L. Martin, Major, USAF

Our mission was slated against the air campaign's number-one target, but not one member of my crew believed we would actually execute the war plan. Our movement to the Forward Operating Base (FOB) of Al Jouf, Saudi Arabia, did little to change that belief. On 16 January the squadron commander stopped by our aircraft and told us there would be a meeting at 1600 hours that day because H-Hour (the day the war would commence) had been set for 0300 hours the next day. At that moment, the silence was deafening.

Our task was to lead four US Army AH64 Apache helicopters from the 1st Battalion of the 101st Airborne Division from Ft. Campbell, Kentucky, to an initial point (IP). An initial point (IP) is the spot just prior to the target area where a combat aircraft begins its attack run, which was approximately eight nautical miles (nm) from an Iraqi early-warning radar site, about 20 nm from the border between Saudi Arabia and Iraq. The Apaches were tasked to destroy the radar site at H-22 minutes or 0238 hours, literally 22 minutes before the war was scheduled to begin. Simultaneously, another formation with the same makeup as ours, two MH-53 Pavelow IIIE helicopters from Hurlburt Field in Florida and four more Apaches, were destroying another radar site 30 nm to the east of our target. This was by no means a "milk run" and everyone involved with the plan knew it. Integrating forces between services—commonly referred to as "joint operations"—is never easy, and when the stakes were as high as they were for this mission, the preparation took on an added measure of urgency.

Planning for this mission began four months earlier at King Fahd air base in eastern Saudi Arabia. Access to the plan was limited to a small number of people for security reasons, and communications discipline was critical from the outset. Without divulging the true nature of our mission we brought the Apache and Pave Low crewmembers together to start training. Challenges that emerged during those first meetings for this historic joint force mission were handled in a professional manner. Even though we were from separate services and different operational cultures, we had the same objective in mind and tried to complement instead of compete with each other—a rare scene in many joint operations. This professional attitude set the tone for our relationship for the rest of our stay in the desert.

The first training flight on 23 October 1990 was little more than a route reconnaissance designed to let each helicopter crew get a feel for the flying procedure and abilities of the rest of their formation. Although the mission was routine, the team aspect was critical. We needed to know what the Apaches were capable of, and how we could best use our technology and talents to get them to their target. I'm certain that they had some similar questions about us. This was the beginning of our disciplined teamwork, learning how to maximize—but not exceed—each other's capabilities.

All of our training flights were flown at night for security reasons and because of the high probability of executing the actual mission at night. Special Forces operators are used to working at night, but the hazards associated with training in the dark with new team members added an increased level of risk to our preparations. We took great care to thoroughly brief all aspects of the training missions, as we didn't want the mission to fail before we had an opportunity to execute it.

The combination of the capabilities of the two radically different helicopters and crews made a very difficult mission seem relatively easy. In truth, either weapon system could probably have accomplished the mission independently, but the risk of failure would have been unnecessarily high. The importance of knocking out the early warning radar sites can not be overstated. Literally hundreds of allied aircraft would already be airborne on their way into Iraq. To fail at our mission would put these Air Force and Navy pilots at much greater risk. But the combined capabilities of the helicopters, and the disciplined teamwork of the crews virtually guaranteed success.

Few of us had forgotten the lessons learned from Desert One, the attempted rescue of American hostages in 1980. That mission had ended in an embarrassing failure, due in large part to interservice rivalry, outside influences, and an extremely complex plan of operations. We wanted no question unanswered, and this time both the Army and the Air Force would be pulling in the same direction. (In

1980, President Carter ordered a military rescue attempt of 53 Americans being held hostage by Iranian terrorists after the Shah of Iran was deposed—Operation DESERT ONE.)

The aircraft

The first Pave Lows were produced during the 1970s, and were built to navigate through adverse weather and over difficult terrain in order to rescue downed aircrew members. The failure of the Iranian rescue mission in 1980 highlighted a serious deficiency in the capabilities of U.S. military helicopters: the ability to fly long distances under adverse conditions. The Pave Lows were quickly acquired by special operations because their navigation system made them an ideal platform for leading other helicopters over long distances under most any conditions. The MH53Ys we flew during Desert Storm were air refuelable, protected by a very good electronic and infrared countermeasures system; they had armor plating, and carried .50 caliber machine guns for self-defense. The navigation system included terrain following/terrain avoidance radar (TF/TA); forward-looking infrared radar (FLIR); projected map display (PMD); Doppler navigation; inertial navigation; and a global positioning system (GPS). All of these systems worked together through a sophisticated data bus to provide a highly accurate navigation solution. In the desert the TF/TA radar and the GPS navigation system turned out to be the most critical pieces of equipment. In a nutshell, we had the world's most sophisticated all-weather, all-terrain aircraft to lead in some of the world's most lethal, Army Apaches. We were well trained and comfortable with our weapons systems.

The AH64 Apache was built to kill tanks. Its navigation system and defensive systems were adequate, but its real strength was targeting and firepower. The Apaches routinely carried eight Hellfire missiles, thirty-eight 2.75 inch rockets with various warheads, and approximately 300 rounds of 30 mm ammunition for the chain gun located in the nose turret. If we could get the Army in close enough to the radar sites undetected, there was little doubt that they would finish the job. We were an ideal match. But the Iraqis knew we would be coming, just not when or from where. We wanted to keep it that way until the lights went out. It would take teamwork and discipline, and a good bit more practice.

Our second training mission, on 25 October, was memorable because of the lessons we learned from the first flight and the new tactics we had started to develop. Although we were primarily "comm out" for security reasons, we had to be able to communicate certain key pieces of information. First, as the flight lead, we learned that we had to plan and fly predetermined ground speeds to keep the formation together, but our second challenge was far more difficult and it resulted in our most critical new tactic. Our Pave Lows had a far more sophisticated navigation capability than the Apaches, but the question was how could we communicate an accurate Initial Point for the Apaches to begin their attack from without compromising communications security. The point had to be an accurate fix close enough to the target to allow their navigation and targeting systems to the best possible update before they had to shoot. Many brain cells were destroyed or damaged pondering the answer to this question. The answer came from a couple of Pave Low gunners—some of the lowest-ranking members of the team. They suggested that we simply drop a bundle of chemical light sticks (the kind you can buy at your local camping supply store) on the ground as we fly over the point and let the Apaches fly to the light. It is amazing that something this simple worked so well, but it did. The Apache gunners began to use night vision goggles (NVG) to assist their pilots in locating the light bundles, as well as keeping track of us. Simple is often better.

The third training flight took place in early November. The significant event on this mission was the Apaches' firing of their 30-mm chain gun on the Dhahran range. We had been firing our .50 caliber machine guns on this range since August because an aerial gunner from the 16th SOS (AC130H Spectre gunships) had convinced the range control officers of the benefits of letting us practice. But the helicopters from the 16th SOW were the only American helicopters allowed to fire on the range. Both our squadron commander and the Apache battalion commander were third-generation Lebanese Arabs. So, they went down to the range and had tea with the officers in charge, and the next thing you know 1st Battalion Apaches were able to fire on the range.

On the fourth training flight on 28 November, the Apaches were allowed to fire one Hellfire missile from each formation for a total of three missiles. The CENTCOM commander wanted to validate our ability to accomplish the mission, so he approved the expenditure of this costly asset. The Apaches destroyed all three targets with ease.

Between the end of November and the 10th of January 1991, we didn't conduct any more training flights. Both units met on an informal basis about once a week. Our professional relationship had developed into a lasting friendship. During December the Iraqis moved their early warning sights north from the border changing their area of radar coverage. This changed our tasking: now we had to destroy only two radar sites. Two additional new tactics were developed during December, both related to the same challenge. The Apaches did not have the fuel required to fly from the forward operating base (FOB) to the target and return to the FOB. The Apache crews developed a configuration never before tested to increase their fuel capacity. They successfully replaced one of their 19 shot rocket pods with an auxiliary fuel tank. Some of the ingenious flight engineers from our squadron invented a way to refuel the Apaches. A fire hose was connected to the Pave Low fuel dump tubes with clamps. This concept was tested and it worked. However, it would never have passed service safety requirements.

During the first week of January everyone was called together for a briefing. Our squadron commander briefed all of the crews on the overall mission. Then, they were briefed on the detailed plan for the mission. Finally, Intel briefed the threat and targeting priorities for each site. Both radar sites were defended by anti-aircraft artillery (AAA). The Apache and Pave Low crews worked together and constructed individual mission packets that fit the personality of their formation. On 10 January we flew our final training mission using exact distances and times. We incorporated the tactics and procedures we had developed over the last three months. The flight was a success. We were a team and we were ready.

On 14 January, our crew led an eight ship formation of Pave Lows to Al Jouf, Saudi Arabia which is located 100 nm south of the Iraqi border. Little did we know that this would be our home for the next two months. On 15 January we flew an operational check flight on our helicopter. We had adjusted the pitch on our rotor blades to give us more lift in case we encountered a situation where we needed more power. One of the older and wiser pilots in our squadron had come up with this idea.

On 16 January when we were checking the gear on our helicopter the squadron commander stopped by and told us our mission was on. At 1600 hours he told us to rest up and write home. Most of us could only manage the latter task. At 2100 hours we had our final crew brief. The commanders offered us words of caution and encouragement. Both of the flight leads went over last minute details with their formations. At 2330 hours we arrived at our helicopter, started our engines at 0030 hours, and took off for Iraq at 0113 hours. We were seven minutes behind the White team, which had the eastern target. En route to the target we flew lights out at 50–75 feet above the ground and 120 knots groundspeed. There wasn't any moon illumination but the sky was clear and the temperature was cold. About 20 nm from the border we spotted a building with lights that seemed to be located on the border. We adjusted our route to avoid the building and to maintain our timing. As we crossed the border, abeam the building, the lights were turned out. Someone had heard our formation, whether they were Saudi or Iraqi we did not know. At 0225 hours we marked the initial point with our chem lights and headed back toward Saudi Arabia. But the excitement was far from over.

En route to our holding point and back in Saudi Arabia our tail gunner and para-rescue specialist called a threat break for a missile launch. Both maneuvers were executed using counter measures at about 50 feet above the ground. We quickly learned that during low level flight at night, reaction swiftly follows the worked "break" as there is no time to wait for further information. It is imperative to have enough situational awareness to know the location of your wingmen, flight lead, and the ground. Again, planning and training were keys to success and survival.

We located a holding area clear of threats in enough time to witness the start of Desert Storm. At 0238 hours we watched explosions occur in the vicinity of both radar sites. A few minutes later we saw the flash of bombs being dropped southeast of White team's target. The flash was quickly followed by AAA tracers racing skyward. In our FLIR we watched blacked out F15E's fly by at 100–200 feet above the ground. By 0250 hours all of the Apaches from our formation had rejoined us at the holding point. We headed for Al Jouf; en route we did another threat break because one of the gunners on our crew inadvertently launched a flare. I'm sure he wasn't the only nervous individual in our formation. We called higher headquarters and relayed the code word that meant our target was destroyed and dropped off the Apaches at a designated waypoint. They proceeded to Al Jouf to refuel and then continued on to the desert south of Rafha, Saudi Arabia where they awaited the ground war for the next month and a half. We climbed up to 500 feet to air refuel with an HC130 from the 9th SOS. Our follow-on mission was combat search and rescue. Flying south we could look up and see what seemed like hundreds of aircraft anti-collision lights in the sky. It looked like rush hour on the Los Angeles freeway at 20,000 feet. Groups of small lights were chasing bigger lights all over the sky as our tanker aircraft earned their flight pay (I guess everyone needed gas that night). Every once in a while a group of lights would head north and then disappear. We landed at 0500 hours and a sense of relief washed over the crew. We debriefed the mission and headed back to our quarters for alert and a change of uniform. We knew the next day would bring a new mission.

This mission has been described in many different ways by different authors—many not participants. The truth changes depending on who you talk to last. We knocked out two early-warning radar sites, which allowed coalition strike packages to ingress safely and undetected to their targets in western Iraq. We were able to accomplish the mission because of training, professionalism, and most importantly teamwork. Teamwork and discipline within our crews, our formations, and our organizations. We flew a successful mission and we brought everybody home.

A final perspective

Flight discipline occurs in many forms. The differences between these two case studies are ones of planning, communication, and team discipline. In the second case, the aircrews and planners clearly understood the need for simplicity and control. It is also noteworthy how often the author mentions teamwork, and even friendship between the team members from the different services. In contrast to the "sit on your hands" mentality evident in Eagle Claw, one of the key solutions in the second case came from a lower ranking enlisted gunner, and was readily accepted and implemented by the flight commanders.

There are several lessons for all aviators in these case studies. First, flight operations in any regime demands thorough and accurate planning, which includes thinking and planning for the unexpected. In the chaos of change, a simple answer is often the best one. Second, communications discipline means knowing when to talk, as well as when not to. Finally, practice like you will play. In the heat of battle—and it doesn't matter whether that battle occurs in a military aircraft, a commercial airliner, or a Piper Cub—aviators tend to fall back upon habit patterns, be they good habits or bad. Fill your kit bag with good habits by practicing sound discipline on every flight.

Chapter review questions

1. Using the information contained in previous chapters, how many components of flight discipline can you find in the case studies presented?

References

Kyle, J. H. 1990. *The Guts to Try*. New York: Orion Books.

Martin, Corby. 1996. *Black Knight: A Mission That Couldn't Afford to Fail*. Unpublished.

NASA ASRS 1993. "A Well-Planned Response," *NASA CALLBACK* No. 173, NASA-Ames Research Center. Moffett Field, California.

17

Flight insurance

A personal program for improving flight discipline

To take no action is to take an undecided action.
Robert McNamara

We began this book with a quote from General George Patton which said the "only discipline is perfect discipline." As an ideal, this is sound guidance, but in the real world of flight, we all make errors of discipline from time to time, although we might not care to admit it, a subject we will deal with forthrightly in a moment. On occasion, I have been guilty of not fully listening to the ATIS broadcast at my local terminal, and I have heard other pilots confess to shortchanging their flight planning because they arrived late to the airfield. But the real point of this book is that we recognize these behaviors as wrong and strive to eliminate them. Improvement, if not perfection, is well within all of our grasps.

Armed with a knowledge of what flight discipline is and how it impacts our day to day operations, you are well on your way towards the Patton ideal, and far surpass many aviators with whom we share the skies. In short, the knowledge we have gained together about the nature of good and bad flight discipline puts us in a select group of aviators who have the vision to see undisciplined behavior and the willpower to correct it.

By reading this book, you have already taken the first step in developing a personal plan for improving flight discipline. Your knowledge of the phenomenon has a framework upon which to grow. But

337

it is only a framework. You must flesh out your understanding and application of flight discipline based on your type of flying and the organizational context in which you operate. The next step to take on your path to better discipline may be the most difficult. You must come to grips with the realization that that you can improve in this most fundamental and critical area of airmanship.

Denial—Not just a river in Egypt

The first thunderstorm we must circumnavigate on our flight to better discipline is to realize our own faults, to overcome the belief that others—but certainly not ourselves—are guilty of practicing poor flight discipline. We must overcome our denial of the possibility that we suffer from some of the same infirmities as others we read about in the accident briefs and safety reports. After all, if we look back at the statistics in Chapter 2, somebody is out there making those 90,280 self-reported deviations from rules and regulations between 1988 and 1994. Isn't it at least possible, if not probable, that we might be part of the problem?

Denial takes several forms, and because it is one of the first and perhaps the largest obstacles we must overcome, it is worth a bit more discussion. One of the easiest ways for us to deny that we need to improve our flight discipline is to ignore the issue altogether.

Withdrawing from the discussion

Tell me if you have ever heard the following phrase, or something similar. "We can talk until we are blue in the face about all these human factors, but the simple fact of the matter is that you can't teach judgment, so let's talk about something else." Sound familiar? As a human-factors instructor for the past ten years, I can always identify the "hard cases" when I walk into a classroom to deliver a lecture or facilitate a human-factors discussion. They are usually displaying some not-so-subtle body language. You know, arms crossed over their chests, slumping in their chairs, looking down at the floor as they shake their heads at the first mention of the topic. I refer to these wonderful students as "grumpy faced nay-sayers." I usually try to get them in the game early by telling them that the class will really need to learn from their insights and experience. Sometimes this thinly veiled pandering works, but the real point here is that these individuals typify denial by withdrawal.

There may be reasons why they are uncomfortable with the discussion on flight discipline. Perhaps it strikes too close to home. Maybe they themselves have a challenge with personal discipline, or perhaps they know a colleague who does, and even considering taking action in either case is disquieting. Better to ignore it altogether. This is the psychological version of one of physics most basic axioms: an object at rest tends to stay at rest—emotional inertia.

A second method of denial is similar, but puts a slightly different spin on the reason for not addressing the challenge: "I don't have time right now."

Procrastination

The second demon of denial is procrastination, and most of us can relate to this one. While we may realize that flight discipline is a challenge for us to improve our overall airmanship, right now we may be working on something else, and we plan to get around to flight discipline just as soon as...

Most aviators are active people, and as such we tend to be involved in multiple activities, all of which take up time, focus, and energy. Maybe to overcome procrastination as it relates to flight discipline, we need to ask ourselves one simple question: Of all our activities, which one can kill us the quickest? The answer may be the incentive we need to re-prioritize and get to work on better flight discipline practices.

Selective perception

Memory is a funny thing. We are not the objective observers we like to think we are. According to experts at *6th Sense,* an organization that specializes in helping people overcome obstacles to improvement:

> *"Our memory does not function like a video camera recording faithfully and objectively what we see and hear. Our mind has a powerful process which actively filters, alters, or reconstructs those memories of ours which are out of line with our perception of ourselves. In a classic study by Marigold Linton, a psychologist at the University of Utah, she found that she recalled the high points—the good times with her friends, the research successes, the new things she had been able to buy. But on examination of her notes she found that her mind had thoughtfully buried a wealth of details*

with one common denominator. These details either were painful such as the feeling of helplessness when her car broke down and there was no one to help, the fight she had with her lover; or inconsistent with her image of herself as a professional person..." (6th Sense 1996).

These findings suggest that our mind literally rewrites reality to fit in with a certain image we have of ourselves. Intuitively, I know this is true of my own recollections of flying. I fondly recall the best crosswind landings, night air refuelings, and defensive countermeasures to deliver weapons with pinpoint accuracy, but have difficulty remembering the circumstances surrounding a less than perfect performance during an annual inspection, or the day I blew a tire with overaggressive braking following a long landing.

The danger here, of course, it that our selective memory may conveniently block long term recollection of bad habits or poor discipline. A partial answer to this dilemma is to self-critique in the form of a journal, a technique we will discuss later in this chapter.

Rationalization

Some of us feel our acts of poor discipline are warranted because of other pressures. Our typical response to this type of behavior is "It's OK because..." This is a tough one, because in some cases, it might well be "OK because..." depending on the criticality of the because. However, in most cases it is rationalization for what you want to do versus what you should do. Let me provide a personal example. When I was a B-1B instructor pilot at our "schoolhouse" in Abilene, Texas (Dyess AFB), we had tight training schedules that required students to graduate on schedule or lots of bad things happened—mostly in the form of additional paperwork. On one particular mission, I needed to get some automatic terrain following (ATF) training done to complete one of my better students, but a combination of factors placed the aircraft just above the maximum gross weight restrictions for the ATF maneuvers. The weight restriction guarantees that the aircraft can provide the required "G" during an automatic fail-safe fly-up to save us if for some reason the automatic system failed or we did not see a mountain we were approaching. The fly-up feature is primarily designed for operations at night or in the weather, but the flight manual does not make any allowances for these factors. As I saw it, I had three choices. I could (1) "reduce our gross weight" by dumping jet fuel over the barren lands of West Texas, (2) fly the ATF out of limits (but just barely), being prepared to disconnect the system if we got a fly-up and hand fly the aircraft out of trouble, or (3) reschedule an additional mission for my student, who would then go past his projected graduation date and cause me paperwork grief. Which option do you suppose I selected? What would you have done?

I flew the ATF out of gross weight limits and accomplished the training, rationalizing that I had, (1) saved the pristine ecology of the western Texas landscape (which ironically sits on a huge oil well), (2) saved the taxpayers thousands of dollars by not requiring another mission to be generated to complete my student, and (3) maintained complete safety by relying on my outstanding judgment and superior skills to intervene should something have gone awry with the technology. But what had I really done?

My student made it clear to me in the debrief. Even though I had spelled out why I was choosing to "bend the rules" on this—and only this—occasion, I had left a confused student, who was prompted to ask, "What other stupid rules do we have in this program?" Through my example I had taught this young man that I operated on a sliding scale of discipline, and I was ashamed. As I reflect now on that decision, and others like it I have made in the name of training, the mission, or whatever, I find myself much more sympathetic to the crew of IFO 21, the CT-43 crew in the case study in Chapter 1. I'm sure that they too rationalized their way into bad decisions, thinking that it was worth the risk. Perhaps only the situation separates the winners from the losers—or the living from the dead—among those of us who rationalize our way into poor flight discipline.

Egotism

The final method of denial that affects our ability to perceive poor discipline as a reality in our flight practices is what I will call "egotism" or "idealization." It is closely related to selective perception in that it is the tendency for us to believe we are already operating at our maximum potential, and that there is no more room for improvement. Idealization is closely related to egotism, in that we falsely assume that we are better than we really are. Often, pilots use successful outcomes to convince themselves, and sometimes others, that they are airmanship personified. It usually goes something like this. "I just landed in a 20 knot crosswind after shooting a back course localizer through moderate turbulence. How could I possibly have a discipline problem? I exercised perfect discipline on that approach."

The logic here is obviously flawed. In the same way that one good approach does not make an excellent instrument pilot, one example of good discipline does not make a disciplined pilot. In fact, it can have just the opposite effect, by inducing complacency through overconfidence.

A final word on denial

The bottom line here is that we should all approach our personal flight discipline as if it can be improved. If we do not feel that we need to work on this area, take a close look to see if any of the natural defense mechanisms are at play. Withdrawing, procrastination, selective perception, rationalizing, or egotism are all dangerous conditions for an aviator, because once the denial "switch" is activated, the people denying are often unaware and quite possibly incapable of realizing they are doing it. How can we ask ourselves to correct something we don't know we are doing? Impossible.

The second realization I want us to make is that many top achievers in other career fields have a different view of disciplined professionalism than most pilots I know (including myself, although I am working on it). Sustained top-drawer performance requires sustained effort, a point I have come to realize more clearly recently.

Professionalism

For over a decade, I have searched the archives and literature in an attempt to come to grips with what it means to be a professional, and then to articulate clearly to others that the flying game is indeed a profession in the truest sense of the word. I have heard the well-meaning (or maybe not) put-downs from my engineering and medical doctor friends. Names like "throttle jockey" and "glorified bus driver" get tossed around in casual social settings. I've heard my airlifter colleagues called "trash haulers," and have been reminded of the quote from the movie *Top Gun*, where the skipper warned Maverick that if "he pulled another stunt like that, he'd be hauling rubber dog... [manure] out'a Hong Kong." Even my commercial pilot friends do not escape the wrath of the mud slingers. "How was the cattle drive today?" "Did you fill up the boxcar?" Frankly, I became sick of it and set out to prove the critics of my profession wrong.

"But is there anything to the criticism?" I asked myself.

In my search for an answer I thought I already knew—namely that pilots were in fact highly trained professionals in every sense of the word—I have found evidence that has given me cause to reconsider. Perhaps the best explanation of my emerging conviction that we can be much more than we are today is captured in this letter I received from a colleague in Australia (hence the difference in the spelling of some words). Read it thoughtfully.

"When asked, how to achieve excellence in flying? 'Be a professional' was my response, squaring my shoulders. This was back when I belatedly got serious about it. I mean, before that, every parrot at the base pet shop had been squawking professionalism slogans for a decade or more. Now, I was going to do it. But what is it? 'Check with a real professional' mutters the parrot, with snide contempt.

Thelma, my first choice, isn't home, business assignment no doubt, so I go around to see my buddy the heart surgeon. 'Follow me for a week,' he says. 'I'll show you professionalism in action, and here are some tranquilizers for the operating theatre sessions.'

He operates three days a week, sometimes for a 12-hour day, on four or five patients a day. I reflect on his skills. On the really tough ones he was engaged for 90 percent of the operation. He was doing extraordinary things. Apart from incisions, any error which would mean the end for the patient, he had to make these incredibly tiny stitches, tying them off using only the fingers of one hand. And then there was, throughout, his quiet control of the entire surgical team and all of its complex logistics.

I came to in the recovery room. The tranquilizers can't have been the right strength.

Hmmm. Food for thought, though, that awesome display of high-level skills. For me, on a normal flight, the maximum skill demand for take-off approach/circuit, and landing would occupy about 1 percent of the total flight time (instruction adds a bit more). To be frank, for the rest of the time, I'm a couch potato.

And as for the really tough skills, the life-saving stuff, well, technological reliability being what it is nowadays, I can be pretty sure I'll never have to use them (except in the occasional simulator ride, but that's different).

I'm sorry I asked this bloke for advice now. He turns up at my home at 7.00 AM, next day, 'C'mon,' he says brightly, 'we have to do rounds.' That's where he goes around the three hospitals he works in, and checks the fifty or so patients he's operated on in recent weeks, sees how they're doing, seven days a week. I go along.

It's a long day, no operations, but patient after patient seen in the consulting rooms. Going on 6.00 PM the magnetic attraction of Rosie's gin palace is palpable. The doctor unexpectedly announces, 'Time for evening rounds.' 'Sure,' I enthuse, 'Love to. I'm coping well with this influenza bug I picked up in Cairo. I can't believe that some experts think it could be fatal for patients in hospital.'

I thought that had shaken loose this ferocious overachiever, but he springs me in Rosie's, at 7.30. 'No thanks," to my offer of a drink. 'We have to get to a continuing professional education (CPE) seminar.' This stuff is getting tedious, but the gin is doing a better job than his drug had. The surgeon explains along the way that his license is dependent on his demonstrating continuing familiarity with all emerging knowledge. CPE seminars occur weekly. They take about two hours each. He is required to attend two a month, 24 a year.

There's a point score, like, if you do four in one month, you can take the next month off. But you can't count holidays as downtime . You have to have enough points in the bank before you take a break. Drop too far behind, hand in your license.

Refreshed by an hour or so of light sleep, I suggest a brief stopover at Rosie's on the way home. 'No, no,' he can't. Today's the day the Medical Journal arrives in his post box. He's got to read it. It would be heresy not to be fully on top of the leading articles by coffee break the next day. All the surgeons subtly test each other out. 'Gotta know your stuff, keep up, it all moves so fast. Tell you what, though,' he says, 'I'm going out for a light jog before rounds in the morning, an hour should be enough. Care to join me?'

'The Cairo flu, Doctor, my little children...OK?'

Enough, you get the picture.

Doctors, lawyers, architects, engineers, every form of professional you can think of, submit to the most rigorous regime of, in the first place, qualification, and then maintenance of their high-level professional skills (and knowledge). The common factors that produce this imperative are public safety and vicarious responsibility for others' lives and/or livelihoods. Pilots are clearly in the same category. However, in my experience, while we replicate the tough entry training/qualification criteria, we don't mirror the rigor of ongoing professional training and CPE.

Why not is pretty plain to see. Even for military pilots, there aren't enough training hours to get anything like sufficient stick time actually practicing the full gamut of physical skills, to compare with, say, the lead violinist in the orchestra. And for pilots like me who pay for their flying, well, forget it.

Simulators? Nope, not enough of them either (and none for me and my lot).

Give up? Well, no. There are in fact a host of opportunities for professional training for pilots; they just don't happen to look like flying. The manipulative and eye–hand co-ordination skills used in many sports have direct cross-over with the physical skills of flying. It s not 100 percent, but it's...far better than nothing. Tennis, squash and golf are examples. So are juggling, and any activity requiring balance, like skiing, riding a unicycle, or 'balance board.'

Flight simulators on PCs are also valuable, provided they are flown with a stick or mouse. Whacking keys on the keyboard is (not yet) equivalent to flying.

Cognitive skills can also be worked up for flying without actually doing it. In a general sense, you can keep your mind in good shape by exercising it, with complex games like bridge or chess, or learn a foreign language. In addition, specific flying exercises and checklist responses, using visualization (chair flying) are clearly beneficial, if done seriously, and with plenty of repetition.

Accident analysis (and real-life experience as a pilot) confirms that a serious emergency, say, an engine failure, will bring with it a certain measure of emotional stress. It's important that pilots can withstand the potentially debilitating effects of gross apprehension. Well, how about parasailing or parachuting? This is how you can assure yourself that you will be able to follow the checklist despite the pressure of the event. In other words, if you want to be sure you can function when scared, you have to practice functioning while scared.

Indeed, that's the message in summary. Skills are only honed by practice...This applies to all skills, inter-personal, cognitive, and physical. And you don't have to be in the plane or simulator.

And how? Well, there's enough here for you to get the general principles. Once you've accepted them, there's no limit to the innovative training routines you can devise for yourself and fellow aircrew (all crew, not just the pilots). Like, there are lots of activities you could imagine that will enable you to exercise your crew cooperation habits. I'll bet there is a lot of time in most pilots' lives when they could do some more towards maintaining their critical skills.

To go back to my mate, the doctor (it's a true story, by the way), folk like him offer us benchmarks. If we aviators want to be regarded as professionals, then we need to put just as much effort into achieving standards of excellence in practice as do other professionals. Have a close look at a top performer yourself: doctor, lawyer, whatever. Then ask if you are putting that amount of real work and training into honing your skill base.

I hope this provokes some useful thought.

Regards,

Doug"

My friend's letter certainly gives us something to think about, doesn't it? While I am quite certain that not all doctors perform at the fever pitch of Doug's acquaintance the heart surgeon (or how else would they get so good at tennis?), the built-in requirements for continuation training coupled with the intrinsic drive for current knowledge may well be far greater in other career fields than in our own. While this may be a slight condemnation of aviation as a whole, it does provide a competitive advantage for individuals who do take the time and make the effort to strive for standards of skill and knowledge beyond what the profession currently demands. But before we can even begin this process, we must first come to the realization that we might have a problem ourselves in the flight discipline arena.

Becoming a more disciplined flyer

The next step in solidifying flight discipline is the same one we all took when we became good stick and rudder pilots—we realized that we were in control. I will never forget that magical moment in my early training when I came to realize that I was no longer riding a bucking bronco, holding on for dear life, but rather was the controller of a very predictable machine. Sometimes even today, nearly 20 years later, when I'm not flying as well as I should, I will make a few firm positive control inputs to the aircraft, just to remind myself who is flying whom.

In the very same manner, we can make a few positive control inputs to improve our flight discipline—or any behavior for that matter. The key is to realize that you are in control of your own destiny. Current self-improvement gurus Tony Robbins and Stephen Covey have made millions pointing out this simple fact to the multitudes who flock to their seminars and buy their books, audio, and video programs in search of some "magic bullet" that will help them to change their lives for the better. These programs, regardless of their origin, usually contain the same message—the power you seek is already inside you. You are the master of your own destiny—or to use a more appropriate metaphor—you are the pilot in command of your life. Although there are many names and special methods for accomplishing self-improvement, and most of the highly successful gurus put their own special spin on it, in the final analysis it is all pretty much the same item—behavioral modification. The steps are relatively simple, and well grounded in the psychology of Carl Rogers and Abraham Maslow in the 1950s. Here is my spin on the fundamentals of self-improvement in aviation.

Modifying human behavior as it relates to flying an aircraft, is a great deal like flying the aircraft itself. First we learn what flight is all about and we obtain the necessary credentials—our license. The next few steps you know by heart. We establish a flight plan, start up the aircraft, fly it in the direction we want to go, carefully monitor our course, correct for unexpected winds, turbulence, or whatever, and land at a new destination. Voila! We are someplace new. Most of you will realize that this is an oversimplification of what it takes to fly an aircraft, but it serves as a useful metaphor. Other steps, such as fueling the plane (motivation to improve) and coordinating with air traffic control (restructuring our priorities) are also involved in the process.

This entire process is best accomplished through keeping a companion journal to your pilot log book. I will refer to the journaling process as we move through the checklist and explain the specific steps involved. But let's keep it simple, and let's get started.

Step 1: Licensing yourself for change

The first step in the checklist is to obtain your license to change. The right to this license was granted to you at birth, and by making your way to the final chapter of this book you are well on your way to establishing the necessary credentials. No aviator worth his salt would begin a flight unless he knew where he was going. This approach is grounded in common sense. My mother used to tell me, when I first got my driver's license, not to "go cruising in the family car with nowhere to go." She pointed out that young men got into trouble that way. Of course I didn't listen, she was right, and I learned my lesson, but in a far more forgiving environment that the one we are discussing today.

Get a firm hold on what airmanship goals you wish to pursue by reading, through hangar talk with other pilots, or through self-assessment, reflection and analysis. You then have your license for change and are ready to file a flight plan.

Step 2: Flight planning for better discipline

A flight plan is nothing more than an organized document that explains how you plan to get from where you are to where you want to go. In the world of behavioral modification psychology, this step is referred to as the goal-setting stage. As in flight planning, be as specific and precise as you can. As you are well aware, a shoddy flight plan often is the first step to a sloppy flight.

Sit down by yourself or with a colleague or instructor, if you are brave enough to bare your flight discipline concerns to another, and specify exactly where you want to go. Use the airmanship model in Chapter 1 as a guide to areas you wish to improve.

Once the destination is determined, you must insure you are equipped to make the flight. What has to be done to reach your goal? What activities will you need to accomplish. These may include further study, more skill development, or proficiency in areas or maneuvers previously ignored or taken for granted, or an outside evaluation with a professional instructor to help you isolate and analyze your areas of need. Decide what your time en route to your new destination will be. That is to say, how many hours per week do you plan to study, practice, or whatever? Be specific; write these numbers down. Additionally, determine what resources you will need, such as updated regulations, self-study guides, flying hours, etc. Make a written list and begin the process of procuring them.

A cautionary note. This is the point at which denial most often sets in. If we fail to be brutally honest with ourselves, it is like beginning a flight from a false location in your GPS. Your flight plan will be of little use, and you may well run out of gas. Since this step is so critical, the third step in the checklist addresses it directly.

Step 3: Recording a baseline

Once you have identified what time and resources you estimate it will take to reach your destination, begin to record what you have available and are doing now in terms of those same resources and time expenditures. This is a very necessary step, because life, like flying, is a resource-constrained environment. We have to maintain situational awareness and control of our daily activities just like we manage the flight path of an aircraft. To do this, I recommend keeping a detailed accounting of your time, actions, and activities for a minimum of two weeks, and most professional programs recommend at least 21 days. It is important that this record is an accurate reflection of your current efforts, before you begin your actual increased study or other improvement activities. This may seem like wasted time, but trust me, it is not. Accurate self-assessment is exceedingly difficult for most people, especially pilots, and even if you never make it past step three, you will have taken a giant step towards self-improvement, because, as Socrates said so insightfully over two millennia ago, the origin of all wisdom is to "Know thyself."

Step 4: Double-check your flight plan

Now that you know where you are, where you are going, and the resources required are on hand, you can make an accurate estimate of the reality of reaching your destination. Do you have the time, energy, and resources to reach the destination you established for yourself in Step 1 in a reasonable time frame? Will you be able to keep your life in balance while making the effort and achieving the goals? If the answer to either of these questions is "no," then reevaluate your destination and make it more realistic—less demanding. Short flights can be as rewarding as long ones, and if you are like me, you may need near-term reinforcements to keep the drive alive. So adjust your goals accordingly.

Since we are half-way through the checklist, let's stop for a mid-mission review, utilizing a hypothetical example.

"On January 1st, I decided to make a New Year's resolution to improve my airmanship, as other priorities had left me with a fuzzy feeling that my skills and knowledge had been slipping. Not knowing exactly where to begin, I pulled out my flight manual and the a copy of last year's AIM and started reading, but I felt this approach was a bit disjointed. I went down to the library and picked up a book on airmanship and talked to a couple of my flying buddies about 'getting back on track.'"

(Step 1: This pilot has licensed herself for change)

"January 7th: From the readings and discussions, I realized that expert flyers are disciplined as well as skilled. I think I'm OK there, but I didn't realize that there is so much to know. I've decided to master the multiple bases of information on weather, the FARs and local regs, my aircraft, and CRM. In addition, I'm going to work these into my training flights to reinforce the book knowledge."

(Step 2: She has set a destination.)

"January 25th: I have kept a record of my study habits and resources available for two and a half weeks now, and I realize that with aerobics and soccer practices/games, I only have about two hours per week to dedicate to this effort without driving myself crazy. I also realized from my last flight that I'm not using my checklist as much as I should.

Maybe my initial assessment and goals were not completely realistic."

(Step 3: She has established a baseline and knows her real capabilities much better.)

"January 27th: I have decided to focus my improvement on mastering the vocabulary in the AIM, reviewing the FARs that apply to IFR operations, and perfecting my checklist discipline. I will strive not to miss a checklist item, by actually referring to my checklist to insure completion. My new goal is to pass the commercial instrument exam this calendar year!"

(Step 4: Revising to meet the demands on your time and resources.)

Step 5: Establishing waypoints

Now that a realistic and obtainable set of goals have been established, we need to set up some milestones, or waypoints along the flight path to our destination. These need not be too detailed and might just be to track your actual contact time with the study material. For example, monthly waypoints for our example above might look something like this.

"February: Eight hours of study time; A through E of the AIM vocabulary list; minimum of two flights, one with some instrument instruction.

March: Eight hours of study time; F through J of the AIM vocabulary list; review FARs relating to TCA procedures on IFR flight plan; two flights."

The purpose of these waypoints is the same as they would be on a normal flight plan, to track our progress and react to any unexpected consequences that Murphy may throw at us. Waypoints can be changed as required, as long as we see how they fit the flight plan towards our ultimate destination. For example:

"April: Last month a complete washout because of Tommy's illness. Repeat March goals and strive for a bit more. Probably need a review flight for basic proficiency."

Here is where we depart from the purity of the flight plan metaphor, because it is not necessary to completely "re-file your plan" if circumstances out of your control throw you off track. If the aviator in our example has a good April, she may well be back on track by May. If not, she may just have to adjust subsequent monthly waypoints, or possibly, revise her end-of-year destination or expected completion date.

The point is simply that structured improvement plans will outperform unstructured ones every time. The amount of pressure you want to place on yourself to "stay on course" is a personal decision, but keep in mind that too much pressure may cause you to abandon the self-improvement plan all together. Any continuous improvement is beneficial, regardless of the pace.

Step 6: Record and reward your progress

In addition to restructuring your future waypoints, it is critical to check your gauges in flight. Just like you record your fuel readings on whatever type flight plan you use in your aircraft, it is important to track your accomplishments towards your self-improvement destination, and to reward yourself when they are met.

> *"April 15th: Paid my taxes, memorized AIM vocabulary A–J and got in two flights already this month...Already put in 9 hours in the books—I'm taking a weekend off to go with Bob and Tommy to the lake!"*

Sometimes you might want to make the reward specified before you reach the waypoint, as an incentive. Maybe a night out, or a purchase of something "impractical" but that you have always wanted (That's how I justified my first set of aviator Ray-Ban sunglasses). Or, if you are more motivated by negative consequences, you might consider an appropriate "punishment" for failing to achieve your goal. Perhaps an hour of listening to music you hate or forcing yourself to watch a rerun of *Heidi*. That would sure keep me going.

In the words of the Nissan ad man, "Life is a journey; enjoy the ride." Make your improvement process fun, not just rewarding. I think one of the unfortunate circumstances of our time is that we all tend to take ourselves too seriously. If the drive for improvement starts to get a bit heavy, lighten up, take a walk, go fishing—you may be surprised how much it helps in all areas of life.

Step 7: Arrive intact, enjoy your success, and set a new goal

Upon reaching your destination, stop and reflect on your success, kind of a postflight debrief. Take a close look at the costs and benefits of your journey. You may well find, as I have, that the process itself is the biggest benefit of all. I have been able to transfer this little seven-step checklist to many other areas of my life with great success. I have also missed a few destinations, for example, I've yet to catch a ten-pound bass, or harvest a bull elk but I'm working new waypoints for those.

You may also find that the process took a bit too much out of you, so your next plan should account for these factors. Most of all, enjoy the accomplishment, bask for a while in the moment of success, remember what it feels like, and set a new goal. Let's look at the final entry of our hypothetical aviator on her journey towards a commercial instrument rating.

"I did it! Mastered the AIM vocabulary, FARs, and passed the quiz! Even better, my improved knowledge and checklist discipline has improved my flying skills dramatically. The CFII I flew with yesterday congratulated me and it felt great. During the debrief, I asked him how he liked his job and how he went about getting into this profession. So far, flying for me has been just a hobby, but my job at the computer firm is getting a bit stale, maybe..."

There are three keys to this entire self-improvement journey. The first is to get past the denial hurdle, realizing that you could be at risk of practicing poor flight discipline. The second key is to make a commitment for action and set upon a plan for self-improvement. The final step is to implement and execute the plan, staying realistic and flexible. The critical item here is accurate record keeping and keeping your journal or notes organized. Self-management comes from action; take the first step today.

In addition to the self-improvement process, there are several other final reminders I would like to reinforce. Although they have all been covered in other places in this book, they are so critical to flight discipline safety that they deserve one more mention.

A final perspective: Some practical reminders

Better flight discipline through personal commitment and growth benefits everyone in the aviation industry, including:

- Designers, who will begin to have more faith in the human element of the human-machine equation.

- Manufacturers, who will be able to spend more time and money on improving the aircraft flight characteristics and less on providing automated backups to eliminate human error.

- Educators and trainers, who can spend their time more productively with well-prepared and mature students.

- Regulators, who can look more to the efficiency of the system than to enforcement of discipline issues.

- Air traffic controllers, who can provide their services better in an environment where pilots know, understand, and follow the rules.

- Pilots, who will learn to have more trust in themselves and each other as they overcome obstacles to reach their peak performance.

Even if you aren't up for the whole self-improvement plan of action, there are a few simple, immediate steps you can take to improve now. If you can just work on the following four items, your flight discipline should improve dramatically to the benefit of all those listed above.

Understand and use your checklist

If we could get all aviators to take this simple step, the accident rate would drop significantly. Although most aviators quickly reach a point in their aircraft where they have much of the checklist memorized and can recite it by heart, the problem with this approach is that the memory does not work the same way in an emergency as it does under normal stress levels. This point is easy to prove to yourself by asking one simple question? How many of the pilots that have landed gear up did not know that lowering the landing gear was part of the "Before Landing" checklist? Of course they all knew it, but complacency ("I have that checklist memorized.") and stress—or some change in their normal habit pattern—short-circuited their normal memory operation. If you always use the checklist, you will not be at the mercy of a fallible memory. The parts to commit to memory are the amplified procedures in your flight manual that explain the whys and hows of each checklist step.

Know your resources and be able to tap into them under stress

The essence—in fact the definition—of good cockpit/crew resource management is to be able to effectively utilize all available resources. Although this was not a book on CRM, a short review of the chapter titles will illustrate how closely flight discipline and CRM are tied. The stress of a situation can quickly overwhelm a single human pilot, and often with little or no warning. We must train ourselves to know where to find all of the help at our disposal on a moment's notice, and then how to integrate them into a plan of action. Many excellent books have been written on CRM for all flight environments, and I strongly recommend further study and work in this area.

Overlearn emergency procedures, regulations, and systems

Closely related to the last two suggestions for improving flight discipline, is the need to "overlearn" some basic information and procedures. Starting with your aircraft's emergency procedures, identify your own "top ten" EPs, regulations, and critical aircraft systems data. Learn these backwards and forwards. Be able to recall and recite them at any time under any conditions. Here's one technique that I have used effectively to overlearn critical flight information. Keep some flash cards handy and challenge yourself by telling a family member or friend that you defy them to stump you at any time on any of the information contained on the flash cards. Make a friendly bet. It's far better to lose a Coke to a friend than your life to a memory lapse under emergency conditions in flight.

Don't bite off more than you can chew

We have to know our limitations. Although we try to tell our family, friends, and even ourselves differently, flying is a dangerous endeavor. Keep as wide a margin of safety around yourself as possible. One simple guide is, if you question your ability to accomplish something in flight, don't do it unless one of two conditions exist: (1) you are with a qualified instructor, (2) you have no other choice. Never, repeat never, attempt something you are not trained and certified to do.

Chapter review questions

1. How does the author modify his earlier stand on perfect flight discipline quoted from General Patton in Chapter 1? Why does he say this approach is necessary?

2. What are the various types of denial which prevent the development of flight discipline? Do you exhibit any of these tendencies?

3. Reread the letter about the heart surgeon. Are aviators professionals? How much study and continuous improvement is necessary in aviation? How much do you do now?

4. Who—besides yourself—benefits from improved flight discipline? Do you have any obligation to these groups?

5. What four steps can you take immediately to improve your flight discipline?

6. What steps are you prepared to take?

References

Baron, R. A., and Byrne, D. 1994. *Social Psychology: Understanding Human Interaction* (7th ed.). Allyn and Bacon. Boston

6th Sense, World Wide Web Site, 1996. "How Our Mind Rewrites History." Internet. http://www.pracpsy.com/rewrite.htm.

Edwards, Doug. Personal correspondence. 1997. 9 July.

Appendix A

Automation-related aircraft accidents and incidents

This appendix contains brief descriptions of automation-related air-. craft mishaps. It is taken from Billings, C. E.: *Human Centered Aviation Automation: Principles and Guidelines*, Ames Research Center, Moffett Field, CA: NASA Technical Memorandum 111830, 1996. It is reprinted by permission of the author.

The occurrences are listed chronologically; each summary is followed by a reference.

6/30/1956: TWA L1049A and United Air Lines DC-7, Grand Canyon, AZ

At approximately 1031 hrs PST, a TWA L-1049A and a United Air Lines DC-7 collided at about 21,000 ft over Grand Canyon, AZ. Both aircraft fell into the Canyon; there were no survivors among the 128 persons aboard the two flights. There were no witnesses to the disaster.

The Civil Aeronautics Board determined that the flights were properly dispatched. In flight, the TWA crew requested 21000 ft, or 1000 ft on top (above cloud tops). 21,000 ft was denied by ATC because of UAL 718. TWA then climbed to and flew at 21,000 ft above clouds. The last position report from each aircraft indicated that both were at 21,000 ft, estimating their next fix at 1031. The aircraft were in uncontrolled airspace and were not receiving traffic control services at the time of the collision.

The Board determined that the probable cause of the collision was that the pilots did not see each other in time to avoid the collision. The Board could not determine why the pilots did not see each other but suggested the following factors: intervening clouds, visual limitations due to cockpit visibility, preoccupation with matters unrelated to cockpit duties such as attempting to provide the passengers with a more scenic view of the Grand Canyon, physiological limits to human vision, or insufficiency of en route air traffic advisory information due to inadequacy of facilities and lack of personnel. (CAB, 1957)

2/3/1959 Pan American World Airways B-707 over the Atlantic Ocean

Pan American flight 115 was en route from London, England to New York when it entered an uncontrolled descent of approximately 29,000 feet. Following recovery from the maneuver, the airplane was flown to Gander, Newfoundland, where a safe landing was made. A few of the 129 persons on board suffered minor injuries; the aircraft incurred extensive structural damage.

The aircraft was at 35,000 ft in smooth air with the autopilot engaged when the captain left the cockpit and entered the main cabin. During his absence the autopilot disengaged and the aircraft smoothly and slowly entered a steep descending spiral. The copilot was not properly monitoring the aircraft instruments and was unaware of the airplane's attitude until considerable speed had been gained and altitude lost. During the rapid descent the copilot was unable to affect recovery. When the captain became aware of the unusual attitude he returned to the cockpit with considerable difficulty. With the aid of the other crew members, he was finally able to regain control of the aircraft at an altitude of about 6000 feet.

The Civil Aeronautics Board determined that the accident resulted from the inattention of the copilot to the flight instruments during the captain's absence from the cockpit, and the involuntary disengagement of the autopilot. Contributing factors were the autopilot disengage warning light in the dim position and the Mach trim switch in the "off" position. During analysis, which was hindered by the flight data recorder having exhausted its supply of metal recording foil, it was indicated that the airplane had reached Mach 0.95 in its abrupt descent. Very high G forces were indicated by the recorder and had been reported by the pilots during their attempts to recover from the spiral dive. After landing at Gander, the lower surface skin of the horizontal stabilizers was found to be buckled; both wing panels and both outboard ailerons were damaged; the wing-to-fuselage fairings were damaged and a three-foot section of the right fairing had separated in flight. Both wing panels suffered a small amount of permanent set. All four wing-to-strut fairing sections of the engine nacelle struts were buckled and other damage was also evident. (CAB, 1959)

6/18/1972: British European Airways Trident, Heathrow Airport, London, England

This aircraft commenced its operation under the command of a very senior BEA captain. The first officer was relatively inexperienced and the second officer was a recent graduate of the airline's training school. The airline was undergoing a difficult labor-management conflict, and the captain had been involved in a heated altercation in the crew room before departure.

Shortly after takeoff, when the first reduction of flaps occurred, it is thought that the first officer inadvertently actuated the wing leading edge slat handle as well, raising the slats at a speed too low to sustain flight. Based on post-mortem evidence, it is believed that the captain had a severe cardiac event at about the same time. Many warning lights and aural signals were actuated by the premature retraction of the slats. The inexperienced first officer was unable to diagnose the problem or to regain control of the airplane, which crashed into a reservoir just west of the airport. There were no survivors. (Department of Trade and Industry, 1973)

12/29/1972: Eastern Air Lines L-1011, Miami, FL

The airplane crashed in the Everglades at night after an undetected autopilot disconnect. The airplane was flying at 2000 ft after declaring and executing a missed approach at Miami because of a suspected landing gear malfunction. Three flight crewmembers and a jumpseat occupant became immersed in diagnosing the malfunction. The accident caused 99 fatalities among the 176 persons on board.

The NTSB believed that the airplane was being flown on manual throttle with the autopilot in control wheel steering mode, and that the altitude hold function was disengaged by light force on the yoke. The crew did not hear the altitude alert departing 2000 ft and did not monitor the flight instruments until the final seconds before impact.

The Board found the probable cause to be the crew's failure to monitor the flight instruments for the final 4 minutes of the flight and to detect an unexpected descent soon enough to prevent impact with the ground. The Captain failed to assure that a pilot was monitoring the progress of the aircraft at all times. The Board discussed overreliance on automatic equipment in its report and pointed out the need for procedures to offset the effect of distractions such as the malfunction during this flight (p. 21). (NTSB, 1974a)

7/31/1973: Delta Air Lines DC9-31, Boston, MA

This airplane struck a seawall bounding Boston's Logan Airport during an approach for landing after a flight from Burlington, VT to Boston, killing all 89 persons on board. The point of impact was 165 ft right of the runway 4R centerline and 3000 ft short of the displaced runway threshold. The weather was sky obscured, 400 ft ceiling, visibility 1-1/2 miles in fog.

The CVR showed that 25 sec before impact, a crewmember had stated, "You better go to raw data; I don't trust that thing." The next airplane on the approach, 4 minutes later, made a missed approach due to visibility below minimums. The accident airplane had been converted from a Northeast Airlines to a Delta Air Lines configuration in April, 1973, at which time the Collins flight director had been replaced with a Sperry device; there had been numerous write-ups for mechanical deficiencies since that time. The flight director command bars were different (see Fig. 11, page 20 for the two presentations), as were the rotary switches controlling the flight director. The crew were former Northeast Airlines pilots. If the crew had been operating in the go-around mode, which required only a slight extra motion of the replacement rotary switch, the crew would have received steering and wing-leveling guidance only, instead of ILS guidance. Required altitude callouts were not made during the approach.

The NTSB found the probable cause to be the failure of the crew to monitor altitude and their passage through decision height during an unstabilized approach in rapidly changing meteorological conditions. The unstabilized approach was due to passage of the outer marker above the glide slope, fast, in part due to nonstandard ATC procedures. This was compounded by the flight crew's preoccupation with questionable information presented by the flight director system.

The Board commented that, "An accumulation of discrepancies, none critical (in themselves), can rapidly deteriorate, without positive flight management, into a high-risk situation...the first officer, who was flying, was preoccupied with the information presented by his flight director system, to the detriment of his attention to altitude, heading and airspeed control..." (NTSB, 1974b)

4/12/77: Delta Air Lines L-1011, Los Angeles, CA

This airplane landed safely at Los Angeles after its left elevator jammed in the full-up position shortly after takeoff from San Diego. The flight crew found themselves unable to control the airplane by any normal or standard procedural means. They were able, after considerable difficulty, to restore a limited degree of pitch and roll control by using differential power on the three engines. Using power from the tail-mounted center engine to adjust pitch and wing engines differentially to maintain directional control, and verifying airplane performance at each successive configuration change during an emergency approach to Los Angeles, the crew succeeded in landing the airplane safely and without damage to the aircraft or injury to its occupants. (McMahon, 1978)

12/18/1977: United Airlines DC-8, near Kaysville, UT

A cargo aircraft encountered electrical problems during its approach to the Salt Lake City Airport. The flight requested and accepted a holding clearance from the approach controller. The flight then requested and received clearance to leave the approach control frequency in order to communicate with Company maintenance (one of the two communications radios had failed due to the electrical problem). Flight 2860 was absent from the approach control frequency for over 7 minutes, during which time the flight entered an area near hazardous terrain. The approach controller recognized the crew's predicament but was unable to contact the flight.

When the crew returned to his frequency, the controller told the flight that it was too close to terrain on its right and to make an immediate left turn. After the controller repeated the instructions, the flight began a left turn. About 15 seconds later, the controller told the flight to climb immediately to 8000 ft. Eleven seconds later, the flight reported that it was climbing from 6000 to 8000 ft. The airplane crashed into a 7665 ft mountain near the 7200 ft level.

The NTSB determined that the probable cause of the accident was the approach controller's issuance and the flight crew's acceptance of an incomplete and ambiguous holding clearance, in combination with the flight crew's failure to adhere to prescribed impairment-of-communications procedures and prescribed holding procedures. The controller's and flight crew's actions were attributed to probable habits of imprecise communication and of imprecise adherence to procedures, developed through years of exposure to operations in a radar environment. A contributing factor was failure of the airplane's no. 1 electrical system for unknown reasons. The Board noted that the GPWS would not have provided a warning until 7.7 to 10.2 sec before impact, which was too late because of the rapidly rising terrain. (NTSB, 1978a)

5/8/1978: National Airlines B727-235, Escambia Bay, Pensacola, FL

Flight 193 crashed into Escambia Bay about 3 miles short of the runway while executing a surveillance radar approach to Pensacola Airport runway 25 at night in limited visibility. The aircraft came to rest in about 12 ft of water. Of 58 persons on board, 3 passengers drowned.

The NTSB determined that the probable cause of the accident was the flight crew's unprofessionally conducted nonprecision instrument approach, in that the captain and crew failed to monitor the descent rate and altitude and the first officer failed to provide the captain with required altitude and approach performance callouts. The crew failed to check and utilize all instruments available for altitude awareness, turned off the ground proximity warning system, and failed to configure the aircraft properly and in a timely manner for the approach. Contributing to the accident were the radar controller's failure to provide advance notice of the start-descent point, which accelerated the pace of the crew's cockpit activities after the passage of the final approach fix.

The Board noted that the approach was rushed, that final flaps were never extended and that the captain was unable to establish a stable descent rate after descending below 1300 ft. The captain either misread or did not read his altimeters during the latter stages of the approach; the first officer did not make any of the required altitude callouts. The flight engineer's inhibition of the GPWS coincided with the captain's raising the nose and decreasing the descent rate. The pilots were misled into believing the problem was solved. (NTSB, 1978b)

12/28/1978: United Airlines DC-8-61, Portland, OR

This airplane crashed into a wooded area during an approach to Portland International Airport. The airplane had delayed southeast of the airport for about an hour while the flight crew coped with a landing gear malfunction and prepared its passengers for a possible emergency landing. After failure of all four engines due to fuel exhaustion, the airplane crashed about 6 miles southeast of the airport, with a loss of 10 persons and injuries to 23.

The NTSB found the probable cause to be the failure of the Captain to monitor the fuel state and to respond properly to a low fuel state and to crewmember advisories regarding the fuel state. His inattention resulted from preoccupation with the landing gear malfunction and preparations for the possible emergency landing. Contributing to the accident was the failure of the other two crew members to fully comprehend the criticality of the fuel state or to successfully communicate their concern to the Captain. The Board discussed crew coordination, management, and teamwork in its report. (NTSB, 1979a)

3/10/1979: Swift Aire Aerospatiale Nord 262, Marina Del Rey, CA

This commuter aircraft was taking off at dusk from Los Angeles en route to Santa Maria, CA, when a crewmember transmitted "Emergency, going down" on tower frequency. Witnesses stated that the right propeller was slowing as the airplane passed the far end of the runway; popping sounds were heard as it passed the shoreline. The airplane turned north parallel to the shoreline, descended, ditched smoothly in shallow water, and sank immediately. The cockpit partially separated from the fuselage at impact. The accident was fatal to the two crewmembers and one passenger.

The flaps were set at 35 degrees, the right propeller was fully feathered and the left propeller was in flight fine position. It was found that the right propeller pitot pressure line had failed; the line was deteriorated and would have been susceptible to spontaneous rupture or a leak. The left engine fuel valve was closed (it is throttle-actuated). Once the fuel valve has been closed, the engine's propeller must be feathered and a normal engine start initiated to reopen the valve. The aircraft operating manual did not state this and the pilots did not know it.

The NTSB found that the right engine had autofeathered when the pitot pressure line had failed; the pilots shut down the left engine shortly thereafter, probably due to improper identification of the engine that had failed. Their attempts to restart the good engine were unsuccessful because of their unawareness of the proper starting sequence after a fuel valve has been closed. Engine failure procedures were revised following this accident. (NTSB, 1979b)

11/11/1979: Aeromexico DC-10-30 over Luxembourg

During an evening climb in good weather to 31,000 ft en route to Miami from Frankfurt, flight 945 entered prestall buffet and a sustained stall at 29,800 ft. Stall recovery was affected at 18,900 ft. The crew performed a functional check of the airplane and after finding that it operated properly they continued to its intended destination. After arrival, it was discovered that parts of both outboard elevators and the lower fuselage tail maintenance access door were missing.

The flight data recorder showed that the airplane slowed to 226 kt during a climb on autopilot, quite possibly in vertical speed mode rather than indicated airspeed mode. Buffet speed was calculated to be 241 kt. After initial buffet, the #3 engine was shut down and the airplane slowed to below stall speed.

The NTSB found the probable cause to be failure of the flight crew to follow standard climb procedures and to adequately monitor the airplane's flight instruments. This resulted in the aircraft entering into prolonged stall buffet which placed it outside the design envelope. (NTSB, 1980)

10/7/1970: Aircraft Separation Incidents at Hartsfield Airport, Atlanta, GA

This episode involved several conflicts among aircraft operating under the direction of air traffic control in the Atlanta terminal area. In at least two cases, evasive action was required to avoid collisions. The conflicts were caused by multiple failures of coordination and execution by several controllers during a very busy period.

The NTSB found that the near collisions were the result of inept traffic handling by control personnel. This ineptness was due in part to inadequacies in training, procedural deficiencies, and some difficulties imposed by the physical layout of the control room. The Board also found that the design of the low altitude/conflict alert system contributed to the controller's not recognizing the conflicts. The report stated that, "The flashing visual conflict alert is not conspicuous when the data tag is also flashing in the handoff status. The low altitude warning and conflict alerts utilize the same audio signal which is audible to all control room personnel rather than being restricted to only those immediately concerned with the aircraft. This results in a 'cry wolf' syndrome in which controllers are psychologically conditioned to disregard the alarms." (NTSB, 1981)

1/13/1982: Air Florida B-737, Washington National Airport, DC

This airplane crashed into the 14th Street bridge over the Potomac River shortly after takeoff from Washington National Airport in snow conditions, killing 74 of 79 persons on board. The airplane had been de-iced 1 hour before departure, but a substantial period of time had elapsed since that operation before it reached takeoff position. The engines developed substantially less than takeoff power during the takeoff and thereafter due to incorrect setting of takeoff power by the pilots. It was believed that the differential pressure probes in both engines were iced over, providing incorrect (too high) EGT indications in the cockpit. This should have been detected by examination of the other engine instruments, but was not.

The NTSB found that the probable cause of the accident was the flight crew's failure to use engine antiice during ground operation and takeoff, their decision to take off with snow/ice on the airfoils, and the captain's failure to reject the takeoff at an early stage when his attention was called to anomalous engine instrument readings. Contributing factors included the prolonged ground delay after deicing, the known inherent pitching characteristics of the B-737 when the wing leading edges are contaminated, and the limited experience of the flight crew in jet transport winter operations. (NTSB, 1982)

9/3/1983: Korean Air Lines B-747 over Sakhalin Island, USSR

The airplane was destroyed in cruise flight by air-to-air missiles fired from a Soviet fighter after it strayed into a forbidden area en route from Anchorage, AK to Soeul, Korea. The airplane had twice violated Soviet airspace during its flight. The flight data and cockpit voice recorders were not recovered from the sea. After extensive investigation by the International Civil Aviation Organization, it was believed that its aberrant flight path had been the result of one or more incorrect sets of waypoints loaded into the INS systems prior to departure from Anchorage.

Many years later, the Russian government made available further information on the flight which supported a finding that the crew had inadvertently left the airplane's autopilot in heading mode rather than INS mode for an extended period of time. As a result, the flight path took the airplane over Soviet territory, where it was destroyed by a Soviet fighter. (Stein, 1985; see also incident of 2/13/90.)

2/28/1984: Scandinavian Airlines DC-10-30, J. F. Kennedy Airport, NY

After crossing the runway threshold at proper height but 50 kt above reference speed, the airplane touched down 4700 ft beyond the threshold of an 8400 ft runway and could not be stopped on the runway. It was steered to the right and came to rest in water 600 ft from the runway end. A few passengers sustained minor injuries during evacuation. The weather was very poor and the runway was wet.

The airplane's autothrottle system had been unreliable for approximately one month and had not reduced speed when commanded during the first (Stockholm-Oslo) leg of this flight. The Captain had deliberately selected 168 kt to compensate for a threatened wind shear. The throttles did not retard passing 50 ft and did not respond to the autothrottle speed control system commands (the flight crew was not required to use the autothrottle speed control system for this approach).

The NTSB cited as the probable cause the flight crew's disregard for prescribed procedures for monitoring and controlling airspeed during the final stages of the approach, its decision to continue the landing rather than to execute a missed approach, and overreliance on the autothrottle speed control system which had a history of recent malfunctions. It noted that "performance was either aberrant or represents a tendency for the crew to be complacent and overrely on automated systems." It also noted that there were three speed indications available to the crew: its airspeed indications, the fast-slow indicators on the attitude director, and an indicated vertical speed of 1840 ft per minute on glide slope. In its report, the Board discussed the issue of overreliance on automated systems at length (report pp. 37–39) and cited several other examples of the phenomenon. (NTSB, 1984)

2/19/1985: China Airlines B747-SP, 300 miles northwest of San Francisco

The airplane, flying at 41,000 ft en route to Los Angeles from Taipei, suffered an inflight upset after an uneventful flight. The airplane was on autopilot when the #4 engine lost power. During attempts to re-light the engine, the airplane rolled to the right, nosed over and be-gan an uncontrollable descent. The Captain was unable to restore the airplane to stable flight until it had descended to 9500 ft.

The autopilot was operating in the performance management system (PMS) mode for pitch guidance and altitude hold. Roll commands were provided by the INS; in this mode, the autopilot uses only the ailerons and spoilers for lateral control; rudder and rudder trim are not used. In light turbulence, that airspeed began to fluctuate; the PMS followed the fluctuations and retarded the throttles when air-speed increased. As the airplane slowed, the PMS moved the throt-tles forward; engines 1, 2, and 3 accelerated but #4 did not. The flight engineer moved the #4 throttle forward but without effect. The INS caused the autopilot to hold the left wing down since it could not correct with rudder. The airplane decelerated due to the lack of power. After attempting to correct the situation with autopilot, the Captain disengaged the autopilot at which time the airplane rolled to the right, yawed, then entered a steep descent in cloud, during which it exceeded maximum operating speed. It was extensively damaged during the descent and recovery; the landing gear de-ployed, 10–11 ft of the left horizontal stabilizer was torn off and the no. 1 hydraulic system lines were severed. The right stabilizer and 3/4 of the right outboard elevator were missing when the airplane landed; the wings were also bent upward.

The NTSB determined that the probable cause was the Captain's pre-occupation with an inflight malfunction and his failure to monitor properly the airplane's flight instruments which resulted in his losing control of the airplane. Contributing to the accident was the Cap-tain's overreliance on the autopilot after a loss of thrust on #4 en-gine. The Board noted that the autopilot effectively masked the approaching onset of loss of control of the airplane. (NTSB, 1986)

3/31/1986: United Airlines B-767, San Francisco, CA

This airplane was passing through 3100 ft on its climb from San Francisco when both engines lost power abruptly. The engines were restarted and the airplane returned to San Francisco, where it landed without incident. The crew reported that engine power was lost when the flight crew attempted to switch from manual operation to the engine electronic control system, a procedure which prior to that time was normally carried out at 3000 ft during the climb. The EEC switches are guarded. It is believed that the crew may have inadvertently shut off fuel to the engines when they intended to engage the EEC, as in the incident cited immediately below. (AWST, 1986)

6/30/1987: Delta Air Lines B-767, Los Angeles, CA

Over water, shortly after takeoff from Los Angeles, this twin-engine airplane suffered a double-engine failure when the captain, attempting to deactivate an electronic engine controller in response to an EEC caution light, shut off the fuel valves instead. The crew was able to restart the engines within one minute after an altitude loss of several hundred feet. The fuel valves were located immediately above the electronic engine control switches on the airplane center console, though the switches were dissimilar in shape.

The FAA thereafter issued an emergency airworthiness directive requiring installation of a guard device between the cockpit fuel control switches. (AWST, 1987)

7/8/1987: Delta Air Lines L-1011/Continental Airlines B-747 over Atlantic Ocean

These two airplanes experienced a near midair collision over the North Atlantic Ocean after the Delta airplane strayed 60 miles off its assigned oceanic route. The incident, which was observed by other aircraft in the area but not, apparently, by the Delta crew, was believed to have been caused by an incorrectly inserted waypoint in the Delta airplane's INS prior to departure. (Preble, 1987)

8/16/1987: Northwest Airlines DC9-82, Detroit Metro Airport, Romulus, MI

The airplane crashed almost immediately after takeoff from runway 3C en route to Phoenix. [Runways are numbered to indicate their magnetic heading to the nearest 10 degrees; 3 = 30 degrees (actually from 26–34 degrees). Parallel runways also have letter designators: L = left, C = center, R = right.] The airplane began its rotation about 1200–1500 feet from the end of the 8500 ft runway and lifted off near the end. After liftoff, the wings rolled to the left and right; it then collided with a light pole located 1/2 mile beyond the end of the runway. 154 persons were killed; one survived.

During the investigation, it was found that the trailing edge flaps and leading edge slats were fully retracted. Cockpit voice recorder (CVR) readout indicated that the takeoff warning system did not function and thus did not warn the flight crew that the airplane was improperly configured for takeoff.

The NTSB attributed the accident to the flight crew's failure to use the taxi checklist to insure that the flaps and slats were extended. The failure of the takeoff warning system was a contributing factor. This airplane has a stall protection system which announces a stall and incorporates a stick pusher, but autoslat extension and post-stall recovery is disabled if the slats are retracted. Its caution and warning system also provides tone and voice warning of a stall, but this is disabled in flight by nose gear extension. (NTSB, 1988b)

6/26/1988: Air France Airbus A320, Mulhouse-Habsheim, France

This airplane crashed into tall trees following a very slow, very low altitude flyover at a general aviation airfield during an air show. Three of 136 persons aboard the aircraft were killed; 36 were injured. The captain, an experienced A320 check pilot, was demonstrating the slow-speed maneuverability of the then-new airplane.

The French Commission of Inquiry found that the flyover was conducted at an altitude lower than the minimum of 170 ft specified by regulations and considerably lower than the intended 100 ft altitude level pass briefed to the crew by the captain prior to flight. It stated that, "The training given to the pilots emphasized all the protections from which the A320 benefits with respect to its lift which could have given them the feeling, which indeed is justified, of increased safety...However, emphasis was perhaps not sufficiently placed on the fact that, if the (angle of attack) limit cannot be exceeded, it nevertheless exists and still affects the performance." (emphasis supplied). The Commission noted that automatic go-around protection had been inhibited and that this decision was compatible with the Captain's objective of maintaining 100 ft. In effect, below 100 ft, this protection was not active.

The Commission attributed the cause of the accident to the very low flyover height, very slow and reducing speed, engine power at flight idle, and a late application of go-around power. It commented on insufficient flight preparation, inadequate task sharing in the cockpit, and possible overconfidence because of the envelope protection features of the A320. (Ministry of Planning, Housing, Transport and Maritime Affairs, 1989)

8/31/1988: Delta Airlines B727-232, Dallas-Fort Worth Airport, TX

The airplane, flight 1141, crashed shortly after takeoff from runway 18L en route to Salt Lake City. The takeoff roll was normal but as the main gear left the ground the crew heard two explosions and the airplane began to roll violently; it struck an ILS antenna 1000 ft past the runway end after being airborne for about 22 sec. 14 persons were killed, 26 injured, 68 uninjured.

The investigation showed that the flaps and slats were fully retracted. Evidence suggested that there was an intermittent fault in the takeoff warning system that was not detected and corrected during the last maintenance action. This problem could have manifested itself during the takeoff.

The NTSB found the probable cause to be the captain's and first officer's inadequate cockpit discipline and failure of the takeoff configuration warning system to alert the crew that the airplane was not properly configured for takeoff. It found as contributing factors certain management and procedural deficiencies and lack of sufficiently aggressive action by FAA to correct known deficiencies in the air carrier. The Board took note of extensive nonduty related conversations and the lengthy presence in the cockpit of a flight attendant which reduced the flight crew's vigilance in insuring that the airplane was properly prepared for flight. (NTSB, 1988a)

3/10/1989: Air Ontario Fokker F-28, Dryden, Ontario, Canada

This airplane was dispatched from Winnipeg, Man. to Thunder Bay, Ont., thence via Dryden, Ont. back to Winnipeg, with an inoperative auxiliary power unit. While preparing for the return trip at Thunder Bay, the crew found more passengers than had been planned for or could be accommodated if enough fuel for the entire flight to Winnipeg was boarded, as it had been. The captain preferred to offload passengers rather than fuel; he was overruled by the company. This action required a delay for defueling at Thunder Bay and a landing at Dryden to take on additional fuel. The company's system operations center did not inform the captain of freezing rain forecast for Dryden.

Upon arriving at Dryden, which had no ground power units with which to start the airplane's engines, the captain was required to take on fuel with one engine running. This was a permitted action, though it was performed with passengers on board, which was not permitted. The airplane could not be de-iced with engines running, however, and freezing rain was falling prior to his takeoff, which was also delayed by a lost aircraft trying to land. The airplane crashed immediately after takeoff; ice was noted on the wings by surviving passengers and cabin crew.

The captain in this accident was placed in a "triple bind." He could not uplift sufficient fuel to fly to Winnipeg with the full passenger load. If he landed at Dryden, he could refuel but could not de-ice if required. The defueling at Thunder Bay had already made his flight over one hour late. He received inadequate information and no guidance from his company.

The subsequent Commission of Inquiry found a large number of latent factors at many levels within the company, its parent, Air Canada, and Transport Canada, the regulatory authority. (Moshansky, 1992)

11/21/1989 British Airways B747, Heathrow Airport, London, England

The aircraft approached London in very bad weather after a flight from Bahrain. Fuel was low due to headwinds; the copilot had been incapacitated for part of the flight due to gastroenteritis and diarrhea. The copilot was not certified for category II or III landings. BA flight operations authorized the approach despite the copilot's lack of qualifications. The approach, to runway 27 instead of 9 as briefed, was hurried. When the aircraft captured the localizer and glide slope, the autopilots failed to stabilize the aircraft, possibly due to late capture of the radio beams. 125 feet above ground, the runway was not in sight and the captain gently began a missed approach. The aircraft sank to 75 feet above ground before gaining altitude. After a second, successful approach, the aircraft landed safely.

An investigation by British Airways disclosed that during the first approach, the aircraft had been seriously to the right of the localizer course and had overflown a hotel to the north of the airport only a few feet above the highest obstacle on its course. The pilot and crew were suspended; legal action was later taken against the captain for endangering the passengers and persons on the ground. (Wilkinson, 1994)

1/25/1990: Avianca B-707-321, Cove Neck, New York

Avianca flight 052 crashed in a wooded residential area during an approach to Kennedy International Airport after all engines failed due to fuel exhaustion. The flight from Medellin, Colombia had been placed in holding patterns three times for a total of about 1.3 hours. During the third period of holding, the crew reported that the airplane could not hold longer than 5 minutes, that it was running out of fuel, and that it could not reach its alternate airport in Boston. Subsequently, the flight executed a missed approach at Kennedy. While trying to return to the airport, the airplane lost power in all four engines and crashed 16 miles from the runway.

The NTSB determined that the probable cause of the accident was the failure of the flight crew to adequately manage the airplane's fuel load, and their failure to communicate an emergency fuel situation to air traffic control before fuel exhaustion occurred. Contributing to the accident was the flight crew's failure to use an airline operational control dispatch system to assist them during the international flight into a high-density airport in poor weather. Also contributing was inadequate traffic flow management by the FAA and the lack of standardized understandable terminology for pilots and controllers for minimum and emergency fuel states. Windshear, crew fatigue and stress were other factors that led to the unsuccessful completion of the first approach and thus contributed to the accident. (NTSB, 1991a)

2/13/1990 El Al B747 and British Airways B747 over the Atlantic Ocean

An El Al B-747 enroute from Tel Aviv to New York almost collided with a British Airways 747 in the Reykjavik Flight Information Region after its crew failed to switch back from heading mode to INS mode after being cleared by Shanwick control to a new oceanic track. The crew deviated 110 nm north of the new track before realizing their error. Upon recognizing the error, the flight crew notified ATC but provided no information on the magnitude of their deviation. ATC cleared them to turn left to reintercept their cleared track, which they did.

The near collision occurred while the crew were navigating back to the correct track without descending 1000 ft below the prevailing traffic flow, as prescribed by North Atlantic Special Procedures for In-flight Contingencies. The El Al 747 passed right-to-left ahead of a westbound British Airways 747 which took evasive action, missing El Al by approximately 600 ft. (Pan American World Airways, 1990)

2/14/1990: Indian Airlines Airbus A320, Bangalore, India

(Official report not available) This airplane crashed short of the runway during an approach to land in good weather, killing 94 of 146 persons aboard including the pilots. The best available data indicates that the airplane had descended at idle power in the "idle open descent" mode until shortly before the accident, when an attempt was made to recover by adding power but too late to permit engine spool-up prior to impact. The airplane was being flown by a Captain undergoing a route check by a check airman.

The crew allowed the speed to decrease to 25 kt below the nominal approach speed late in the descent. The recovery from this condition was started at an altitude of only 140 ft, while flying at minimum speed and maximum angle of attack. The check captain noted that the flight director should be off, and the trainee responded that it was off. The check captain corrected him by stating, "But you did not put off mine." If either flight director is engaged, the selected autothrust mode will remain operative, in this case, the idle open descent mode. The alpha floor mode was automatically activated by the declining speed and increasing angle of attack; it caused the autothrust system to advance the power, but this occurred too late for recovery to be affected before the airplane impacted the ground. (Lenorovitz, 1990)

12/3/1990: Northwest Airlines B-727 and DC-9, Detroit Metro Airport, MI

These two aircraft collided while the 727 was taking off and the DC-9 had just inadvertently taxiied onto the active runway. The DC-9 was lost on the airport in severely restricted visibility. Both aircraft were on the ground. The accident site was not visible from the tower due to fog; ASDE was not available.

The Board determined that the probable cause of the accident was a lack of proper crew coordination, including a reversal of roles, on the part of the DC-9 pilots. This led to their failure to stop taxiing and alert the ground controller of their positional uncertainty in a timely manner before and after intruding onto the active runway. A number of contributing factors were also cited. (NTSB, 1991b)

2/1/1991: US Air B-737 and Skywest Fairchild Metro, Los Angeles, CA

This accident occurred after the US Air airplane was cleared to land on runway 24L at Los Angeles while the Skywest Metro was positioned on the runway at an intersection awaiting takeoff clearance. There were 34 fatalities and 67 survivors. The Metro may not have been easily visible from the control tower; airport surface detection radar equipment (ASDE) was available but was being used for surveillance of the south side of the airport. The controller was very busy just prior to the time of the accident.

The NTSB investigation indicated that the controller cleared the Metro into position at an intersection on runway 24L, 2400 ft from the threshold, two minutes before the accident. One minute later, the 737 was given a clearance to land on runway 24L. The Board determined that the probable cause of the accident was the failure of Los Angeles Air Traffic Facility management to implement procedures that provided adequate redundancy and the failure of FAA's Air Traffic Management to provide adequate policy direction and oversight. These failures ultimately led to the failure of the local controller to maintain awareness of the traffic situation. (NTSB, 1991c)

5/26/1991 Lauda Air (Austria) B767-300ER over Thailand

This airplane was climbing to altitude on a flight between Bangkok and Vienna when its right engine reverser actuated because of a mechanical failure. The flight crew was unable to control the airplane due to the high level of reverse thrust coming from the right engine. The airplane crashed after an uncontrolled descent. Simulation studies indicated that recovery from such an event was not possible for pilots without advance knowledge of the event. (Ministry of Transport and Communications, Thailand, 1993)

8/12/1991: Ansett Australia A320 and Thai Airways DC-10: Sydney, Australia

During simultaneous crossing runway operations at Kingsford Smith Airport, a Thai DC-10 was landing on runway 34 and an Ansett A320 was on short final approach for intersecting runway 25. Landing instructions for the DC-10 included a requirement for the aircraft to hold short of the runway 25 intersection. While observing the DC-10's landing roll during his landing, the A320 captain judged that the DC-10 might not stop before the runway intersection. He elected to initiate a missed approach from a low height above the runway. The go-around was successful; the A320 passed the centerline of runway 34 at a radio altitude of 52 ft. Under heavy braking, the DC-10 slowed to about 2 kts ground speed when it reached the edge of runway 25.

During the A320 go-around, differing attitude command inputs were recorded from the left and right sidesticks for a period of 12 seconds. Neither the captain, who had taken over control, or the copilot, was aware of control stick inputs from the copilot during this period. Activation of the "takeover button" on the control stick was not a part of Ansett's standard operating procedures. The incident analysis noted that "Although the A320 successfully avoided the DC-10, under different circumstances the cross controlling between the pilots could have jeopardized a safe go-around...This simultaneous input situation would almost certainly have been immediately apparent, and corrected rapidly had there been a sense of movement between the two sidesticks." (Bureau of Air Safety Investigation, 1993)

12/12/1991: Evergreen International Airways B-747, Nakina, Ontario, Canada

While in cruise flight at 31,000 ft, a cargo aircraft entered a steep right bank (greater than 90°) and descended more than 10,000 feet at speeds approaching Mach 1. During the recovery, with vertical accelerations greater than 3g, the right wing was damaged. About 20 feet of honeycomb structure from the underside of the wing was missing; a small honeycomb panel on the upper portion of the wing was damaged and some structure was protruding into the airstream. Upon recovery from the dive, the aircraft was experiencing control difficulties; the crew successfully diverted to Duluth, MN. During the approach and landing, the left and right flaps, as well as the right horizontal stabilizer, were damaged by debris from the damaged right wing. There were no injuries.

The Transportation Safety Board of Canada determined that the flight upset was caused by an uncommanded, insidious roll input by the channel A autopilot roll computer; the roll went undetected by the crew until the aircraft had reached an excessive bank angle and consequential high rate of descent. The recovery action was delayed slightly because of the time required by the crew to determine the aircraft attitude. (NTSB, 1992a)

1/1992: Air Inter Airbus A320 on approach to Strasbourg, France

The airplane was being given radar vectors to a nonprecision (VOR-DME) approach to the airport at Strasbourg. It was given vectors that left little time for cockpit setup prior to intercepting the final approach course. It is believed that the pilots intended to make an automatic approach using a flight path angle of $-33°$ from the final approach fix; this maneuver would have placed them at approximately the correct point for visual descent when they reached minimum descent altitude.

The pilots, however, appear to have executed the approach in heading/vertical speed mode instead of track/flight path angle mode. The Flight Control Unit setting of "-33" yields a vertical descent rate of -3300 ft/min in this mode, and this is almost precisely the rate of descent the airplane realized until it crashed into mountainous terrain several miles short of the airport. A push button on the FCU panel cycles the automation between H/VSI and T/FPA mode.

Modifications to A320 vertical speed/flightpath angle displays (in vertical speed mode, four digits are shown; in flight path angle mode, only two digits are visible) were subsequently made available by the manufacturer to avoid this error. New production A320s have been modified in this manner since November 1993 (Aerospace, 1994a).

12/8/1992: United Airlines B737-291, Colorado Springs, CO

United Airlines flight 585 was on final approach course following a flight from Denver, CO to Colorado Springs, CO under visual meteorological conditions when it was observed by numerous eyewitnesses to roll steadily to the right and pitch nose down, reaching a nearly vertical attitude when it impacted the ground, killing all 25 occupants.

Despite an exhaustive investigation which is continuing, the NTSB has thus far been unable to identify conclusive evidence to explain the loss of this aircraft. It is surmised by the Board that either a rudder control anomaly, or a "rotor," a horizontal axis wind vortex, may have precipitated the loss of control, but this is not certain. (NTSB, 1992b)

9/14/1993: Lufthansa A320, Warsaw, Poland

The aircraft, carrying 70 persons, landed at Warsaw in a downpour with strong, gusty winds. The pilot carried extra airspeed because of the wind conditions; a probable wind shear late in the approach made its ground speed still faster at touchdown. The airplane landed gently despite the gusts. It continued for approximately 8 sec after touchdown before being able to activate ground spoilers and reverse thrust. The airplane overran the runway end, traversed an embankment beyond the departure end and caught fire. Two persons, including the copilot, were killed; 55 were injured.

"Preliminary findings of the Polish inquiry...suggest that the crew, having been advised of wind shear and a wet runway, correctly added 20 kt to the approach speed. When the forecast crosswind unexpectedly became a tailwind, making ground speed about 170 kt, the wheel spinup and oleo squat switches did not (activate). For a critical 9 sec (during which the aircraft may have been aquaplaning) thrust reverse, wheelbraking and lift dumping (full spoiler deployment) remained disarmed...Although the A320 was...still to have the softer landing double-oleo modification, which might have 'made' the switches, the priority question raised by the accident is whether pilots should have manual override of safety locks..." (Aerospace, 1994b; AWST, 1994a)

4/26/1994: China Airlines A-300-600R, Nagoya, Japan

During a normal approach to landing at Nagoya runway 34 in visual meteorological conditions, the captain indicated he was going around but did not indicate why. Within the next 30 seconds, witnesses saw the aircraft in a nose-up attitude, rolling to its right before crashing tail-first 300 ft to the right of the approach end of the runway.

During the approach, the copilot flying apparently triggered the autopilot TOGA (takeoff-go-around) switch, whereupon the automation added power and commanded a pitch-up. The captain warned the copilot of the mode change, but the copilot continued to attempt to guide the aircraft down the glide slope while the automation countered his inputs with nose-up elevator trim. Ultimately, with stabilizer trim in an extreme nose-up position, the copilot was unable to counteract the trim with nose-down elevator. The aircraft nosed up to an attitude in excess of 50°, stalled, and slid backwards to the ground. 264 people were killed in the crash.

This accident is still under civil and criminal investigation. It is presently thought that the pilots failed "to realize that their decision (to continue the approach) contradicted the logic of the airplane's automated safety systems. In February 1991, an Interflug A310 at Moscow experienced a sudden, steep pitch-up similar to the one observed in this accident." (*Aviation Week & Space Technology*, 5/2/94, p. 26; 5/9/94, pp. 31–32; 12/5/94, p. 29)

On 8/31/94, The NTSB issued Safety Recommendations A-94-164 through –166 to the FAA. Its Recommendation stated, "the Safety Board is concerned that the possibility still exists for a pilot-induced 'runaway trim' situation at low altitude and that...such a situation could result in a stall or the airplane landing in a nose-down attitude..." Referring to other transport category aircraft autopilot systems, the Board said, "It is noted that the (autopilot) disconnect and warning systems are fully functional, regardless of altitude, and with or without the autopilot in the land or go-around modes. The Safety Board believes that the autopilot disconnect systems in the Airbus A-300 and A-310 are significantly different...additionally, the lack of a stabilizer-in-motion warning appears to be unique to (these aircraft). The accident in Nagoya and the incident in Moscow indicate that pilots may not be aware that under some circumstances the autopilot will work against them if they try to manually control the airplane."

The Board recommended that these autopilot systems be modified to ensure that the autopilot would disconnect if the pilot applies a specified input to the flight controls or trim system, regardless of the altitude or operating mode of the autopilot, and also to provide a sufficient perceptual alert when the trimmable horizontal stabilizer is in motion, irrespective of the source of the trim command. (NTSB, 1994)

6/6/1994: Dragonair A-320, Kai Tak Airport, Hong Kong

The airplane was attempting a landing at Kai Tak Airport during a severe storm. As the aircraft banked at about 1000 feet, it encountered a wind shear that registered –1.6g. It lost 12 kt of airspeed in 1 second. The buffeting triggered its automatic flap locking safety mechanism, which is set if there is more than a 40 mm difference between the positions of the flaps to prevent them from becoming asymmetrical. The flaps locked at a full setting of 40°, or "flaps 4" (the landing position). The airplane's (leading edge) slats were in the no. 3 position of 22°. Sensing an anomaly, the electronic centralized aircraft monitoring system (ECAM) flashed a warning message for the pilot to correct it by moving the flaps lever to Flaps 3.

Unable to do so, the pilot aborted the landing. On the fourth try, he landed on runway 31, which allowed an approach without a banking maneuver. Two passengers were slightly injured after the aircraft ran off the runway. The incident is still under investigation. The article notes that a similar incident apparently occurred to an Indian

Airlines A320 in November, 1993. Airbus Industrie has recommended since this incident that pilots disregard the ECAM warning message. The software is being rewritten to eliminate the message; changes are also to be made in the flight control computers to prevent discrepancies between the flap lever position and the position of the flaps. (AWST, 1994)

6/21/1994: Brittania Airways B757-200, Manchester, United Kingdom

The aircraft was at light weight and was conducting a full-power takeoff. An altitude of 5000 ft had been selected. The autopilot went to altitude acquisition mode passing 2200 ft because of the rapid climb speed. Power was reduced by the autothrust system and the airplane's speed began to drop rapidly toward takeoff safety speed because of the high pitch angle. Flight director bars continued to command pitch up, then disappeared from view. The pilot reduced the pitch attitude to 10° nose-up and normal acceleration resumed. This incident resembles in many respects the more serious occurrence of the A330 at Toulouse (6/30/94, below), which also involved a rapid switch to altitude acquisition mode after takeoff. (Civil Aviation Authority, UK, 1994)

6/30/1994: Airbus A330-322 test flight, Toulouse Blagnac Airport, France:

This airplane was on a Category III certification test flight to study various pitch transition control laws in the autopilot Speed Reference System mode during engine failure at low altitude, rearward center of gravity, and light aircraft weight. The flight crew included an experienced test pilot flying as captain, a copilot from a customer company, a flight test engineer, and three passengers. The copilot was handling the aircraft.

During the takeoff, the copilot rotated the airplane slightly rapidly; the landing gear was retracted. The autopilot was engaged 6 sec after takeoff at a speed of 150 kt and a pitch angle of almost 25° nose-up. Immediately thereafter, the left engine was brought to idle power and one hydraulic system was shut down, as planned for the test.

When the airplane reached 25° pitch angle, autopilot and flight director mode information were automatically removed from the PFD. A maximum pitch angle of 29° was reached 8 sec after takeoff; the airplane was decelerating. The angle of attack reached 14°, which activated the alpha protection mode of the flight controls. The captain disconnected the autopilot 19 sec after takeoff. Subsequent control actions by the captain, which included reducing power on the right engine to regain control, deactivated alpha floor protection on the left engine. The airplane slowed to 100 kt, appreciably below minimum single-engine control speed of 118 kt, and yawed to the left. The left wing then stalled; speed reached 77 kt with an increasing left bank. Pitch angle reached 43° nose down and the airplane crashed 36 sec after takeoff.

During investigation, it was found that the aircraft autopilot had gone into altitude acquisition (ALT*) mode. In this mode, there was no maximum pitch limitation in the autoflight system software. As a consequence, at low speed, if a major thrust change occurs (as it did here), the autopilot can induce irrelevant pitch attitudes since it is still trying to follow an altitude acquisition path which it cannot achieve.

The investigating committee believed that the accident was caused by the conjunction of several factors, none of which taken separately would have produced the accident. The committee cited the planned and inadvertent conditions under which the flight test was undertaken (high thrust, very aft center of gravity, trim within limits but nose-up, a selected altitude of 2000 feet, late and imprecise definition of respective tasks between the pilot and copilot regarding the test to be performed, firm and quick rotation by the copilot, captain busy with the test actions, taking him out of the piloting loop). They also noted that the lack of pitch protection in the ALT* mode of the autopilot played a key role. Contributing factors included the inability of the flight crew to identify the active autopilot mode (due to the FMA declutter action at 25° nose-up), crew confidence in the anticipated aircraft reactions, late reaction of the flight test engineer to the rapid evolution of flight parameters (particularly the airspeed), and a late captain reaction to an abnormal situation.

A subsequent published article noted that "Contradictory autopilot requirements appear as a key factor that contributed to the loss of control: the 2000 ft altitude was selected while the autopilot also had to simultaneously manage the combination of very low speed, an extremely high angle of attack, and asymmetrical engine thrust." (Director General of Armaments (France), 1994)

7/2/1994: US Air DC-9-31, Charlotte, NC

The airplane was returning from Columbia, SC to Charlotte, NC, when it encountered a wind shear during a very heavy rainstorm while on final approach to the Charlotte-Douglas Airport. A wind shear alert had been received and the crew had briefed a missed approach if necessary. The captain flying ordered a missed approach at 200 feet because of poor visibility and strong, gusty winds. The first officer initiated the missed approach; the landing gear was retracted and flaps reduced from 40° (landing position) to 15°. At 350 ft the crew felt a severe sink developing; full throttles were applied, but full thrust occurred only about 3 sec before impact, too late to arrest the descent and impact about 0.2 nm to the right of runway 18R. 37 occupants were killed.

The crewmembers were unable to recall whether they had heard an aural warning from the wind shear detection system; investigation later revealed that the system's sensitivity is sharply reduced while wing flaps are in transit, to minimize the likelihood of false or nuisance warnings when airflow over the wing is disturbed during the change of configuration. Data provided to the NTSB by the system's manufacturer indicated that an alert would have been furnished 12 sec after a wind shear was detected if flaps were in transit, whereas an alert would have been generated in the presence of a severe shear within 5 sec under other circumstances. As a result, the time lag "rendered the system useless" because the warning "would have occurred too late" for the pilots to perform a successful escape maneuver, according to the NTSB. It is not known whether the pilots were aware of this automatic reduction in sensitivity during flap transit.

The NTSB recommended that the FAA issue a flight standards bulletin informing pilots that wind shear warnings will be unavailable when flaps are in transit, and require modifications in the standard wind shear alert system to delete the delay feature, thereby ensuring "prompt warning activation" when flaps are transitioning between settings. The Board did not speak to the fact that this delay was incorporated in the system's software specifically to avoid nuisance warnings caused by temporary airflow disturbances. Honeywell had stated that such false alarms could cause pilots to "overreact or lose confidence" in the system's detection capabilities. (Phillips, 1994c)

9/8/1994: US Air B737-300, Pittsburgh, PA

During a routine approach to Great Pittsburgh International Airport, US Air flight 427 was cleared to turn left to a heading of 100°, reduce speed to 190 kt and descend to 6000 ft in preparation for a right downwind on a visual approach to runway 28R. The pilots extended their slats and flaps to the "Flaps 1" position. As the airplane began its turn, it rolled left, then decreased its bank angle, then increased it again to at least 100° as the nose pitched downward. The airplane struck the ground 23 sec later at an angle of about 80° and an airspeed in excess of 260 kt. The accident was not survivable.

The NTSB has undertaken extensive investigations of this accident, which thus far remains unexplained. Data collection is continuing. The similarity between certain aspects of this accident and a B737-291 accident at Colorado Springs, CO on 12/8/92, also unexplained, has prompted intensive studies of rudder control and other aircraft mechanisms by the Board, the Boeing Company, and component manufacturers. In both cases, the Board has been hampered by the availability of only limited data from the flight data recorders, which were older models with limited parameter recording capability. (Phillips, 1994b; see also page 69)

9/24/1994: Tarom (Romanian) Airlines A310-300, Orly Airport, Paris, France

The airplane, carrying 182 persons on a flight from Bucharest to Paris, was on final approach to Orly Airport under visual meteorological conditions when it suddenly assumed a steep, nose-high attitude, then rolled into a dive before the pilots regained control at 800 feet above ground. No one was seriously injured and the airplane landed safely. A videotape taken by a witness showed the airplane in a steep nose-up attitude, then rolling off on one wing and descending in a nose-down attitude for several seconds before recovery. The digital flight data recorder was apparently inoperative during the incident, but data were obtained from the cockpit voice recorder and a direct access recorder used for maintenance purposes.

It is believed that the autopilot "suddenly went into the 'level change' mode" because flap limit speed was exceeded by 2 knots during the approach; this resulted in the pitch-up. "According to one report, the electric trim countered the pilot's action" during the attempt to recover from the pitch-up. (AWST, 1994b; Aerospace, 1994c; see also AWST, 1995b)

10/31/1994: American Eagle Airlines ATR72, Roselawn, IN

The airplane went out of control and crashed after flying at 10,000 ft at relatively low airspeed in a holding pattern for an extended period under icing conditions. The airplane carried a highly capable digital flight data recorder, whose data indicated that severe lateral control instability occurred, due, it is presently thought, to an accretion of ice ahead of the ailerons but aft of the wing leading edge de-icer boots. The airplane was being flown on autopilot when control was first lost.

The accident is still under investigation, but the NTSB has issued urgent safety recommendations. The FAA has warned ATR42/72 pilots to avoid prolonged flight under icing conditions and to avoid high angles of attack if lateral instability occurs. Autopilot use under such conditions is proscribed, because autopilot corrective actions can mask the onset of the controllability problem. NTSB was aware of "similar, uncommanded autopilot disengagements and uncommanded lateral excursions" that have occurred on ATR42 aircraft in the past six years. (Phillips, 1994a)

Appendix B

Controlled flight into terrain (CFIT) checklist

Flight Safety Foundation
CFIT Checklist
Evaluate the Risk and Take Action

Flight Safety Foundation (FSF) designed this controlled-flight-into-terrain (CFIT) risk-assessment safety tool as part of its international program to reduce CFIT accidents, which present the greatest risks to aircraft, crews and passengers. The FSF CFIT Checklist is likely to undergo further developments, but the Foundation believes that the checklist is sufficiently developed to warrant distribution to the worldwide aviation community.

Use the checklist to evaluate specific flight operations and to enhance pilot awareness of the CFIT risk. The checklist is divided into three parts. In each part, numerical values are assigned to a variety of factors that the pilot/operator will use to score his/her own situation and to calculate a numerical total.

In *Part I: CFIT Risk Assessment,* the level of CFIT risk is calculated for each flight, sector or leg. In *Part II: CFIT Risk-reduction Factors,* Company Culture, Flight Standards, Hazard Awareness and Training, and Aircraft Equipment are factors, which are calculated in separate sections. In *Part III: Your CFIT Risk,* the totals of the four sections in *Part II* are combined into a single value (a positive number) and compared with the total (a negative number) in *Part I: CFIT Risk Assessment* to determine your CFIT Risk Score. To score the checklist, use a nonpermanent marker (do not use a ballpoint pen or pencil) and erase with a soft cloth.

Part I: CFIT Risk Assessment

Section 1 – Destination CFIT Risk Factors	Value	Score
Airport and Approach Control Capabilities:		
ATC approach radar with MSAWS	0	____
ATC minimum radar vectoring charts	0	____
ATC radar only	-10	____
ATC radar coverage limited by terrain masking	-15	____
No radar coverage available (out of service/not installed)	-30	____
No ATC service	-30	____
Expected Approach:		
Airport located in or near mountainous terrain	-20	____
ILS	0	____
VOR/DME	-15	____
Nonprecision approach with the approach slope from the FAF to the airport TD shallower than 2 ³⁄₄ degrees	-20	____
NDB	-30	____
Visual night "black-hole" approach	-30	____
Runway Lighting:		
Complete approach lighting system	0	____
Limited lighting system	-30	____
Controller/Pilot Language Skills:		
Controllers and pilots speak different primary languages	-20	____
Controllers' spoken English or ICAO phraseology poor	-20	____
Pilots' spoken English poor	-20	____
Departure:		
No published departure procedure	-10	____
Destination CFIT Risk Factors Total	(–)	____

Section 2 – Risk Multiplier

	Value	Score
Your Company's Type of Operation (select only one value):		
Scheduled	1.0	_____
Nonscheduled	1.2	_____
Corporate	1.3	_____
Charter	1.5	_____
Business owner/pilot	2.0	_____
Regional	2.0	_____
Freight	2.5	_____
Domestic	1.0	_____
International	3.0	_____
Departure/Arrival Airport (select single highest applicable value):		
Australia/New Zealand	1.0	_____
United States/Canada	1.0	_____
Western Europe	1.3	_____
Middle East	1.1	_____
Southeast Asia	3.0	_____
Euro-Asia (Eastern Europe and Commonwealth of Independent States)	3.0	_____
South America/Caribbean	5.0	_____
Africa	8.0	_____
Weather/Night Conditions (select only one value):		
Night — no moon	2.0	_____
IMC	3.0	_____
Night and IMC	5.0	_____
Crew (select only one value):		
Single-pilot flight crew	1.5	_____
Flight crew duty day at maximum and ending with a night nonprecision approach	1.2	_____
Flight crew crosses five or more time zones	1.2	_____
Third day of multiple time-zone crossings	1.2	_____

Add Multiplier Values to Calculate Risk Multiplier Total _____

Destination CFIT Risk Factors Total × Risk Multiplier Total = CFIT Risk Factors Total (–) _____

Part II: CFIT Risk-reduction Factors

Section 1 – Company Culture

	Value	Score
Corporate/company management:		
Places safety before schedule	20	_____
CEO signs off on flight operations manual	20	_____
Maintains a centralized safety function	20	_____
Fosters reporting of all CFIT incidents without threat of discipline	20	_____
Fosters communication of hazards to others	15	_____
Requires standards for IFR currency and CRM training	15	_____
Places no negative connotation on a diversion or missed approach	20	_____

115-130 points	Tops in company culture			
105-115 points	Good, but not the best	**Company Culture Total (+)**	_____	*
80-105 points	Improvement needed			
Less than 80 points	High CFIT risk			

Section 2 – Flight Standards

	Value	Score
Specific procedures are written for:		
Reviewing approach or departure procedures charts	10	
Reviewing significant terrain along intended approach or departure course	20	
Maximizing the use of ATC radar monitoring	10	
Ensuring pilot(s) understand that ATC is using radar or radar coverage exists	20	
Altitude changes	10	
Ensuring checklist is complete before initiation of approach	10	
Abbreviated checklist for missed approach	10	
Briefing and observing MSA circles on approach charts as part of plate review	10	
Checking crossing altitudes at IAF positions	10	
Checking crossing altitudes at FAF and glideslope centering	10	
Independent verification by PNF of minimum altitude during stepdown DME (VOR/DME or LOC/DME) approach	20	
Requiring approach/departure procedure charts with terrain in color, shaded contour formats	20	
Radio-altitude setting and light-aural (below MDA) for backup on approach	10	
Independent charts for both pilots, with adequate lighting and holders	10	
Use of 500-foot altitude call and other enhanced procedures for NPA	10	
Ensuring a sterile (free from distraction) cockpit, especially during IMC/night approach or departure	10	
Crew rest, duty times and other considerations especially for multiple-time-zone operation	20	
Periodic third-party or independent audit of procedures	10	
Route and familiarization checks for new pilots		
Domestic	10	
International	20	
Airport familiarization aids, such as audiovisual aids	10	
First officer to fly night or IMC approaches and the captain to monitor the approach	20	
Jump-seat pilot (or engineer or mechanic) to help monitor terrain clearance and the approach in IMC or night conditions	20	
Insisting that you fly the way that you train	25	

300-335 points	Tops in CFIT flight standards	
270-300 points	Good, but not the best	**Flight Standards Total** (+) _____ *
200-270 points	Improvement needed	
Less than 200	High CFIT risk	

Section 3 – Hazard Awareness and Training

	Value	Score
Your company reviews training with the training department or training contractor	10	
Your company's pilots are reviewed annually about the following:		
Flight standards operating procedures	20	
Reasons for and examples of how the procedures can detect a CFIT "trap"	30	
Recent and past CFIT incidents/accidents	50	
Audiovisual aids to illustrate CFIT traps	50	
Minimum altitude definitions for MORA, MOCA, MSA, MEA, etc.	15	
You have a trained flight safety officer who rides the jump seat occasionally	25	
You have flight safety periodicals that describe and analyze CFIT incidents	10	
You have an incident/exceedance review and reporting program	20	
Your organization investigates every instance in which minimum terrain clearance has been compromised	20	

You annually practice recoveries from terrain with GPWS in the simulator .... 40 _____

You train the way that you fly .... 25 _____

285-315 points	Tops in CFIT training		
250-285 points	Good, but not the best	Hazard Awareness and Training Total (+) ____ '	
190-250 points	Improvement needed		
Less than 190	High CFIT risk		

Section 4 – Aircraft Equipment

	Value	Score

Aircraft includes:

Radio altimeter with cockpit display of full 2,500-foot range — captain only .... 20 _____

Radio altimeter with cockpit display of full 2,500-foot range — copilot .... 10 _____

First-generation GPWS .... 20 _____

Second-generation GPWS or better .... 30 _____

GPWS with all approved modifications, data tables and service

bulletins to reduce false warnings .... 10 _____

Navigation display and FMS .... 10 _____

Limited number of automated altitude callouts .... 10 _____

Radio-altitude automated callouts for nonprecision

approach (not heard on ILS approach) and procedure .... 10 _____

Preselected radio altitudes to provide automated callouts that

would not be heard during normal nonprecision approach .... 10 _____

Barometric altitudes and radio altitudes to give automated

"decision" or "minimums" callouts .... 10 _____

An automated excessive "bank angle" callout .... 10 _____

Auto flight/vertical speed mode ....-10 _____

Auto flight/vertical speed mode with no GPWS ....-20 _____

GPS or other long-range navigation equipment to supplement

NDB-only approach .... 15 _____

Terrain-navigation display .... 20 _____

Ground-mapping radar .... 10 _____

175-195 points	Excellent equipment to minimize CFIT risk		
155-175 points	Good, but not the best	Aircraft Equipment Total (+) ____ ⁴	
115-155 points	Improvement needed		
Less than 115	High CFIT risk		

Company Culture _____ + Flight Standards _____ + Hazard Awareness and Training _____

+ Aircraft Equipment _____ = CFIT Risk-reduction Factors Total (+) _____

*** If any section in Part II scores less than "Good," a thorough review is warranted
of that aspect of the company's operation.**

Part III: Your CFIT Risk

Part I CFIT Risk Factors Total (–) ____ + Part II CFIT Risk-reduction Factors Total (+) ____

= CFIT Risk Score (±) ____

**A negative CFIT Risk Score indicates a significant threat; review the sections in Part II and
determine what changes and improvements can be made to reduce CFIT risk.**

FSF CFIT Checklist © 1994 Flight Safety Foundation

Glossary

AAA Anti-Aircraft Artillery
ACC Air Combat Command
AFI Air Force Instruction
AFR Air Force Regulation
AGL Above Gound Level
AIM Airman's Information Manual
AOA Angle of Attack
AOPA Aircraft Owner's and Pilot's Association
Armament Weapons carried upon an aircraft
ARSA Airport Radar Service Area
ARTCC Air Route Traffic Control Center
ASRS Aviation Safety Reporting System
ATC Air Traffic Control
ATIS Automatic Terminal Information System. Provides current, routine information to arriving and departing aircraft by means of continuous repetitive broadcasts
ATP Airline Transport Pilot
BOLDFACE Critical action emergency procedures that must be memorized
CAA Civil Aviation Authority
CAS Close Air Support
CEVG Combat Evaluation Group
CFI Certified Flight Instructor
CFIT Controlled Flight Into Terrain
CG Center of Gravity
CND Could Not Duplicate
CNN Cable News Network
CPT Cockpit Procedural Trainer
CRM Cockpit/Crew Resource Management
CVR Cockpit Voice Recorder
DME Distance Measuring Equipment

DO Deputy Commander for Operations
EMS Emergency Medical Service
EP Emergency Procedure
FAA Federal Aviation Administration
FAC Forward Air Controller
FAR Federal Aviation Regulations
Field grade Military ranks between major and colonel
Finis flight Last flight in a given type of aircraft
FL Flight Level
Flyby Airshow Overflight
FO First Officer
FSF Flight Safety Foundation
GA General Aviation
GPS Global Positioning System
GPWS Ground Proximity Warning System
Groupthink A phenomenon where the group prefers harmony over finding the best solution
HF High Frequency
HUD Heads Up Display
IAS Indicated Airspeed. What the airspeed dial normally reads
IAW In Accordance With
ICAO International Civil Aviation Organization
IFR Instrument Flight Rules
IG Inspector General
ILS Instrument Landing System. A system which allows appropriately equipped aircraft to land in bad weather
IMC Instrument Meteorological Conditions. Flying in conditions below those required for VFR flight
INS Inertial Navagation System
IP Instructor Pilot
IP/AC Instructor Pilot/Aircraft Commander
KIAS Knots Indicated Airspeed
LOFT Line Oriented Flight Training
MAC Military Airlift Command
MAW Medical Airworthiness
MEA Minimum Enroute Altitude
MOA Military Operating Area
MP Mishap Pilot
MSL Mean Sea Level. Altitude above the ocean
NASA National Aeronautics and Space Administration
NDB Non-Directional Radio Beacon
NOTAM Notices to Airmen. Regular updates on rules and regulations
NTSB National Transportation Safety Board

OI Operating Instructions
ORI Operational Readiness Inspection
PF Pilot Flying
PIC Pilot in Command
PIREP Pilot Report
PIT Pilot Instructor Training
PNF Pilot Not Flying
RAF Royal Air Force
RCO Range Control Office
ROE Rules of Engagement
RN Radar Navigator
RVR Runway Visual Range. The range over which the pilot of an aircraft on the center line of a runway can expect to see the runway markings or lights
SAM Surface to Air Missile
SAR Search and Rescue
SD Spatial Disorientation
SID Standard Instrument Departure
SIGMET Significant information concerning en route weather phenomena which may affect the safety of aircraft operations
SOLL Special Operations Low Level
SOP Standard Operating Procedures
Sortie A combat mission made by one aircraft
SPINS Special Mission Instructions
STAN-EVAL Standardization and Evaluation
TA Transition Areas
TACAN Tactical Air Navigation
TCA Terminal Control Radar Areas
TCAS Terminal Collision Avoidance System
TERPS Terminal Instrument Procedures Specialist
TFR Terrain Folllowing Radar
TRW Thunderstorm
TUC Time of Useful Consciousness
UAL United Airlines
USAF United States Air Force
UHF Ultra-High frequency
VFR Visual Flight Rules
VHF Very High Frequency (30–300 MHz)
VMC Visual Meteorolgical Conditions. Flying in conditions at least as good as the minimums required for VFR flight
VOR VHF Omnidirectional Radio Range
VVI Vertical Velocity Indicator
WSO Weapon Systems Officer

Bibliography

Abbott, K., S. Slotte and D. Stimson. 1996. *The Interfaces Between Flight Crews and Modern Flight Deck Systems.* A Federal Aviation Administration Human Factors Team Report. June 18, 1996.

AeroKnowledge ASRS CD-ROM. 1995. Accession Number: 81432.

AeroKnowledge ASRS CD-ROM. 1995. Accession number: 136756.

Alluisi, E. A., (1972). Influence of Work-Rest Scheduling and Sleep Loss on Sustained Performance. In W. P. Colquhoun (ed.), *Aspects of Human Efficiency* (pp. 199–214). London: The English Universities Press.

Alluisi, E. A., (1967). Methodology in the Use of Synthetic Tasks to Assess Complex Performance. *Human Factors.* 9, 375–384.

Alter, Jonathan. 1996. "National Affairs: Maiden Flight." *Newsweek.* 22 April.

André, Babette. 1996. "Captain Jepp, the Gentile Air-mapping Pioneer, Passes Away." Internet. General Aviation News and Flyer at http://www1.drive.net/evird.acgi$pass*222.../ganflyer/dec20-1996/capt._jepp_dies.html

ASRS Callback 217. 1997. *Going Global with GPS.* NASA Aviation Safety Reporting System, Moffet Field, CA. July.

ASRS Directline No. 8. 1995. Internet: http://www-afo.arc.nasa.gov/ASRS/dl8.

Aviation Monthly Safety Summary and Report, 1996. "Accident Statistics," April.

Aviation Safety Journal, 1996. "Low Altitude Turn Causes Baron to Stall, Crash," November.

Baron, R. A., and D. Byrne. 1994. *Social Psychology: Understanding Human Interaction (7th Ed.)* Boston: Allyn and Bacon.

Beaty, David. 1995. *The Naked Pilot: The Human Factor in Aircraft Accidents.* Shrewsbury, UK: Airlife Publishing Ltd.

Beyerchen, A. 1992. "Clausewitz, Nonlinearity and the Unpredictability of War." *International Security 17:3* August.

Billings, C. E. 1996. "Human-Centered Aviation Automation: Principles and Guidelines," *NASA Technical Memorandum 110381*.

Bluecoat newsgroup. WWWeb, http://www.neosoft.com/~sky/BLUE-COAT/intro.html. The Bluecoat Project was created by Bill Bulfer as a means of opening an ongoing discussion between the engineers who build flight deck automation systems and the pilots who use them.

Brannigan, M. 1996. "Captain WOW: When Is Mental State of a Pilot Grounds for Grounding Him?" *The Wall Street Journal*. March 7. New York.

Bringle, Donald. 1996. "The Topgun Mentality." *Proceedings*. April. p. 8–9.

Clausewitz, C. 1976. *On War.* Princeton, New Jersey: Princeton University Press.

Colorado Springs Gazette-Telegraph. 1996. "Pilot Qualifications Questioned in Crash." November 20. pp. B-1, B-5.

Coolidge, C. H. Jr. 1996. *AFI 51-503 Report of Aircraft Accident Investigation on USAF CT-43 73-1149*, vol 1., United States Air Force (USAF).

Degani, A., and E. Wiener, 1994. "Philosophy, Policies, Procedures and Practices: The Four 'P's of Flight Deck Operations," *Aviation Psychology in Practice*. 45–67.

Deitz, Shelia R, and William E. Thoms, eds. 1991. *Pilots, Personality, and Performance: Human Behavior and Stress in the Skies,* New York: Quorum Books.

Diehl, A. 1992. "Does Cockpit Management Training Reduce Aircrew Error?" *ISASI Forum,* 24 (4).

Drew, Charles, Andrew Scott and Robert Matchette. 1993. *ASRS Directline*. Lost Com. Issue 6: August 1993. Internet. http//www-afo.arc.nasa.gov/ASRS/dl6_lost.html

Dreyfus, H. L., and S. E. Dreyfus. 1986. *Mind Over Machine: The Power of Human Intuition and Expertise in the Era of the Computer*. New York: The Free Press.

Edens, E. 1991. "Individual Differences Underlying Cockpit Error." Unpublished doctoral dissertation. National Technical Information Service, Springfield, VA.

Edwards, Doug. Personal correspondence. 1997. 9 July.

Foushee, H. C. 1982. "The Role of Communications, Socio-psychological, and Personality Factors in the Maintenance of Crew Coordination." *Aviation Psychology II*. November. pp. 1062–1066.

Foushee, H. C. 1982. "The Role of Communication, Socio-psychological and Personality Factors in the Maintenance of Crew Coordination." *Aviation, Space and Environmental Medicine,* 53: 1062–1066.

Fry, G. E., and R. F. Reinhardt. 1969. Personality Characteristics of Jet Pilots As Measured by the Edwards Personal Preference Schedule. *Aerospace Medicine, 40,* pp. 484–486.

Ganse, T. 1996. *Approach.* vol 41, no. 5. September–October pp. 24–25. Naval Safety Center, Norfolk, VA.

General Accounting Office (GAO). 1996. Report to Committee on National Security (Honorable Ike Skelton). 27 pages. 1 February 1996.

Gleick, B. 1987. *Chaos: Making a New Science.* New York: Penguin Books.

Graeber, C., February 1990. The Tired Pilot. *Aerospace.* pp. 45–49.

Gregorich, S., Robert L. Helmreich, John A. Wilhelm, and Thomas Chidester. 1989. "Personality-Based Clusters as Predictors of Aviator Attitudes and Performance. In R. S. Jensen's (ed.) *Proceedings of the 5th International Symposium on Aviation Psychology, vol. II.* Columbus, Ohio. pp. 686–691.

Hawkins, F. H. 1987. *Human Factors in Flight. 2d ed.* Brookfield, VT: Ashgate Publishers.

Helmreich, R. 1982. "Pilot Selection and Training." Paper presented to the annual meeting of the American Psychological Association, Washington D.C. August.

Helmreich, R. L. 1980. Social Psychology on the Flight Deck. In proceedings of a NASA/industry workshop, *Resource Management on the Flight Deck,* San Francisco, 26–28 June 1979.

Hoey, R. (1992, November). Fit to Fly? Fatigue in the Cockpit. *Flying Safety.* pp. 8–13.

Hughes, D, M. Dornheim, D. North, and B. Nordwall, 1995. "Automated Cockpits Special Report, Part 2," *Aviation Week & Space Technology.* 48–55. February 6, 1995.

Hughes, D., M. Dornheim, W. Scott, E. Phillips, B. Henderson, and P. Sparaco, 1995. "Automated Cockpits Special Report, Part 1," *Aviation Week & Space Technology.* 52–65. January 30, 1995.

Hughes, D., M. Dornheim, W. Scott, E. Phillips, and B. Henderson, 1992. "Automated Cockpits: Keeping Pilots in the Loop," a special report found in, *Aviation Week & Space Technology.* 50–70. March 23, 1992.

Inkso, C. A. 1985. Balance Theory, the Jordan Paradigm, and the West Tetrahedron. In L. Berkowitz (ed.), *Advances in Experimental Social Psychology.* New York: Academic Press.

Inkso, C. A., R. H. Hoyle, R. L. Pinkley, G. Y. Hong, R. M. Slim, B. Dalton, Y. H. Lin, P. P. Ruffin, G. J. Dardis, P. R. Brenthal, and J. Schloper, 1988. Individual-group Discontinuity: The Role of a Consensus Rule. *Journal of Experimental Social Psychology, 24,* 505–519.

International Civil Aviation Organization. 1989. *Human Factors Digest no. 1,* Circular 216-AN/131, Montreal.

Jensen, R. S. 1995. *Pilot Judgment and Crew Resource Management,* Aldershot, UK: Avebury Aviation.

Kanki, B. G., M. T. Palmer and E. Veinott. 1991. "Communication Variations Related to Leader Personality." NASA-Ames Research Center, Moffett Field, CA. Publication #231.

Kelly, Bill. 1996. NDB's High Margin of Error. *Aviation Safety.* June 15.

Kelly, Bill. 1994. *USAF Flying Safety Magazine,* "Regulations Are Made to Be Broken, Right? November. pp. 17–18

Kern, Anthony T. 1995. A Historical Analysis of US Air Force Tactical Aircrew Error in Operations Desert Shield/Storm. US Army Command and General Staff College monograph. Fort Leavenworth, KS.

Kern, T. 1997. *Redefining Airmanship.* New York: McGraw-Hill.

Kern, T. 1994. *The Human Factor Newsletter.* U.S. Air Force Air Education and Training Command, Randolph AFB, TX.

Lewis, C. S. 1961. *The Screwtape Letters.* New York: Touchstone.

McConnell, Michael G. 1994. AFR 110-14 USAF Accident Investigation Report, vol. 1, June 1994.

Merry, U. 1995. *Coping with Uncertainty: Insights from the New Sciences of Chaos, Self-Organization and Complexity.* Westport CT: Praeger.

Metzger, N. 1996. "My Hail-Lacious Low-Level." *Approach.* vol 41, no. 5. September–October pp. 12–13. Naval Safety Center, Norfolk, VA.

Monan, B. 1991. *ASRS Directline. Readback-Hearback.* Issue 1, March 1991. Internet. http//www-afo.arc.nasa.gov/ASRS/dl6_read.html

Monan, W. P. 1978. Distraction—A Human Factor in Air Carrier Hazard Events. *NASA Aviation Safety Reporting System: Ninth Quarterly Report,* 2–23. Moffett Field, CA: National Aeronautics and Space Administration.

Murphy, D. 1993. "Simuflite CRM Targets Human Aspects of Flight," *Aviation International News.* 38–39. January 1, 1993.

Nance, John J. 1986. *Blind Trust: How Deregulation Has Jeopardized Airline Safety and What You Can Do about It.* New York: William Morrow and Company.

NASA Aircrew Safety Reporting System (ASRS) 1995. *Callback #192: It's Almost Summertime,* p. 1. Moffet Field, CA: NASA.

NASA ASRS. 1994. AeroKnowledge ASRS CD-ROM. NASA ASRS. 1995. *The AeroKnowledge CD-ROM.* Trenton, New Jersey.

National Transportation Safety Board. 1994. *A Review of Flightcrew-Involved, Major Accidents of U.S. Air Carriers, 1978 through 1990. Safety Study.* (Report No. PB94-917001 NTSB/SS-94-01.) Washington DC: National Transportation Safety Board.

National Transportation Safety Board, accident identification number SEA96FA079. Internet at http://www.ntsb.gov.Aviation/SEA/96A079.html

National Transportation Safety Board. November. 1996 Aviation Accidents. NTSB Identification: FTW97FA042, Internet. Found at http://www.ntsb.gov/aviation/9611.html

National Transportation Safety Board, 1995. NTSB Report Brief #46413. Internet. http://www.ntsb.gov/Aviation/FTW/95A244.html

Novello, J. R., and Z. I. Youssef. 1974. Psycho-Social Studies in General Aviation: Personality Profiles of Male Pilots. *Aerospace Medicine*, 45, 185–188.

NTSB ASRS Directline. 1996. Statistics. Issue 6, p. 33.

Olcott, J. 1996. "Complacency: The Silent Killer." *National Business Pilots Association Magazine*. June.

Orlady, H.W. 1992. "Advanced Cockpit Technology in the Real World," *Journal of ATC*. p. 47–51. January–March.

Osland, John J., and J. Christensen. 1994. "Hibbing Air Crash Blamed on Captain." *Minneapolis Star-Tribune*, 25 May 1994.

Palmer, M. T., W. H. Rogers, K. A. Latorella, and T. S. Abbott. 1995. *A Crew-Centered Flight Deck Design Philosophy for High-Speed Civil Transport (HSCT) Aircraft*. NASA Technical Memorandum 109171. January.

Parasuraman, R., R. Molly and I. Singh, 1993. "Performance Consequences of Automation-Induced Complacency," *The International Journal of Aviation Psychology. 3(1)*. 1–23.

Phillips, R. J. 1996. "A Principle Within: Ethical Military Leadership." Presented at the Joint Services Conference on Professional Ethics XVII. Washington, DC. January 25–26, 1996.

Rawnsley, Judith H., 1995. *Total Risk: Nick Leeson and the Fall of Barings Bank*. New York: HarperCollins.

Reed, Robert C. 1968. *Train Wrecks: A Pictorial History of Accidents on the Main Line*. New York: Bonanza Books.

Rippon, T. S., and E. G. Manuel. 1918. "The Characteristics of Successful and Unsuccessful Aviators. With Special Reference to Temperament." *The Lancet*. September 28, 1918. pp. 411–415. London.

Rouse, D. M. 1991. Report of Aircraft Accident Investigation B-52G #59-2593. United States Air Force. February 3.

Schmitt, J. 1995. "Chaos, Complexity and War: What the New Nonlinear Dynamical Sciences May Tell Us about Armed Conflict." Paper written for Marine Corps Combat Development Command. Quantico, VA.

Sheehan, Neil. 1971. *The Arnheiter Affair*. New York: Random House.

Smith, Perry M. 1986. *Taking Charge: A Practical Guide for Leaders*. Washington, DC: National Defense University Press.

Snyder, C. R., and H. L. Fromkin. 1980. *Uniqueness: The Human Pursuit of Difference*. New York: Plenum.

Stewart, J. 1989. *Avoiding Common Pilot Errors: An Air Traffic Controller's View.* Blue Ridge Summit, PA: Tab Books.

Sumwalt, R. L., and A. W. Watson. 1995. What ASRS Incident Data Tell about Flight Crew Performance during Aircraft Malfunctions. *The Ohio State University Eighth International Symposium on Aviation Psychology,* 758–764. Columbus, OH: The Ohio State University.

Thomma, Steven. 1994. "Pilot in Crash near Hibbing Had Checkered Flight Record." *The Minneapolis Star Tribune.* December 17, 1993. B-1.

Transport Canada. 1997. "Crew Resource Management: Stress Management." Internet: www.caar.db.erau.edu/crm/resources/misc/transcan/transcan6.html

United States Air Force. 1994. *Aircrew Awareness and Attention Management Workbook.* Langley Air Force Base, VA.

United States Air Force. 1996. 51-503 Accident Investigation Report on C-130H Mishap at Jackson Hole, Wyoming.

United States Air Force. 1994. *Human Performance Enhancement Workbook.* Langley Air Force Base, VA.

United States Air Force. 1997. Multi Command Instruction (MCI) 11–217, vol. 5. 15 February.

United States Air Force Regulation 110-14, 1992. *Accident Investigation Report,* "B-1B Mishap, 30 November 92." Dyess AFB, Abilene, TX.

Wickens, C.D. 1992. *Engineering Psychology and Human Performance* (2d ed.). New York: Harper Collins Publishers Inc.

Wickens, C. D., and J. M. Flach, 1988. "Information Processing" a chapter in E. L. Wiener and D. C. Nagel, (eds.), *Human Factors In Aviation.* 111–155. San Diego, CA: Academic Press, Inc.

Wiener, E.L. 1981. "Complacency: Is the Term Useful for Air Safety?" Proceedings of the 26th Corporate Aviation Safety Seminar. 116–125. Denver: Flight Safety Foundation, Inc.

Wiener, E. L. 1987. Application of Vigilance Research: Rare, Medium, or Well Done? *Human Factors,* 29(6), 725–736.

Wilkinson, R. T. 1964. Effects of up to 60 Hours of Sleep Deprivation on Different Types of Work. *Ergonomics.* 1, 175–186.

Wilkinson, R. T. 1965. Sleep Deprivation. In O. G. Edholm and A. Bacharach (eds.), *The Physiology of Human Survival* (pp. 399–430). New York: Academic Press.

Wolfe, Sidney, Mary Gabay et. al. 1996. *Questionable Doctors: Disciplined by the States or the Federal Government. 1996 edition.* Washington, DC: Public Citizen Health Research Group. Yeager, Chuck. 1994. Personal Interview. Edwards Air Force Base, CA. March.

6th Sense World Wide Web Site. 1996. "How Our Mind Rewrites History." Internet. Found at http://www.pracpsy.com/rewrite.html

Index

325th Bomb Squadron (BMS), 51–52

A

A-7K practice engagement and
    accident, 43–45
Accelerated Copilot Enrichment
    (ACE), 66, 203
accidents/crashes, 39, 45–46, 263–278
    aerobatics at low altitude, 270–272
    chain of events leading to, 276–277
    clearance changes, 273–276
    Confess-Climb-Conserve-Consult
        checklist, 276
    Eastern Flight 111 and Epps Air
        100, 81–85
    flight plan changes, 272–273
    hail, thunderstorms, 264–266
    IFR to VFR problems, 267–270
    low-altitude flight, 263, 270–272
    midmission changes, 263, 272–276,
        316–317
    poor visibility, clouds, 266–267
    somatogravic illusions, 267–270
    spatial disorientation, 263, 267–270
    vertigo, 269
    weather-related problems, 263,
        264–267
Addison, Joseph, 74–75
advocacy, active, and peer pressure,
    166–169
aerobatics at low altitude, 270–272
age, pilot, 125
aggressive behavior, 129–130, 143, 172
air shows/airshow syndrome
    (*see also* pushing the limits),
    51, 52, 53, 54, 151–153
air taxis, 45–46
air traffic control (ATC), 83–85, 354
    altitudes, 208–209
    callsigns, 206, 209

air traffic control (ATC) *continued*
    causes of communication
        breakdown, 205–206
    clearing, 213–214
    communication with pilot, 203–214
    frequency for ATC communications,
        206, 208
    hear back problems, 206–207
    lost communications, 209–212
    mindset/expectancy, 206
    peer pressure from, 167–168
    phraseology, 212–213
    read back problems, 206
    read back, hear back technique, 205
    slip of mind/tongue, 206
    standard communications
        procedures, 208
    verifying ATC instructions, 208
aircraft check out, 287–288
Aircraft Owners and Pilots
    Association (AOPA), 99
aircrew/aviation safety reporting
    system (ASRS), 42–43, 59, 201,
    205, 240–241
Airman's Information Manual (AIM),
    212, 220
airmanship, 4, 6–9, 10
airspace, 291–292
airworthiness directives, 102
alcohol/drug use, 125
altitudes, 208–209
American Airlines crash, Columbia,
    Dec. 1995, 39
Andre, Babette, 36
antiauthority attitudes, 145–146
anticipating problems, 251–252,
    254–255
approaches, 17–20, 28–29
    Jeppesen manual, 18, 30
    mapping, 36

approaches *continued*
  missed approach, 29
  procedure, 80, 86–89
  separation minimums, 83–85
Aristotle, (q)59
Arnheiter Affair, The, 111
Arnheiter, Marcus A., 110–111
assertiveness, 154–155
attention management (*see also*
  situational awareness (SA), 89–90
  case study—single-focus pilot,
    91–93
  channelized attention, 90–91
  prioritizing tasks, 91
  task saturation, 90–91
automated communications, 213
automated terminal information
  service (ATIS), 213
automatic direction finder (ADF), 13
Automatic Flight Control System
  (AFCS), 237
automation, 223–248, 353–354,
  357–383
  accidents associated with
    malfunctions in, 357–383
  Automatic Flight Control System
    (AFCS), 237
  autopilot misuse, 233–234
  Aviation Safety Reporting System
    (ASRS), 240, 241
  building personal discipline for, 246
  captain's role challenged, 239–240
  case study—Pyote 14 crash,
    Nov. 1992, 224–225
  case study—Scandinavian Air crash,
    Feb. 1984, 225–226
  Cathay Pacific Airways policy, 235
  complacency from use of, 240
  control-management continuum
    for pilots, 239
  controlled flight into terrain (CFIT),
    385–389
  crew coordination errors, 242
  database errors, 242
  Delta Airlines policy, 234–235
  design philosophy of systems, 232
  errors, 256–257, 303–308
  fixation on, 233
  flight management systems (FMS),
    240
  Global Positioning System (GPS)
    limitations, 243–246

automation *continued*
  human factors and, 232
  human ingenuity and, Sioux City
    crash, July 1989, 227
  human-centered, 231–232
  insufficient understanding of, 242
  malfunctions, 256–257, 303–308,
    357–383
  mode confusion, 240–243
  mode transition problems, 242
  navigation errors, 227, 228–229
  overautomation problems, 226
  philosophy of organization on,
    229–230, 233–237
  pilot decisions and, 232–233
  pilot views on, 232
  pitfalls of, 237–238
  policies for, 230
  problems of using, 233
  procedures for, 230–231
  programming errors, 242
  situational awareness (SA), 225
  standard operating procedures
    (SOP), 231
  strengths and weaknesses of, 227
  training, 243
  U.S. Air Force C-17 Operations
    policy, 236–237
  United Airlines policy, 235–236
  unknown failures, 242
  workload reduction or increase, 238
autopilots (*see also* automation),
  233–234
Avoiding Common Pilot Errors, 204

B

Baring, Peter, 113–114
Barker, Pat, 279–300, 317
baseline record for self, 349
Bates, Stacy, 133
Beaukieu, Roderic, 123
becoming a more disciplined flyer,
  347–353
Billings, Charles E., 223, 231–232,
  239, 243, 357
Black Knight mission, 329–335
Blind Trust, 173
boredom, 254
Bragg, Bob, 48
briefings, 289–290
Bringle, Donald, 133–134
Brown, Ronald H. (*see* IF0 21 crash)

C

callsigns, 206, 209
capabilities management, 316
Cathay Pacific Airways, automation
   policy, 235
CFIT Risk Assessment Checklist, 299
change management, 301–318
   capabilities management, 316
   chaos theory, 301–302, 316–317
   checklists, 316
   coincidence, chaos, both?, 303–308
   collective errors, 311
   communication lacking, 303–308
   delaying reaction to change, 314
   errors made by changing
      conditions, 310–311
   errors of commission, 311–312
   errors of omission, 311, 312
   four basic steps to control, 315–316
   help from experts, 314
   human errors, 311
   human reaction to change, 302–303
   hurry-up syndrome, 309, 312–313
   immediate reaction to change, 313
   midmission change, 316–317
   Murphy's Law, 301
   reacting to change, 308
   small events, big consequences,
      303, 307
   stress, 312–313
   tempo or pace of actions, 315–316
   temporal distortion, 308, 312–313
   time, the fourth dimension of
      change, 313–315
   worst-case scenarios, 314
channelized attention, 90–91
chaos theory, 301–302, 316–317
character (*see also* personality
   factors), 125
checklists, 46–48, 78, 91–93, 95–96,
   96–97, 217–218, 258, 260–261,
   276, 282–285, 299, 316, 320, 354
clearance changes, 273–276
climate versus culture (organizational),
   100–104
clouds, ceiling, 266–267
cockpit procedural trainer (CPT), 189
cockpit/crew resource management
   (CRM), 16–17, 101, 288–291,
   328, 354–355
   briefings, 289
   conflict can be good, 291

cockpit/crew resource management
   continued
   interpersonal relationships, 290–291
   peer pressure affects, 161–163
   team work, team formation, 289
collisions and near-misses
   (*see* accidents/crashes)
communications discipline, 91, 117,
   118, 201–221, 249, 304
   air traffic control to pilot, 203–214
   altitudes, 208–209
   automated communications, 213
   automated terminal information
      service (ATIS), 213
   callsigns, 206, 209
   catastrophic effects of
      noncommunication, 201–202
   causes of communication
      breakdown, 205–206
   causes of lost communications,
      211–212
   checklists, 217–218
   clearing, 213–214
   crew communication, 203
   crew leadership, 214
   debriefing, 218
   differences resolved, 215–216
   enroute flight advisory service
      (EFAS), 213
   error chains, 218
   example of miscommunication,
      207–208, 217–218
   experience versus risk of losing
      communication, 211
   frequency for ATC
      communications, 206, 208, 211
   ground communications, 216
   hear back problems, 206–207
   lack of, 303–308
   lost communications, 209–212
   mindset/expectancy, 206
   National Airspace System (NAS),
      203–204
   NOTAMs, 213
   Operation Eagle Claw, breakdown,
      325–326, 328
   opinions stated, 215
   personality and, 128–131
   phase of flight and lost
      communications, 210–211
   phraseology, 212–213
   portable transceivers as backup, 212

communications discipline *continued*
  precision in communication,
    218–220
  preflight planning, 211, 213
  questions and answers, 215
  read back problems, 206
  read back, hear back technique, 205
  recommendations of ASRS study,
    208–209
  slip of mind/tongue, 206
  standard communications
    procedures, 208
  sterile cockpit concept, 212, 220
  subtle messages, 214
  transcribed weather briefing
    (TWEB), 213
  verifying ATC instructions, 208
  written communications, 213
competing with yourself, 169–171
complacency, 150–151, 240
compliance/non-compliance, 15–16,
  21–22
Confess-Climb-Conserve-Consult
  checklist, 276
confidence, 187–188
conflict among crewmembers, 291
confrontation as discipline, 196–197
confusion over procedure, 95–96,
  104–106
consistency as discipline, 197–198
contingency planning, 252
controlled flight into terrain (CFIT),
  385–389
control-management continuum for
  pilots, 239
cost of flying, 4
cost of safety, 106–108
costs of poor flight discipline, 35–57
  A-7K practice engagement and
    accident, 43–45
  accident rate, 39
  air taxi near-miss, 45–46
  aircrew/aviation safety reporting
    system (ASRS), 42–43
  checklist oversight example, 46–48
  commercial costs, 39–40
  cost-benefit equation, 38
  deaths from mishaps, 39
  devolving discipline, 48–49
  final perspective, 55–56
  incident reports, 42–43
  insured and uninsured costs, 41
  military costs, 40

costs of poor flight discipline
  continued
  mishap data (Aviation Monthly
    Safety Summary), 39
  Operation Eagle Claw failure,
    322–329
  patterns of poor discipline, 49–55
  poor role models, 49–55
  scope of discipline problem, 38–46
  self-reported statistics, 42–43
  statistical evidence of cost, 39
Covey, Stephen, 347
creativity versus rogue behavior, 115
crew awareness, 258–260
crew discipline, 20–21
crew management (*see* cockpit/crew
    resource management)
crosschecks, 27, 29
curiosity and dangerous situations, 145
Czar 52 crash, 54

D

Dash 11 Tech Order, 51–52
deaths from mishaps, 39
debriefing, 218
decision-making, 6, 13, 16, 187–188
  peer pressure and, 174–175
  personality and, 132
decisiveness, 187–188
deferential attitude, excessive, 154–155
defining flight discipline, 9
Delta Airlines, automation policy,
  234–235
denial of discipline problems, 338–342
Desert Storm, 143
  Black Knight mission, 329–335
designers, 353
Directline, 205
disciplined attention (*see* situational
    awareness (SA)
discipline problems, 3–33
distractions, 27–28, 168–169, 251,
  254, 255, 257–258, 286–287
domineering personalities, 129–130
Downeast Airlines, peer pressure
    case study, 172–174
Dubroff, Jessica, 163–165
Dunne, John, 201 (q)

E

Eastern Flight 111 and Epps Air 100
    collision, 81–85
economic pressures on organizations,
  106–108

Edwards, Doug, 308
egotism, 341–342
emergency procedures, 79, 91–93,
   187–188, 290, 319–322, 355
   impulsiveness, impetuous
     behavior, 148
   lost communications, 209–212
   resignation, giving up, 149–150
   water rescue using GPS, 245–246
Emerson, Ralph W., 118 (q)
"emotional jet lag", 153–154
enroute flight advisory service
   (EFAS), 213
Erickson, Chad, 171–172
error chains, 78, 141, 218, 276–277,
   303–308
errors of commission, 311–312
errors of omission, 311, 312
ethics (see also peer pressure), 175
evaluating situations, 251
expertise and discipline, 6

F
Falitz, Marvin, 171–172
false horizon (see somatogravic
   illusion)
FAR Part 91, 51
fatigue, 20–21, 23–26, 29
fear or apprehensions, 287
Federal Aviation Administration
   (FAA), 10
flight crew information files (FCIF),
   102
flight management systems (FMS), 240
flight plans, 20–21, 66–69, 86–89,
   211, 213, 252, 272–273, 279–300,
   348–349
   aircraft check out, 287–288
   airspace, 291–292
   art of, 281
   briefings, 289–290
   changing the plan, 23–24
   checklists, 282–285
   cockpit/crew resource
     management (CRM), 288–291
   double checking, 350–351
   environment of flight, 291–292
   failure to prepare, case study,
     292–296
   familiar routines can need
     planning, 280–281
   fundamental planning criteria, 288
   insufficient planning, 282–285

flight plans continued
   knowns and unknowns, 279–280
   pilot self-check, 285–287
   risk management, 296–299
   workload management, 292
Flight Safety Foundation, 299
fly overs, 51–52, 153
focusing attention, 250–251, 253
formation flight, 53–54
Foundations of Aviation Excellence,
   317
frequency for ATC communications,
   206, 208, 211

G
Ganse, Tom, 267–270
General Accounting Office (GAO), 40
Gibb, Randall W., 192–198
Gleick, James, 302
global fixes, 102
Global Positioning System (GPS),
   43–246
   hand-held GPS limits, 244–245
   limitations of, 243–246
   water rescue, 245–246
Global Power Mission, 52
goal setting, 351–353
Graeber, Curt, 25
Graff, Harold, 48
Gray, Alfred, 316–317
ground communications, 216
Gunther, John, 263 (q)

H
habit patterns and consistency, 93, 254
hail, thunderstorms, 264–266
Hansman, R. John, 241–243
Hawkins, Frank, 201, 213, 216
hazardous attitudes (see also
   personality factors), 139–156, 286
   aggressiveness, 143
   air-show syndrome, 142, 151–153
   antiauthority attitudes, 145–146
   assertiveness, 154–155
   competing with yourself, 169–171
   complacency, 150–151
   curiosity and dangerous situations,
     145
   deferential attitude, excessive,
     154–155
   "emotional jet lag", 153–154
   fear or apprehensions, 287
   hurry-up syndrome, 309, 312–313

hazardous attitudes *continued*
impact on flight discipline, 142–143
impulsiveness, impetuous
behavior, 148
invulnerability syndrome, 147–148
machismo, 142, 146–147
medical airworthiness, 285
mistakes and bouncing back,
153–154
pressing situations, 142, 143–145
psychological airworthiness,
285–286
pushing the limits, 139–142
resignation, giving up, 149–150
self-assessment of, 155–156,
286–287
stress, 286, 308, 309, 312–313
"take a look" syndrome, 145
temporal distortion, 308, 312–313
hazardous situations, 9
Helmreich, Robert, 132
high-risk activities, 257
*Human Centered Aviation
Automation*, 357
human errors, 311
Human Factors in Flight, 201
human factors, 21–24
hurry-up syndrome, 309, 312–313
Huxley, Thomas, 250

I

ICAO Journal, 99
idealization, 341–342
IFO 21 crash in Croatia (USAF CT-43),
11–31
analysis of crew actions, 29
approach safety problems, 18–20
crew discipline problems, 20–21
crucial navaids missing, 12–13
decision-making process and
players, 13, 15
distractions, 27–28
external pressure on crew
performance, 21–24
failed discipline, 30
failing to take no for answer, 18–20
fast-paced environment, 21–22
fatigue, 20–21, 23–26, 29
final approach, 28–29
flight planning problems, 20–21
good intentions and discipline,
30–31
human factors, 21–24

IFO 21 crash in Croatia (USAF CT-43)
continued
internal stress factors on crew,
24–28
Jeppesen approach information,
18, 30
missed approach procedure, 29
multiple mission changes, 23–24
non-compliance, 15–16, 21–22
pressing attitude, 26–27
training problems, 16–17
unapproved instrument approach,
17–18
VIP passengers, 22–23
IFR to VFR problems, 267–270
ignorance of regulations, 63–65
impulsiveness, impetuous behavior,
148
incident reporting, 42–43
self-reports of nonadherence, 59
instinctive cues to problems,
253–254
instructors, 70–72, 181–199, 354
character and integrity, 197
classroom study, 193
confrontation as discipline, 196–197
consistency as discipline, 197–198
expecting the unexpected,
188–191
hours of operation, 193–194
human stops, 191–192
importance of discipline in, 194–195
indecisiveness, 187
lack of confidence, 187–188
logbooks, 195
losing a student, 182–185
organizational pressures, 195
overzealous, 165–166
peer pressure from, 165–166
"preventative medicine" in
discipline, 196
pride and proficiency, 185–187
proficiency challenge, 185–187
riding the controls, 191–192
role models, 192–198
setting an example, 185
stress, 187–188
instrument landing systems (ILS), 13
interpersonal relationships, 290–291
interruptions in routine, 255
invulnerability syndrome, 147–148
Iranian hostage rescue (*see* Operation
Eagle Claw)

J
Jeppesen, 18, 30
Jeppesen, Elrey B., 35–38
judgment, 3, 7–8, 38

K
Kelly, Bill, 59–62
killing conditions (*see*
accidents/crashes), 263–278
Knutson, Don, 47–48
Kompus, Larry, 112

L
labor disputes, 106
Lancet, 123, 128
Lauber, John, 171–172
Leeson, Nick, 113–114
Lewis, C.S., 31
Lewis, Eugene, 47
licensing, 348
limitations, personal, 6
logbooks, 195
lost communications, 209–212
low altitude parachute extraction
system (LAPES), 139–142
low-altitude flight, 263, 270–272
aerobatics, 270–272

M
MAC Flyer, The, 80
machismo, 142, 146–147
malfunctions, 255–258, 303–308,
326–327
manufacturers, 354
Market Time operation, Vietnam, 111
Martin, Corby L., 329–335
Maslow, Abraham, 347
McNamara, Robert, 337 (q)
mechanical knowledge, 126
medical airworthiness, 285
mental characteristics (*see also*
personality factors), 126
Merry, Uri, 302
Mesa Club, peer pressure, 176
METRO weather system, 36
Metzer, Norman, 264–266
midmission changes, 263, 272–276,
316–317
clearance changes, 273–276
flight plan changes, 272–273
military
costs of mishaps, 40
good discipline example, Black
Knight, 329–335
poor discipline example,

Operation Eagle Claw, 322–329, 325
mindset versus personality, 132–135
minimum altitudes, 53–54
mishap data (Aviation Monthly Safety
Summary), 39
missed approach, 29
mission weighting, 163–166
mistakes and bouncing back, 153–154
Monan, William P., 205, 257
monitoring surroundings, 250–251, 259
Moshansky, Virgil P., 103
Murphy's Law, 301

N
Nance, John, 173–174
Napier, William, 249 (q)
Napoleon, 99(q), 313
NASA, 42
NASA-Ames Research Center, 129
National Airspace System (NAS), 201,
203–204
National Transportation Safety Board
(NTSB), 39
navaids, 12–13
Naval Proceedings, 134
navigation, 91, 249
automation errors, 227, 228–229
negative transfer of training, 93
noncompliance, 82–83
antiauthority attitudes, 145–146
combating, 72–74
machismo, 146–147
willful, 69–72
nondirectional radiobeacon (NDB), 13
Norris, Peter, 114
NOTAMs, 213

O
Olcott, John W., 151
operating instructions (OI), 102
Operation Eagle Claw, 322–329
analysis of failure, 327–329
casualties, 327
communications breakdown,
325–326, 328
end of plan, 327–329
execution details, 324–327
malfunctions, 326–327, 326
plan by joint military ops, 323–324
weather-related problems, 324–325,
326
organizational discipline, 10–11,
99–119
acting against rogue behavior,
116–117

automation policies, 229–230,
233–237
bureucratic responses to unsafe
conditions, 102
change, chaos, confusion, 104–106
climate versus culture, 100–104
communication, 117, 118
conflict resolution, 117–118
creativity versus rogue behavior,
115
cultures where rogues flourish,
116
denial responses, 102
Downeast Airlines case study,
172–174
economic pressures, 106–108
example setting, 117
fix problems, 118
generative responses to unsafe
conditions, 102
global fixes, 102
impact of organizations on safety,
99–100
information flow, 102
inquiries, 103
intuition, 118
labor disputes, 106
latent unsafe conditions, 102–104,
103, 118
management's role in safety,
107–108
mergers and rapid growth cause
stress, 104–106
pathological responses to unsafe
conditions, 102
peer pressure from seniors, 171–174
peer pressure, 157–158
personnel hiring, 117
pilot deviation rate at time of
merger, 105
pressure on instructors, 195
pressure to perform, 172–174
response to unsafe conditions, 102
rogues, 108–117
seven principles for organizational
flight discipline, 117–118
shaping of organizational climate,
102–104
tolerance of poor flight discipline,
101–102
workload and confusion, 104–106
Orlady, Harry W., 237–238
overreaching ability, 355

P
passengers, 22–23, 60–62
patterns of poor discipline, 49–55
Patton, George, 337
peer pressure, 70, 157–177
active advocacy, 166–169
aggressiveness in others, 172
air traffic controllers and, 167–168
case study, bad weather and,
163–165
competing with yourself, 169–171
competition from, 159
crew coordination affected by,
161–163
decision making in spite of, 174–175
defining peer pressure, 157–158
distractions, 168–169
Downeast Airlines case study,
172–174
ethical behavior and, 175
fear of retribution, 159
how it works, 159
influencing judgment through, 159
Jessica Dubroff incident, 163–165
mission weighting, 163–166
organizational pressure to perform,
172–174
overzealous instructors, 165–166
positive aspects of, 175–176
pressure from seniors, 171–174
pride and, 159–163
quota systems, 166–169
resisting, 174–175
sources of, 158
personal improvement plan (see also
personality factors), 337–356
baseline record for self, 349
becoming a more disciplined flyer,
347–353
denial of discipline problems,
338–342
egotism, 341–342
flight planning, 348–351
idealization, 341–342
license yourself, 348
procrastination, 339
professionalism, 342–347
rationalization, 340–341
recording, rewarding progress, 352
reminders, 353–354
selective perceptions, 339–340
waypoints or goal setting, 351–353
withdrawing from discussion of
discipline, 338–339

personal nature of discipline, 5–6
personality factors
personality factors (*see also*
  hazardous attitudes), 123–138
age, 125
aggression, 129–130
alcohol use, 125
becoming a more disciplined flyer,
  347–353
character, 125
communication and, 128–131
conflict can be good, 291
correcting personality flaws,
  130–131
decision making ability, 132
denial of discipline problems,
  338–342
domineering personalities, 129–130
egotism, 341–342
fear or apprehensions, 287
hazardous attitudes, 286
high-spiritedness, 125
human reaction to change, 302–303
hurry-up syndrome, 309, 312–313
idealization, 341–342
interpersonal relationships, 290–291
intuitive feel for aircraft, 126
mechanical knowledge, 126
mental characteristics, 126
mindset versus personality, 132–135
modern demands on pilots, 127–128
occupation or career, 125
performance shaped by, 128
permanence of personality, 128
personal improvement plan,
  337–356
pilot personalities, 123–127
procrastination, 339
professionalism, 342–347
psychological airworthiness,
  285–286
rationalization, 340–341
recreation, 125
selective perceptions, 339–340
self-assessment of, 135, 286–287
sportsmanlike attitude, 125
stereotype of pilot, 124–127
stress, 131, 286, 308, 309, 312–313
temporal distortion, 312–313
"top-gun" mentality, 132–135
unsuccessful pilot characterstics,
  126–127
withdrawing from discussion of
  discipline, 338–339

Phillips, Don, 228–229
Phillips, R.J., 175
Phormio of Athens, 77 (q)
phraseology, ATC communications,
  212–213
pilot self-check, 285–287
baseline record for self, 349
dangerous mind set, 286
distractions, 286–287
fear or apprehensions, 287
hazardous attitudes, 286
medical airworthiness, 285
psychological airworthiness,
  285–286
self-evaluation, 286–287
stress, 286, 308, 309
pitch up maneuver, 52–53
plane, path, people checklist, 258
preflight planning, 213
pressing situations, 26–27, 142,
  143–145
pride and peer pressure, 159–163,
  185–187, 314
prioritizing tasks, 91
procedural discipline, 77–98
air traffic control (ATC), 83–85
analyzing error chain, 82–83
approach procedure, 80, 86–89
attention management, 89–90
automation use, 230–231
case study in procedural errors,
  81–85, 86–89, 319–322
case study—single-focus pilot,
  91–93
case study—too many procedures,
  93–94
channelized attention, 90–91
checklists, 78, 91–93, 95–97
communications skills, 91
conflicting procedures, procedure
  overload, 93–94
confusion, 95–96
emergency procedures, 79, 91–93,
  319–322, 355
error chains, 78
habit patterns and consistency,
  93–97
keys to procedural excellence, 78
knowledge of procedures, 79
mandatory procedures, 79
navigation skills, 91
negative transfer of training, 93
noncompliance, 82–83
pattern interference, habits, 94–95

procedural discipline *continued*
performance circle of procedure,
89, 89
prioritizing tasks, 91
procedural errors, 78
self-assessment of procedural
proficiency, 97
separation minimums, 83–85
skill and proficiency, 85–86
standard operating procedures
(SOP), 231
task saturation, 90–91
technique versus procedures, 79–85
training, 85–86, 97
unlearning poor procedures, 93
willful disregard for, 85
procrastination, 339
professionalism, 342–347
proficiency, 7–8, 185–187
psychological airworthiness, 285–286
psychological study of disciplined
pilots, 6
pushing the limits, 74, 139–142
Pyote 14 crash, Nov. 1992, 224–225

Q
Questionable Doctors, 112
quota systems, peer pressure, 166–169

R
radio failure, 209–212
rationalization, 72–74, 340–341
Rawnsley, Judith, 113
read back, hear back technique, 205
Reason, James, 102
recording, rewarding progress, 352
recurrent training (refresher rides), 97
Redefining Airmanship, 6, 49, 68,
86, 297
Reed, Robert, 109
regulatory discipline, 10–11, 59–75
deviations from regulations, 63–72
"getting away with it", 73
"good flyers don't need rules",
73–74
good pilots follow rules, 74–75
ignorance of regulations, 63–65
making excuses for deviation, 66–69
noncompliance, combating, 72–74
noncompliance, willful, 69–72
"overregulation", 74
poor planning, 66–69
"pushing the envelope", 74

regulatory discipline *continued*
rationalizing poor flight discipline,
72–74
Regulations Are Made to be
Broken article, 59–62
"safety margins", 73
self-assessment of regs knowledge,
64–65
situation awareness (SA), 68–69
Reichert, S. William, 236
reminders, 255, 260–261, 353–354
resignation, giving up, 149–150
risk management, 5, 296–299
CFIT Risk Assessment Checklist, 299
controlling risk, 298
identifying risk, 297–298
Robbins, Tony, 347
Rogers, Carl, 347
rogue behavior, 108–117
acting against rogue behavior,
116–117
business professional, 113–114
common characteristics, 114–115
creativity versus rogue behavior,
115
cultures where rogues flourish, 116
medical profession, 111–113
naval captain, 110–111
railroad engineer, 109–110
role models of poor discipline, 49–55
role models, instructors as, 192–198
Roosevelt, Teddy, 319(q)
Roswell AFB, air show, 53
Rouse, 303–308

S
safety, 4
Scandinavian Air crash, Feb. 1984,
225–226
scanning, 260
scissors maneuver, 70–71
scope of discipline problem, 38–46
Screwtape Letters, The, 31
scud running (*see also* aerobatics), 143
selective perceptions, 339–340
separation minimums, 83–85
Sheehan, Neil, 111
showing off (*see* airshow syndrome)
simulators, 189
Sioux City crash, July 1989, 227
situational awareness (SA), 68–69,
225, 249–262
absence of cues, detecting, 256

situational awareness (SA) *continued*
anticipating problems, 251–252, 254–255
automation errors, 256–257
boredom, 254
building a crystal ball, 258–261
checklists, 258, 260–261
contingency planning, 252
crew awareness, 258–260
distractions, 251, 254, 255, 257–258
evaluating situation, 251
expectations, 254–255
flight plans, 252
focusing attention, 250–251, 253
habits versus, 254
high-risk activities, 257
instinctive cues to problems, 253–254
interruptions in routine, 255
malfunctions, 255–258
monitoring surroundings, 250–251, 259
plane, path, people checklist, 258
reminders, 255, 260–261
scanning, 260
what-if scenarios, 252
skill and proficiency, 7–8, 85–86, 126
Smith, Perry, 118
somatogravic illusions, 267–270
spatial disorientation, 263, 267–270
IFR to VFR problems, 267–270
somatogravic illusions, 267–270
vertigo, 269
sportsmanlike attitudes, 125
standard operating procedures (SOP), 231
standards and guidelines, 4–5
statistical evidence of poor discipline, 39
sterile cockpit concept, 212, 220
Stewart, John, 204, 220
Strauch, Barry A., 241–243
stress, 24–28, 70, 187–188, 308, 309, 312–313
personality and, 131
pressing situations, 26–27, 142–145
studying to be a better flyer, 4–5
Sumwalt, R.L., 258
Sun-Tzu, 157
Syrus, Pubilius, 75 (q)

T
"take a look" situations, 145
Taking Charge, 118

task saturation, 90–91
team work, team formation with crew, 289
technique versus procedures, 79–85
tempo or pace of actions, 315–316
temporal distortion, 308, 312–313
Terminal Instrument Procedures (TERPS), 17
Thompson, D'Arcy W., 157 (q)
thunderstorms, 264–266
time, the fourth dimension of change, 313–315
Toffler, Alvin, 301 (q)
Top Gun, 342
"top-gun" mentality (*see also* air show syndrome), 132–135
Total Risk, 113
training, 16–17, 64–65, 85–86
automation systems use, 243
Cockpit/Crew Resource Management (CRM), 16–17
habit patterns and consistency, 93–97
mistakes and bouncing back, 153–154
negative transfer, 93
recurrent training (refresher rides), 97
unlearning poor procedure, 93
transceivers, 212
transcribed weather briefing (TWEB), 213

U
U.S. Air Force C-17 Operations policy, automation, 236–237
United Airlines, automation policy, 235–236
USAF Flying Safety Magazine, 59

V
Valle, Christopher, automation discipline, 223–248
vertigo, 269
violations to flight discipline, 38
visibility, 266–267
von Clausewitz, Carl, 301
von Moltke, Helmuth, 279(q), 279
VOR, 13

W
Wall Street Journal, 130
water rescue using GPS, 245–246
Watson, A.W., 258

waypoints or goals, 351–353
weather-related problems, 263,
    264–267, 324–326
  case study in peer pressure,
    163–165
  hail, thunderstorms, 264–266
  Jessica Dubroff incident, 163–165
  poor visibility, clouds, 266–267
  pressing situations, 143–144
  thunderstorms, 264–266
weighting, mission, 163–166
Westrum, Ron, 99
"what-if" scenarios, 252, 292, 314
Wickens, Christopher, 227
Wiener, Earl L., 238

wingover maneuver, 52
withdrawing from discussion of
    discipline, 338–339
Witter, Wayne O., 130, 131
workload management, 292
Wright brothers, 36
Wright, Wilbur, 3
written communications, 213

Y
Yakima Bombing Range, 53–54
Yeager, Chuck, 108–109, 150

Z
Zend, Robert, 123 (q)

About the Author

Dr. Tony Kern is a Lieutenant Colonel in the U.S. Air Force. He has commanded KC-135 tankers, piloted B-1B bombers, and served in various operational and training capacities in his Air Force career. He has designed numerous aviation education & training programs that have been implemented across the spectrum of aviation. He lectures and consults internationally on aviation human factors and pilot accountability.